R00069 82331

CHICAGO PUBLIC LIBRARY
HAROLD WASHINGTON LIBRARY CENTER
Y0-DJO-333
R0006982331

REF
QC
176.84
.R3
I57
1975
Vol.2
Cop.1

FORM 125 M
Business/Science/Technology
Division

The Chicago Public Library

Received _____ JAN 19 1978

ION BEAM SURFACE LAYER ANALYSIS
Volume 2

ION BEAM SURFACE LAYER ANALYSIS
Volume 2

Edited by
O. Meyer, G. Linker, and F. Käppeler
Nuclear Research Center
Karlsruhe, Germany

PLENUM PRESS · NEW YORK AND LONDON

Library of Congress Cataloging in Publication Data

International Conference on Ion Beam Surface Layer Analysis, Karlsruhe, 1975.
  Ion beam surface layer analysis.

Proceedings of the International Conference on Ion Beam Surface Layer Analysis held at Karlsruhe, Ger., Sept. 15-19, 1975.
  Includes bibliographical references and indexes.
  1. Thin films, Effect of radiation on—Congresses. 2. Ion bombardment—Congresses. 3. Straggling (Nuclear physics)—Congresses. 4. Backscattering—Congresses. 5. Surfaces (Technology)—Congresses. I. Meyer, Otto. II. Linker, Gerhard. III. Käppeler, Franz. IV. Title.
QC176.84.R3I57 1975   530.4′1   76-2606
ISBN 0-306-35046-7

Proceedings of the second half of the International Conference on Ion Beam Surface Layer Analysis held at Karlsruhe, Germany, September 15-19, 1975

©1976 Plenum Press, New York
A Division of Plenum Publishing Corporation
227 West 17th Street, New York, N.Y. 10011

United Kingdom edition published by Plenum Press, London
A Division of Plenum Publishing Company, Ltd.
Davis House (4th Floor), 8 Scrubs Lane, Harlesden, London, NW10 6SE, England

All rights reserved

No part of this book may be reproduced, stored in a retrieval system, or transmitted, in any form or by any means, electronic, mechanical, photocopying, microfilming, recording, or otherwise, without written permission from the Publisher

Printed in the United States of America

Preface

The II. International Conference on Ion Beam Surface Layer Analysis was held on September 15-19, 1975 at the Nuclear Research Center, Karlsruhe, Germany. The date fell between two related conferences: "Application of Ion-Beams to Materials" at Warwick, England and "Atomic Collisions in Solids" at Amsterdam, the Netherlands.

The first conference on Ion Beam Surface Layer Analysis was held at Yorktown Heights, New York, 1973. The major topic of that and the present conference was the material analysis with ion beams including backscattering and channeling, nuclear reactions and ion induced X-rays with emphasis on technical problems and novel applications. The increasing interest in this field was documented by 7 invited papers and 85 contributions which were presented at the meeting in Karlsruhe to about 150 participants from 21 countries. The oral presentations were followed by parallel sessions on "Fundamental Aspects", "Analytical Problems" and "Applications" encouraging detailed discussions on the topics of most current interest. Summaries of these sessions were presented by the discussion leaders to the whole conference. All invited and contributed papers are included in these proceedings; summaries of the discussion sessions will appear in a separate booklet and are availble from the editors.

The application of ion beams to material analysis is now well established. The results of the conference in Karlsruhe have shown the progress in analytical problems, such as depth resolution and depth profiling and the necessity of using combined techniques for a complete material analysis. The backscattering technique has been extended to new materials such as biological samples, to more complicated layered structures and to interfacial reactions. Progress has been achieved in the application of backscattering and channeling to investigate bubble and blister formation in metals as well as in lattice location determination of light atoms in heavy mass host materials by nuclear reactions. The advantages of ion-induced X-ray analysis have been pointed out and this will certainly lead to specific applications and to the combination with other techniques.

The sponsorship and the financial support of the Gesellschaft für Kernforschung mbH was vital to the success of the Conference. We particularly wish to acknowledge help and advice from the International Advisory Committee consisting of:
G. Amsel (Paris), R. Behrisch (Garching), L.C. Feldman (Bell Labs), J.W. Mayer (Caltech), S.T. Picraux (Sandia), E. Wolicki (Naval Research) and J.F. Ziegler (IBM).

The operation and success of the discussion sessions is due to G. Amsel (Paris), J. Biersack (HMI, Berlin), J. Cairns (Harwell), W.K. Chu (IBM), L.C. Feldman (Bell Labs), F. Folkmann (Darmstadt), S. Kalbitzer (MPI Heidelberg), J. Poate (Bell Labs) and B. Scherzer (Garching).
The Committee responsible for organizing this conference consisted of W. Gläser, F. Käppeler, G. Linker, O. Meyer and G. Schatz.
Special thanks are due to Mrs. E. Maaß and Mr. P. Emmerich for their help and excellent skill in organization.

Karlsruhe, Germany
September 1975

O. Meyer

G. Linker

F. Käppeler

Contents of Volume 2

V. LATTICE LOCATION AND DECHANNELING IN DISORDERED CRYSTALS

Analysis Problems in Lattice Location Studies 497
    H.D. Carstanjen

Multi-String Statistical Equilibrium Calculation
of Scattering Yield for Foreign Atom Location Using
Channeling Effect 517
    K. Komaki

Ion Channeling Studies of the Lattice Location of Interstitial Impurities: Hydrogen in Metals 527
    S.T. Picraux

Depth Distribution of Damage Obtained by Rutherford
Backscattering Combined with Channeling 539
    R. Behrisch, J. Roth

Helium Trapping in Aluminum and Sintered Aluminum Powders . . 567
    S.K. Das, M. Kaminsky, T. Rossing

Helium Re-Emission and Surface Deformation in Niobium
During Multiple Temperature Helium Implantation 575
    W. Bauer, G.J. Thomas

Studies on Compound Thin Film Semiconductors by Ion
Beam and Electron Microscopy Techniques 585
    S.U. Campisano, G. Foti, E. Rimini, G. Vitali,
    C. Corsi

Energy Dependence of Channeling Analysis in Implantation
Damaged Al. 597
    E. Rimini, S.U. Campisano, G. Foti, P. Baeri,
    S.T. Picraux

Characterization of Reordered (001) Au Surfaces by the
Combined Techniques of Positive Ion Channeling Spectroscopy (PICS), LEED-AES, and Computer Simulation 607
    B.R. Appleton, D.M. Zehner, T.S. Noggle,
    J.W. Miller, O.E. Schow III, L.H. Jenkins,
    J.H. Barrett

Proton Channeling Applied to the Study of Thermal Disorder
in AgBr 627
    M. Roulet, C. Jaccard, H. Huber

Neon-Ion Implantation Damage Gettering of Heavy
Metal Impurities in Silicon 635
    Shanghai Institute of Metallurgy, Chinese Academy
    of Sciences, Shanghai Institute of Nuclear Research

VI. RELATED TECHNIQUES

A New Scanning Ion Microscope for Surface and
In-Depth Analysis 649
    K. Wittmaack

Sputtering of Thin Films in an Ion Microprobe 659
    W.O. Hofer, H. Liebl

Xe^+ Ion Beam Induced Secondary Ion (Si^+) Yield from
Si-Metal Interfaces 665
    T. Narusawa, T. Satake, S. Komiya,
    A. Shimizu, M. Iwami, A. Hiraki

Surface Analysis of Ion Bombarded Metal Foils by XPS 675
    E. Henrich, H.J. Schmidt

CONTENTS OF VOLUME 2

Chemical Reaction Enhancement and Damage Rate of Surface
Layer Bombarded with Inert Ion Beams 685
    T. Tsurushima, H. Tanoue

VII. ION INDUCED X-RAY SPECTROSCOPY

Progress in the Description of Ion Induced X-Ray Production;
Theory and Implication for Analysis 695
    F. Folkmann

K-Shell Ionization of Boron Induced by Light Ions
Bombardment 719
    K. Kawatsura, K. Ozawa, F. Fujimoto,
    M. Terasawa

Effect of Channeling on Impurity Analysis by
Charged Particle Induced X-Rays 727
    P.B. Price, B.E. Cooke, G.T. Ewan,
    J.L. Whitton

Depth Profiling with Ion Induced X-Rays 735
    L.C. Feldman, P.J. Silverman

Sensitivity in Trace Element Analysis of Thick
Samples Using Proton Induced X-Rays 747
    F. Folkmann

Review of Trace Analysis by Ion Induced X-Rays 759
    J.F. Ziegler

The Use of Proton Induced X-Rays to Monitor the
Near Surface Composition of Catalysts 773
    J.A. Cairns, A. Lurio, J.F. Ziegler,
    D.F. Holloway, J. Cookson

Application of Proton-Induced X-Ray Emission to
Elemental Analysis of Oligo-Elements in Human
Lymphocytes 785
    M.C. Bonnet, J.P. Thomas, H. Betuel,
    M. Fallavier, S. Marsaud

Elemental Analysis of Biological Samples Using
Deuteron Induced X-Rays and Charged Particles 795
    L. Amtén, L. Glantz, B. Morenius,
    J. Pihl, B. Sundqvist

Suppression of Radioactive Background in Ion
Induced X-Ray Analysis 803
    H. Sobiesiak, D. Heck, F. Käppeler

VIII. NUCLEAR REACTIONS

Depth Profiling of Hydrogen and Helium Isotopes
in Solids by Nuclear Reaction Analysis 811
    J. Bøttiger, S.T. Picraux, N. Rud

Achievable Depth Resolution in Profiling Light
Atoms by Nuclear Reactions 821
    W. Eckstein, R. Behrisch, J. Roth

Depth Profiling of Deuterons in Metals at Large
Implantation Depths Using the Nuclear Reaction Technique . . . 831
    M. Hufschmidt, W. Möller, V. Heintze,
    D. Kamke

Gas Reemission and Blister Formation on Nickel Surfaces
During High Energy Deuteron Bombardment 841
    W. Möller, Th. Pfeiffer, D. Kamke

Unfolding Techiques for the Determination of
Distribution Profiles from Resonance Reaction Gamma-Ray
Yields 851
    D.J. Land, D.G. Simons, J.G. Brennan,
    M.D. Brown

Z_2 Dependence of the Electronic Stopping Power
of 800 keV $^{14}N^+$ Ions in Targets from Carbon
Through Molybdenum 863
    D.G. Simons, D.J. Land, J.G. Brennan,
    M.D. Brown

Sensitivity of Fluorine Detection in Different Matrices
and at Different Depths Through the $^{19}F(p,\alpha\gamma)^{16}O$
Reaction 873
    M.A. Chaudhri, G. Burns, J.L. Rouse,
    B.M. Spicer

Ion Beam Analysis Techniques in Corrosion Science 885
    G. Dearnaley

CONTENTS OF VOLUME 2

Quantitative Measurement of Light Element Profiles in
Thick Corrosion Films on Steels, Using the Harwell
Nuclear Microbeam 901
    C.R. Allen, G. Dearnaley, N.E.W. Hartley

Nuclear Microprobe Analysis of Reactor Materials 913
    J.W. McMillan, T.B. Pierce

Analysis or Microgram Quantities of Aluminum in Germanium . 925
    I.V. Mitchell, W.N. Lennard

Analysis of Fluorine by Nuclear Reactions and Application
to Human Dental Enamel 933
    J. Stroobants, F. Bodart, G. Deconninck,
    G. Demoriter, G. Nicolas

Determination of Nitrogen Depth Distributions in Cereals
Using the $^{14}N(d,p_0)^{15}N$ Reaction 945
    B. Sundqvist, L. Gönczi, I. Koersner,
    R. Bergman, U. Lindh

Experimental Measurements, Mathematical Analysis and Partial
Deconvolution of the Asymmetrical Response of Surface
Barrier Detectors to MeV ^{4}He, ^{12}C, ^{14}N, ^{16}O Ions 953
    G. Amsel, C. Cohen, A. L'Hoir

Experimental Study of the Stopping Power and Energy
Straggling of MeV ^{4}He, ^{12}C, ^{14}N, ^{16}O Ions in Amorphous
Aluminium Oxide 965
    A. L'Hoir, C. Cohen, G. Amsel

AUTHOR INDEX 977

SUBJECT INDEX 981

Contents of Volume 1

I. ENERGY LOSS AND STRAGGLING

The Treatment of Energy-Loss Fluctuations in
Surface-Layer Analysis by Ion Beams 3
    J.W. Butler

Evidence of Solid State Effects in the Energy Loss
of ^{4}He Ions in Matter 15
    J.F. Ziegler, W.K. Chu, J.S.-Y. Feng

Empirical Stopping Cross Sections for ^{4}He Ions 29
    R.A. Baragiola, J.C. Eckardt

Determination of Stopping Cross Sections by Rutherford
Backscattering 33
    B.M.U. Scherzer, P. Børgesen, M.-A. Nicolet, J.W. Mayer

Depth Profiling of Implanted ^{3}He in Solids by Nuclear
Reaction and Rutherford Backscattering 47
    J. Roth, R. Behrisch, W. Eckstein, B.M.U. Scherzer

Energy Loss Straggling of Protons in Thick Absorbers . . . 55
    B.H. Armitage, P.N. Trehan

Energy Dependence of Proton Straggling in Carbon 65
    D. Olmos, F. Aldape, J. Calvillo, A. Chi, S. Romero,
    J. Rickards

Energy Straggling of ^{4}He Ions in Al and Cu in the
Backscattering Geometry 75
    M. Luomajärvi, A. Fontell, M. Bister

Energy Spreading Calculations and Consequences 87
    G. Deconninck, Y. Fouilhe

Analysis of Nuclear Scattering Cross Sections
by Means of Molecular Ions 99
    G. Thieme

II. BACKSCATTERING ANALYSIS

Determining Concentration vs. Depth Profiles from
Backscattering Spectra without Using Energy Loss Values . . 111
    R.F. Lever

Comparative Analysis of Surface Layers by Backscattering and by Auger Electron Spectroscopy 125
    J.K. Howard, W.K. Chu, R.F. Lever

Analyzing the Formation of a Thin Compound Film
by Taking Moments on Backscattering Spectra 149
    R.F. Lever, W.K. Chu

Computer Analysis of Nuclear Backscattering 163
    J.F. Ziegler, R.F. Lever, J.K. Hirvonen

Some Practical Aspects of Depth Profiling Gases in
Metals by Proton Backscattering: Application to
Helium and Hydrogen Isotopes 185
    R.S. Blewer

Depth Profiling of Deuterium and Helium in Metals by
Elastic Proton Scattering: A Measurement of the En-
hancement of the Elastic Scattering Cross Section over
Rutherford Scattering Cross Section 201
    R.A. Langley

Near-Surface Investigation by Backscattering of N^+ Ions
and Grazing Angle Beam Incidence 211
    W. Pabst

The Application of Low Angle Rutherford Backscattering
to Surface Layer Analysis 223
    J.S. Williams

CONTENTS OF VOLUME 1

Measurement of Projected and Lateral Range Parameters for Low Energy Heavy Ions in Silicon by Rutherford Backscattering 235
    W.A. Grant, J.S. Williams, D. Dodds

Range Parameters of Heavy Ions in Silicon and Germanium with Reduced Energies from $0.001 \leq \varepsilon \leq 10$ 245
    H. Oetzmann, A. Feuerstein, H. Grahmann, S. Kalbitzer

On Problems of Resolving Power in Rutherford Backscattering 255
    J. Schou, S. Steenstrup, A. Johansen, L.T. Chadderton

Studies of Surface Contaminations, Composition and Formation of Superconducting Layers of V, Nb_3Sn and of Tunneling Elements Using High Energetic Protons Combined with Heavy Ions 265
    P. Müller, G. Ischenko, F. Gabler

Determination of Implanted Carbon Profiles in NbC Single Crystals from Random Backscattering Spectra 273
    K.G. Langguth, G. Linker, J. Geerk

Pore Size from Resonant Charged Particle Backscattering . . 281
    C.D. Mackenzie, B.H. Armitage

Measurement of Thermal Diffusion Profiles of Gold Electrodes on Amorphous Semiconductor Devices by Deconvolution of Ion Backscattering Spectra 293
    J.P. Thomas, S. Marsaud, M. Fallavier

Enhanced Sensitivity of Oxygen Detection by the 3.05 MeV (α,α) Elastic Scattering 303
    G. Mezey, J. Gyulai, T. Nagy, E. Kotai, A. Manuaba

Progress Report on the Backscattering Standards Project (Abstract) 313
    J.E.E. Baglin

III. APPLICATIONS OF BACKSCATTERING AND COMBINED TECHNIQUES

Ion Beam Studies of Thin Films and Interfacial Reactions . . 317
    J.M. Poate

Studies of Tantalum Nitride Thin Film Resistors 337
    R.A. Langley

Investigation of CVD Tungsten Metallizations on
Silicon by Backscattering 353
    P. Eichinger, H. Sauermann, M. Wahl

Ion Beam Analysis of Aluminium Profiles in Heteroepitaxial
$Ga_{1-x}Al_x$As-Layers 363
    P. Bayerl, W. Pabst, P. Eichinger

Analysis of $Ga_{1-x}Al_x$As-GaAs Heteroepitaxial Layers
by Proton Backscattering 375
    K. Gamo, T. Inada, I. Samid, C.P. Lee, J.W. Mayer

Interdiffusion Kinetics in Thin Film Couples 385
    J.E.E. Baglin, F.M. d'Heurle

Backscattering and T.E.M. Studies of Grain Boundary
Diffusion in Thin Metal Films 397
    S.U. Campisano, E. Costanzo, G. Foti, E. Rimini

The Analysis of Nickel and Chromium Migration
Through Gold Layers 407
    A. Barcz, A. Turos, L. Wieluński

Applications of Ion Beam Analysis to Insulators 415
    J.A. Borders, G.W. Arnold

Lithium Ion Backscattering as a Novel Tool for the Characterization of Oxidized Phases of Aluminum Obtained
from Industrial Anodization Procedures 425
    J.P. Thomas, A. Cachard, M. Fallavier,
    J. Tardy, S. Marsaud

Investigation of an Amino Suger-Like Compound from the Cell
Walls of Bacteria Using Backscattering of MeV Particles . . . 437
    T.G. Finstad, T. Olsen, R. Reistad

IV. EQUIPMENT

Versatile Apparatus for Real-Time Profiling of Interacting
Thin Films Deposited in Situ 447
    J.E.E. Baglin, W.N. Hammer

Application of a High-Resolution Magnetic Spectrometer
to Near-Surface Materials Analysis 457
    J.K. Hirvonen, G.K. Hubler

Rutherford Backscattering Analysis with Very High Depth
Resolution Using an Electrostatic Analysing System 471
    A. Feuerstein, H. Grahmann, S. Kalbitzer,
    H. Oetzmann

An Apparatus for the Study of Ion and Photon Emission
from Ion Bombarded Surfaces: I. Some Preliminary Results . . 483
    R.J. MacDonald, A.R. Bayly, P.J. Martin

AUTHOR INDEX xix

SUBJECT INDEX xxii

Lattice Location and Dechanneling

in Disordered Crystals

ANALYSIS PROBLEMS IN LATTICE LOCATION STUDIES

Heinz-Dieter Carstanjen

Sektion Physik, University of Munich

D8000 München, Amalienstr. 54, Germany

1 INTRODUCTION

In the last ten years the channeling of fast ions has found an increasing application in the lattice location of impurities in crystals, in particular at low concentrations. Up to now more than a hundred systems have been investigated with this technique; a detailed list may be found in the review article by Picraux /1/. At the moment the most extensive investigations have been made for impurities in the metals Al, Fe and Cu and the semiconductors Si and Ge. The increasing popularity of the method may be due to the fact that in simple cases an assignment of the lattice location of an impurity in question can be achieved by elementary considerations only. In complicated cases however, such as in cases of simultaneous occupation of more than one interstitial site by the same impurity detailed profile measurements and flux distribution calculations are necessary. The method was not sufficiently elaborated until very recently and some of the interpretations given can not be considered reliable. In certain cases - in particular in ion implantation - the method has been used primarily to determine the substitutional fraction of the implanted atoms. The method has improved and even complicated systems, e.g. Br in Fe /7/, were lately analysed successfully.

In Section 2 the underlying principles of the method are outlined, in Section 3 problems connected with the interpretation of experimental data are discussed and finally, in Section 4, various experimental techniques are presented. There, as an example (in 4.2), the location of an impurity (O in Nb) is shown in detail.

Comprehensive reviews are given by Davies /2/, Mayer et al. /3/, also including impurities in semiconductors, and de Waard et al. /4/, including hyperfine interaction. Reviews on channeling can be found as a book edited by Morgan /5/ which includes theory and application and lately by Gemmell /6/.

2 PRINCIPLES

When a well collimated beam of fast ions enters a single crystal along a low index crystal axis the penetrating ions experience strong steering by the (repulsive) potentials of the atom rows along this axis (axial channeling). Fig. 1 shows the projection of calculated trajectories of channeled ions on the transverse plane of the incidence of the projectile. The motion of an ion in the transverse plane is at parallel incidence restricted to the area inside that equipotential contour line $U(\vec{r})$ at which the ion started at the crystal surface. Therefore, all ions can pass through the channel center, however only the projectiles with large $U(\vec{r})$ traverse the region inside the atom rows. The consequence is an enhanced ion flux density in the channel center (flux peaking) and a reduced flux density at the atom rows. In Fig. 2 this is shown by contours of constant relative flux density (relative to a uniform distribution). The flux density is strongly peaked in the channel center at M and almost zero inside the atom rows.

If an impurity is located at the channel center, it experiences a high flux density. If it is located inside the atom rows which we call in the following 'substitutional' it experiences a low flux density. When measuring the yield of a reaction between the incident projectile and the impurity (e.g., production of X-rays, of back-scattered ions or of nuclear reaction products) a high yield indi-

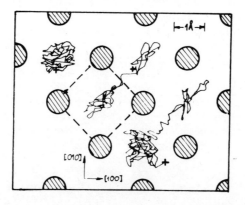

Fig. 1 Projections of some [001] channel trajectories onto the (001) face of bcc "Cu". The starting points of the 5 keV Cu ions are marked by crosses. The hatched circles represent the [001] atom rows which form the channel (dashed line). After Robinson and Oen /8/.

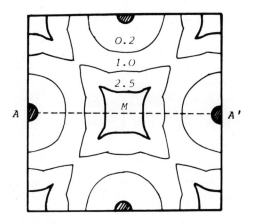

Fig. 2 Contours of constant relative deuteron flux in the {100} transverse plane. The hatched circles represent the <100> atom rows which form the <100> channel. 300 keV deuterons in niobium, depth = 4125 Å. After Carstanjen and Sizmann /9/.

cates impurities in the channel center and a low yield indicates 'substitutional' impurities. When the crystal is tilted towards the ion beam in a 'random' direction the flux distribution changes into the usual uniform flux distribution. Accordingly the reaction yield from impurities in the channel center decreases whereas the reaction yield from 'substitutional' impurities increases. Thus monitoring the reaction yield during such an angular scan through a low index crystal axis, the measured yield profile provides the information about the lattice location of the impurity species. Similar considerations hold for planar channeling. In this case the channel 'center' is the center plane between two neighbouring atom planes. The flux density peaks along the center plane, if the ion beam is incident parallel to the lattice plane.

Because of the lattice symmetry operations, a certain interstitial site can have several different projections onto the transverse plane. In such a case it is necessary to measure the reaction yield profiles for angular scans across various low index axes or planes in order to obtain sufficient information about the site for an unambiguous analysis. A proper choice of the axial or planar channel (channels with lower symmetry are the most favourable) often reduces the number of projections of a particular site to just one (e.g., only in the channel center). In such a case an interpretation can be given rather easily (cf. Fig. 9 and Section 3). In case of simultaneous occupation of more than one not equivalent site the analysis requires a comparison with yield profiles calculated for each of the various sites exclusively occupied.

3 ELEMENTS OF INTERPRETATION

In this Section various problems connected with the interpretation of experimental yield profiles are discussed. The reaction yield Y emitted from an volume element ΔV at ion incidence with angle ψ and energy E_o is given by:

(1) $\quad Y(\psi,E_o) = \int\limits_{\Delta V,\Omega_o,E_o,V'} W(\vec{r}')\ p_{\vec{r}'}(E)\ \sigma(E,\theta)\ b(\vec{r}-\vec{r}')\ f(\vec{r})\ d\vec{r}'\,dE\,d\Omega\,d\vec{r}$

Here $W(\vec{r})$ is the ion flux density, $p_{\vec{r}}(E)$ the probability to find an ion with energy E at position $\vec{r}$. In both functions enters the dependence on the incidence angle. $\sigma(E,\theta)$ is the cross section of the reaction, depending on the ion energy E and the emission angle θ. $b(\vec{r}-\vec{r}')$ describes the impact parameter dependence of the reaction. If there is a considerable contribution from impact parameters greater than the Thomas Fermi screening radius, a reaction is not appropiate for atom location. $f(\vec{r})$ gives the distribution of the impurities inthe crystal, depending on their concentration profile, on the occupied interstitial sites and on the probability distribution around the sites (due to thermal vibrations). It should be noted that Eq. (1) does not include absorption or energy loss of the emitted signal. Further information on yield calculations, in particular on calculations of $W(\vec{r})$ and $p_{\vec{r}}(E)$ by Monte Carlo methods or analytical methods is given in Refs. /7,9,10/.

If the signal is emitted from a small depth interval near the surface the angular yield profile Y from a certain position in the transverse plane is approximately proportional to the angular profile of the flux density $W(\vec{r})$ at this position. Hence, for a first interpretation, it is sufficient to know the angular dependence of the flux density $W(\vec{r})$. In the following the yield profiles which are produced by impurity positions at certain lattice positions will be discussed and compared with flux density profiles from these positions.

3.1 Dips

Impurities at substitutional sites and, for a given incidence direction of the projectiles, at sites with projections which lie inside atom rows or planes (both will be called 'substitutional' positions in the following) show dips in the angular yield profile. If the thermal vibration amplitudes of the host and the impurity atoms are comparable, then the _relative_ (normalised to 1 at random) yield profiles from the _impurities and_ from the host atoms coincide, provided both signals are taken from the same depth. For illustration Fig. 3 shows backscattering profiles in 2 at. % Au in Cu from the Au impurities and from the Cu host lattice /11/. The Au atoms are to 100 % located at substitutional sites, since both profiles coincide exactly (note that in this experiment, the Au atoms were already known to occupy substitutional sites).

ANALYSIS PROBLEMS IN LATTICE LOCATION STUDIES

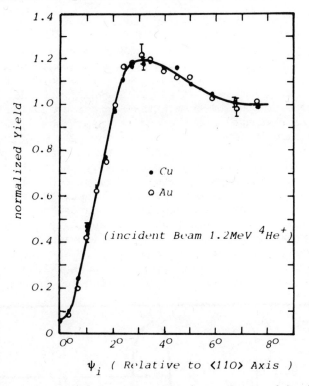

Fig. 3 Channeling angular distributions observed by backscattering from Cu-2 at. % Au for 1.2-MeV He incident along the <110> axis. After Alexander and Poate /11/.

The host yield profile from a small depth interval is also a measure of the relative flux density profile at 'substitutional' positions in this depth. It is, therefore, useful to measure the signal from the host lattice which can be obtained from the backscattering yield during crystal orientation. It also allows to subtract the contributions from 'substitutional' positions in a mixed occupation (e.g., the octahedral positions in bcc crystals along <100>, Fig. 9). When only a part of the impurities is 'substitutional' and the rest is located on random positions (i.e., they give on the average the random yield) then the observed dips have the same widths as the host lattice dips, but a different minimum yield y. This is found for instance when a fraction of impurities is located on low symmetry sites (cf. Section 3.3) or when precipitation occurs. In such a case the 'substitutional' fraction f_s can be determined from the minimum yields of the host atoms y_h and the impurities y_i:

(2) $$f_s = (1 - y_i)/(1 - y_h)$$

Impurities slightly displaced from 'substitutional' positions generally give narrower dips than the host atoms and higher minimum yields. A detailed discussion of this is given in Refs. /1/ and /12/.

3.2 Center Peaks

Impurities located at or near a channel center yield peaks at perfect alignment (cf. Fig. 4). However, peaks are also obtained from atoms located at saddle points with low (axial) potential (cf. Fig. 9). Experimentally the peak height is strongly dependent on the angular divergence of the ion beam, the surface condition of the crystal (oxide layer, damage) and the crystal depth where the signal comes from. A quantitative discussion of the peak height has been given by Van Vliet /13/. These factors are particularly important, since the half widths of the peaks are typically one third or less of the half widths of the dips from the host atoms. Mostly, the experimental conditions are not known exactly. It becomes even more complicated when also other positions in the channel are occupied like 'substitutional' positions (they would essentially lower the peak height) or off center positions (see the following section. If they are not resolved, they lead to a broadening of the center peak). Therefore, when measuring a relatively low peak (≤1.3), it is difficult (without any other evidence) to distinguish between a position

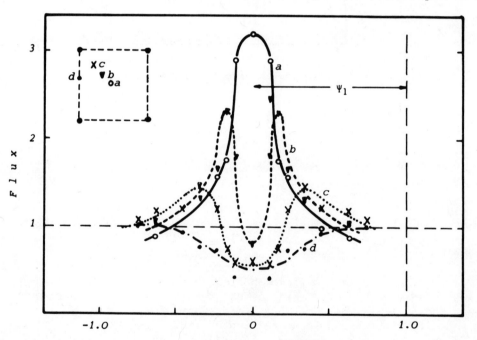

Fig. 4 Calculated particle flux as a function of the angle of incidence ψ_o. The calculations are for 1.5-MeV alpha particles in the <100> axial channel of Cu. The four curves correspond to points $\vec{r}$ in the transverse plane as shown in the insert. The angle of tilt ψ_o is such that the beam is incident in a (110) planar direction. After Morgan and Van Vliet /14/.

at a channel center, near the channel center or at a saddle point (compare the situation for octahedral sites in Nb along <111> and <110> in Fig. 9). Also it is not simple to determine the fraction of atoms which occupy such positions. However, a sharp and high peak (>1.5) indicates the occupation of positions at or very close to the channel center (cf. Fig. 9).

3.3 Off Center Peaks

An important element of the location analysis are off center peaks. They occur when off center positions in an axial or planar channel are occupied (cf. Fig. 4). It should be noted that positions close to the channel center give narrow peaks, positions further away give broad off center peaks. This is due to the different gradients of the planar or axial potential at these positions. The angular position ψ_i of such a peak allows to determine the potential contour where the particular site is located by observing Equ. (3):

$$(3) \quad E \psi_i^2 = U(r_i) - U_c$$

Here E is the ion energy, $U(r)$ the potential of the axial or planar channel and U_c the potential at the center. For a planar channel the contours are parallel planes with distance r_i from the center. For an axial channel they constitute cylinders. With the symmetry of the crystal structure additional information on the lattice site can be obtained. For instance an off center peak in a {100} planar channel in a simple cubic structure indicates that the particular site lies on the intersection of the three equivalent equipotential planes. It should be noted that the narrow peaks near 0° do not allow a better location of the equipotential contours because of the flat potential near the channel center. A most successful application of off center peaks has been made by Alexander at al. (/7/, cf. Fig.5).

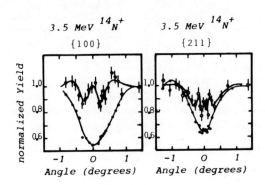

Fig. 5 Angular scans across {100} and {211} planes in Br-implanted Fe. ^{14}N backscattering yields from Fe (lower curves) and from Br (upper curves) are shown. After Alexander et al. /7/.

For a channel with high symmetry pronounced off center peaks are obtained only, if the impurity atoms lie on or close to a site with high symmetry. For off center positions with low symmetry in a high symmetry channel there exist many equivalent positions which give on the average almost the random yield. There is often a strong interaction between impurity atoms and defects which may shift the impurity atoms towards positions of lower symmetry. This interaction between defects induced by the irradiation during the location experiment and the impurity atoms is probably the reason why at low impurity concentrations often a considerable change of the yield from 'substitutional' impurities (e.g. an increase of the minimum yield) is observed while the yield from the host lattice remains almost unchanged.

3.4 Depth Dependence

The flux density distribution is depth dependent. At high ion energies flux peaks and dips are damped with increasing penetration depth due to an increase of transverse energy of the ions. This is known for dips from backscattering measurements (e.g. /15/) and for peaks in model calculations (e.g. /13/,/16/). Therefore, for atom location preferentially only the signals produced near the surface should be used. For low energies, however, flux peaking may be enhanced with increasing depth due to electronic stopping (e.g./16/, /17/). Compare also the flux peaking calculations for 19 keV Ne^+ in Cu with a flux peak up to 25 times the random flux /18/. In moderate depths the flux density as well at 'substitutional' positions as in the channel center oscillates (cf. /19/,/13/). The oscillation lengths for flux peaking vary between a few hundred and a few thousand Å with increasing energy. In general the depth interval where the signal is taken from is large compared with the oscillation length and an average flux density is seen. However, for low energy heavy ions the projected range is only a few times the oscillation length. This is important in particular in heavy ion X-ray excitation.

4 EXPERIMENTAL TECHNIQUES

The impurity concentrations in location experiments are typically ≥ 0.01 at. % for backscattering and nuclear reactions and ≥ 0.1 at. % for X-rays as the observed reaction yield. The lower limit is set by a sufficient high number of counts (≥ 1000), a minimum of radiation damage, and the background signal to impurity signal ratio. An estimate of the number of counts is obtained from the number of counts at random incidence N_r from an energy interval ΔE (Equ. (4)):

(4) $$N_r = N_i \int_{\Delta E} C(x(E))\, \sigma(E) / (dE/dx)\, dE$$

Here N_i is the number of incident ions, $C(x)$ the depth dependent concentration profile of the impurity, $\sigma(E)$ the energy dependent reaction cross section, dE/dx the stopping power of the material. Again no absorption of the emitted signal is included. The separation of the impurity signal from the host signal is a main problem in atom location experiments and is usually done by energy discrimination. Depth resolution is not a serious problem (except for X-rays), since the signal may be taken from depth intervals up to 1 µm (i.e. as long as channeling effects are sufficiently pronounced). However, contributions from the crystal surface (e.g., O-, N-, C-contaminations, amorphous or enriched surface layers) should be eliminated.

Ion implantation is a versatile method of introducing impurities. It usually gives a well defined depth interval which contributes to the signal. This is of importance when reactions with relatively small energy dependence are used. By implantation impurity positions can be obtained which by diffusion or by growing from the melt are usually not obtained by simple means. The disadvantage, however, is radiation damage which then influences location experiments. An annealing of the defects can lead to precipitation of the impurities.

4.1 Backscattering

Ion backscattering is the most commonly used technique in atom location. The cross sections are of the order of 10 barn. Because of the typically low impurity concentrations the method is restricted to high mass impurities in low mass crystals, since then an energy discrimination due to the different recoil energies is easily possible. In addition this favors the yield from the impurities, since Rutherford scattering is proportional to the square of the atomic number of the impurity atom. The usual projectile ions are He^+, C^+, N^+, Ne^+ with energies from 1 to 3 MeV. Heavier ions have the advantage of giving a better energy discrimination, however with the disadvantage of causing more radiation damage. A detailed discussion on this subject is found in Ref. /3/. Background problems may arise from pulse pile-up (an anti pile-up device is given in Ref. /7/) and other impurities present (e.g., at the surface or inside the sample material).

Since backscattering is a well known technique, only one application for defect investigations shall be given here. Fig. 6 shows typical backscattering spectra of 1 MeV He^+ from 0.08 at. % Ag in Al for random incidence and aligned incidence along <110> for annealed and radiation damaged crystals /20/. The rather high Ag peak - even in the random spectrum - arises from an Ag enrichment at the surface due to electropolishing. The increase in the minimum Ag yield during irradiation at 60 K is caused by the trapping of Al atoms by the Ag impurities which are thought to form a kind of <100> dumbbell. In an isochronal annealing experiment almost complete recovery of the fraction of displaced Ag atoms (obtained from the minimum yields) is found (cf. Fig. 7) with an annealing stage at ~180 K.

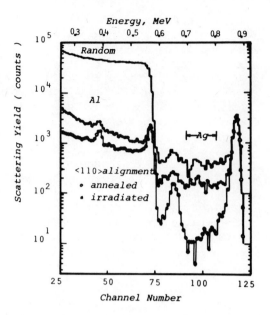

Fig. 6 Typical energy spectra of backscattered 1 MeV He$^+$ ions from Al crystals at 35 K, showing the random spectrum and the <110> aligned spectra for annealed and irradiated crystals. Al-0.08 at. % Ag, irradiation dose 10^{15} 0.3 MeV He$^+$ ions per cm^2 at 60 K. /20/.

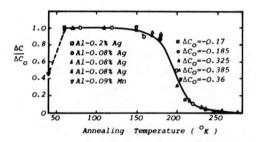

Fig. 7 Recovery of the concentration of displaced impurity atoms during isochronal annealing (10 min. pulses). 1 MeV He$^+$ backscattering at 35 K along <110> (Ag) and <111> (Mn). After Swanson et al. /20/

4.2 Nuclear Reactions

The cross sections of nuclear reactions required in atom location experiments are > 1 mbarn/steradian in order to give sufficient high count rates at a minimum of radiation damage (maximum dose: ≤ 10^{15} ions/cm^2). This reduces the projectile ions used in the analysing beam to H$^+$, D$^+$, ^{3}He$^+$ and the impurities to be investigated to low Z atoms. In Table 1 the most commonly used nuclear reactions

ANALYSIS PROBLEMS IN LATTICE LOCATION STUDIES

Table 1

reaction	Q (MeV)	reaction	Q (MeV)	reaction	Q (MeV)
$D(d,p)T$	4.032	$^{11}B(p,\alpha)^8Be$	8.582	$^{15}N(p,\alpha)^{12}C$	4.964
$^3He(d,p)^4He$	18.3	$^{12}C(d,p)^{13}C$	2.719	$^{16}O(d,p_1)^{17}O$	1.048
$^7Li(p,\alpha)^4He$	17.347	$^{14}N(d,\alpha_0)^{12}C$	13.579	$^{18}O(p,\alpha)^{15}N$	3.97
$^9Be(d,p)^{10}Be$	4.585	$^{14}N(d,\alpha_1)^{12}C$	9.146		

and their Q-values are listed. A detailed study of nuclear reactions is found in a review article by Amsel et al. /21/. The energies of the incident ions range from 300 keV to 2 MeV. Higher energies are not favorable because of the small critical angles for channeling and the increasing number of background reactions. An estimate for the widths of axial channeling dips is obtained from Lindhard's critical angle $\psi_1 = (2Z_1 Z_2 e^2/(d\,E))^{1/2}$ where d is the spacing of the atoms in the atom rows of this axis /22/. For 1 MeV protons ψ_1 is of the order of 1° which makes a beam collimation $< 0.05^\circ$ necessary. For detection an annular surface barrier detector in close distance is used. It is usually shielded by a thin mylar or aluminum window against backscattered projectile ions which may lead to pile up and damage of the detector. The shielding already rules out low Q-value (p,α) or (d,α) reactions because of the large stopping power for alpha particles. An estimate for the energy E_3 of the emitted particles with mass m_3 is obtained for 90° emission from:

(5) $\quad E_3 = Q\, m_4/(m_3 + m_4) + E_1\, (m_4 - m_1)/(m_3 + m_4)$

Here m_1 and E_1 are the mass and energy of the incident ion, m_4 the mass of the remaining nucleus. According to Eq. (5) the best energy discrimination is obtained for ion energies as low as possible. A depth analysis can be made based on the energy loss which the emitted particles experience when traversing the crystal. This is particularly necessary when the impurity atoms contribute also to surface contamination or when the reaction cross section does not vanish for large penetration depths.

In the following we illustrate a case where a nuclear reaction is used to locate O in Nb. Two reactions are possible: $^{18}O(p,\alpha)^{15}N$ with 700 keV protons as applied by a Russian group /23/ with about 0.1 at.% ^{18}O in Nb and $^{16}O(d,p_1)^{17}O$ with 1.1 MeV deuterons as applied by us /24/ with about 0.05 at.% ^{16}O in Nb. Fig.8 shows energy spectra of alpha particles emitted from $^{18}O(p,\alpha)^{15}N$ for <100> and random incidence as obtained by the Russian group.

For interstitials in bcc metals two positions are likely: tetrahedral sites (coordinates: 1/2,1/4,0) and octahedral sites (coordinates: 1/2,0,0 and 1/2,1/2,0). In Fig. 9a and Fig. 9b their projections along various crystal axes and planes are shown. In <111> and <110> for both interstitial sites peaks at perfect alignment should appear

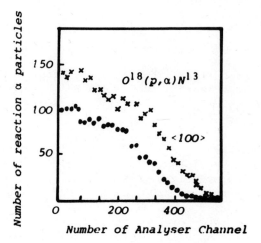

Fig. 8 Energy spectra of α particles from $^{18}O(p,\alpha)^{15}N$ for <100> (crosses) and random incidence of 700 keV protons on Nb containing about 0.1 at. % ^{18}O. After Matyash et al. /23/.

and, in fact, have been found (Fig. 9d and 9e). In these directions a decision with simple arguments is not possible. In <100> octahedral sites show 'substitutional' positions (because of the symmetry of the site with a total weight 4) and positions in the channel center (with weight 8). They yield a sharp maximum in comparison to a broad maximum (with poor resolution) expected for tetrahedral positions. More decisive yield profile measurements are required for a reliable location. They are provided by planar scans. In an angular scan across a {100} plane octahedral sites are shadowed by the host atoms ('substitutional' positions). This produces a dip comparable with the dip obtained from the host lattice. For tetrahedral sites a small peak in a minimum is expected which, however, is hardly resolved. In fact experimentally a simple dip has been obtained (Fig. 9f and Fig. 10c). An unambiguous decision is obtained from a scan across a {111} plane. This plane, however, is a very weak steering plane for channeled projectiles. Here, tetrahedral sites have positions only in planar center, octahedral sites have only 'substitutional' positions. The experiment shows a clear dip for oxygen in Nb. For comparison a yield profile from a D(d,p)T reaction with about 2 at. % D in Nb is shown in Fig. 9h. D in Nb is known to occupy tetrahedral sites. The experiment shows a clear peak which however is rather small due to the long depth range where the signal has been taken from. It should be noted that also a series of other (middle Z) impurities are located on octahedral sites in bcc metals, e.g., C in Fe and B in W (cf. /25/, /26/ and Fig. 10). A comparison with the host lattice yield shows that almost 100 % of the impurities occupy octahedral sites as does O in Nb (Fig. 10c).

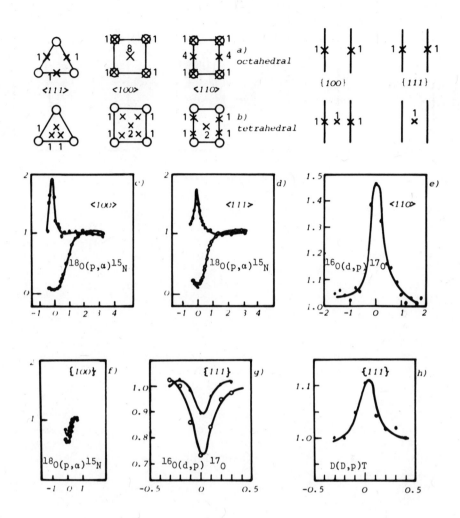

Fig. 9 Oxygen in niobium: Projections of a) octahedral and b) tetrahedral sites on the transverse planes of <100>, <111>, <110>, {100} and {111}. Circles: host lattice sites, crosses: interstitial sites. The numbers give the weight of each projection. $^{18}O(p,\alpha)^{15}N$ yield profiles are shown for angular scans across c) <100>, d) <111> and f) {100}. Impurity yield: upper curves, host lattice yield: lower curves. After Matyash et al. /23/. $^{16}O(d,p)^{17}O$ yield profiles are shown for angular scans across e) <110> and g) {111}. Impurity yield: dots, host lattice yield from near the surface: circles. h) For comparison an angular scan across {111} is shown for 2 at.% deuterium diffused into Nb from D(d,p)T. /24/.

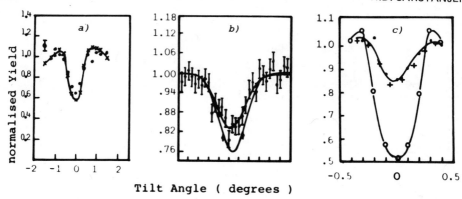

Fig. 10 Angular scans across {100} planar channels in bcc metals.
a) ^{11}B implanted in W. Yields are shown from ^{11}B(p,α)^{8}Be (dots) and
W(p,p) (crosses). After Andersen et al./26/. b) ^{12}C implanted in Fe.
Yields are shown from ^{12}C(d,p)^{13}C (upper line) and Fe(d,d) (lower
line). After Feldman et al./25/. c) ^{16}O diffused in Nb. Yields are
shown from ^{16}O(d,p)^{17}O (dots) and Nb(d,d) (circles: near surface,
crosses: depth corresponding to ^{16}O(d,p)^{17}O) /24/.

The two preceeding methods are quasi unified in (α,α')reactions.
This reaction recently was proposed by a Russian group for locating
C and O in Si. A 22 MeV alpha beam is incident on a thin (~10μm)
silicon crystal. The reactions have cross sections of the order of
barns whereas Rutherford backscattering is strongly reduced due to
the high alpha energy. The alpha yield from Si, C and O consists in
three well separated peaks /27/. Due to the high energy of the alphas the target thickness is of no problem. The main disadvantage of
this method is the very small channeling angle ($\psi_1 < 0.2°$) which requires high precision beam alignment and crystal orientation.

4.3 X-Ray Excitation

The excitation of characteristic X-rays has not been used very
often, although it can be applied quite generally. A main problem
at the typically low impurity concentrations has been how to separate the impurity X-rays from the host lattice X-rays and from the
background (mainly from bremsstrahlung due to secondary electrons).
By the use of high resolution Li-drifted Si-detectors the separation from the host lattice X-rays could be improved. With X-ray excitation no depth analysis can be made except by the choice of the
ion energy and an appropiate implantation profile. In particular
surface effects can not be eliminated. In addition (in particular
for M X-rays) large impact parameters (about the size of the Thomas-Fermi screening radius) may contribute to the yield /28/.

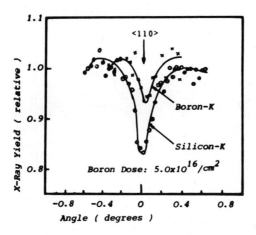

Fig. 11 B-K and Si-K X-ray yield from an angular scan across a <110> axial channel of B-implanted Si with 100 keV protons. After Cairns et al. /29/.

For Coulombic excitation usually protons with energies up to 500 keV are used. One problem connected with Coulombic excitation is that the X-rays are obtained from a relatively large depth interval. An other point is the self absorption by the target.
Fig. 11 shows the B-K and Si-K X-ray yield profiles obtained from B-implanted ($5 \cdot 10^{16}$ B/cm^2) silicon crystals with 100 keV protons. From the dip along <110> it is concluded that boron tends to be located along <110> rows /29/.

The use of heavy ions for producing characteristic X-rays through Pauli-excitation up to a few hundred keV (cf. /30/) has several advantages compared with Coulomb-excitation. (1) The cross section increases rapidly once a certain threshould enrgy has been surpassed and reaches a saturation value at higher energies (cf. /30/). Thus by a proper choice of the projectile energy the depth interval where the X-rays are produced can be limited to the near surface region. (2) The cross sections are orders of magnitude larger than for protons, in particular when the incident analysing projectile ion is identical with the impurity atom investigated (cf. /31/). When the same ion is used the problem arises how to separate the X-rays due to the excitation of the incident ions from the excitation by the host lattice atoms. This can, however, be corrected by measuring the X-ray yield emitted from the crystal prior to implantation.

4.4 Blocking

In various location experiments radioactive impurities have been used, e.g., radon /32/. The alpha particles emitted in the radioactive decay show blocking pattern, if the radon is located at a

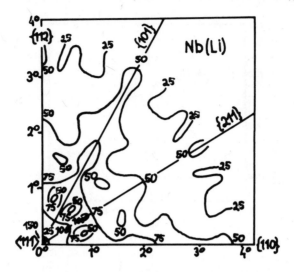

Fig. 12 Blocking pattern of Nb(Li) around around a <111> direction. ^{6}Li was implanted to a concentration of a few at. %. After Biersack et al. /33/.

substitutional site. Recently the emission of light, charged particles from (n,p), (n,t) or (n,α) reactions between thermal neutrons and impurities is used /33/. Thermal neutrons have the advantage of usually not causing radiation damage; the damage produced by the small number of the emitted ions is negligible. Suitable elements with cross sections up to kbarns are: ^{3}He, ^{6}Li, ^{10}B, ^{14}N, ^{17}O. Fig. 12 gives an example. Here a Nb crystal is implanted with 220 keV ^{6}Li up to a maximum concentration of a few at. % and is exposed to reactor neutrons. The emitted alpha particles (2 MeV) of the ^{6}Li(n,α)T reaction are recorded on plastic films as recorders. The figure shows contour lines of constant particle density in the vicinity of the <111> crystal axis. The peak along <111> indicates that the Li atoms are located on interstitial sites; however no definite location is so far made.

4.5 Double Alignment

Finally the double alignment technique - a combination of channeling and blocking - is mentioned, since it is so far the only channeling method which permits the location of low concentration self interstitials (0.1 at. %). In double alignment the ion beam is incident along a fixed low index channeling direction and the ions scattered by the crystal are recorded along other low index blocking directions or planes. In each separate case, blocking or chan-

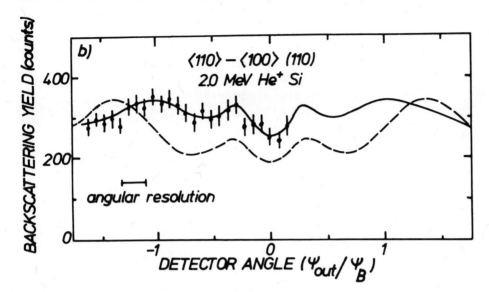

Fig. 13 Backscattering of 2 MeV He$^+$ ions from radiation damaged silicon in double alignment. Ion incidence <110>, direction of observation <010> with detector tilt in {110}. After Carstanjen et al. /34/.

neling, the signal from the lattice is reduced by about 30 - 50. Compared to random incidence defect concentrations of 2 - 3 at. % (without fluxpeaking) and of 0.4 - 1 (including flux enhancement of a factor of 3 - 5) can at most be detected. Double alignment, however, gives (again due to flux peaking) an other reduction by a factor of 3 - 5. This increases the sensitivity to approximately 0.1 at.%.

In Fig. 13 a yield profile of 2 MeV He ions scattered from a radiation damaged Si crystal in a double align ment arrangement is shown /34/ which was obtained on a nuclear track film. The incidence direction is [110], a detector scan is made across a [100] direction in a (1$\bar{1}$0) plane. The profile shows pronounced shoulders which agree well with the yield profile calculated /35/ for $6 \cdot 10^{-3}$ defects per lattice atom and an occupation of 60 % bond centered and 40 % hexahedral positions.

There are two disadvantages of both the double align ment and the blocking technique: (1) The measuring times are by a factor of 100 - 1000 longer than in channeling experiments. (2) So far a good energy analysis of the emitted particles is not possible which is necessary for eliminating suface effects. Also the evaluation of the obtained pattern with electronic devices is rather time consuming. Both methods require two dimensional position sensitive detectors.

Acknowledgement

The author would like to thank Prof. R. Sizmann for many helpful discussions, Mr. W. Schölkopf for his assistance in preparing this publication and the members of the accelerator laboratory of GSF, Neuherberg, for their kind cooperation.

References

1. S. T. Picraux, New Uses of Low-Energy Accelerators (Plenum Press, N. Y.) in press.
2. J. A. Davies, Channeling: Theory, Observation and Applications, ed. by D. V. Morgan (John Wiley, 1973).
3. J. W. Mayer, L. Eriksson, and J. A. Davies, Ion Implantation in Semiconductors (Academic Press, N. Y., 1970).
4. H. de Waard and L. C. Feldman, Applications of Ion Beams to Metals, ed. by S. T. Picraux, E. P. EerNisse, and F. L. Vook (Plenum Press, N. Y., 1974), p. 317.
5. Channeling: Theory, Observation and Applications, ed. by D. V. Morgan (John Wiley, 1973).
6. D. S. Gemmell, Rev. Mod. Phys. 46, 129 (1974).
7. R. B. Alexander, P. T. Callaghan, and J. M. Poate, Phys. Rev. B9, 3022 (1974).
8. M. T. Robinson and O. S. Oen, Phys. Rev. 132, 2385 (1963).
9. H. D. Carstanjen and R. Sizmann, Radiation Effects 12, 225 (1972).
10. R. B. Alexander, G. Dearnaley, D. V. Morgan, and J. M. Poate, Phys. Lett. A32, 365 (1970).
11. R. B. Alexander and J. M. Poate, RAdiation Effects 12, 211 (1972).
12. S. T. Picraux, W. L. Brown, and W. M. Gibson, Phys. Rev. B6, 1382 (1972).
13. D. Van Vliet, Radiation Effects 10, 137 (1971).
14. D. V. Morgan and D. Van Vliet, Radiation Effects 12, 203 (1972).
15. L. M. Howe and S. Schmid, Can. J. Phys. 49, 2321 (1971).
16. M. A. Kumakhov, Radiation Effects 15, 85 (1972).
17. V. V. Beloshitsky, M. A. Kumakhov, and V. A. Muralev, Radiation Effects 13, 9 (1972).
18. D. J. Bierman and D. Van Vliet, Physica 57, 221 (1972).
19. J. H. Barrett, Phys. Rev. B3, 1527 (1971).
20. M. L. Swanson, F. Maury, and A. F. Quenneville, Application of Ion Beams to Metals, ed. by S. T. Picraux, E. P. EerNisse, and F. L. Vook (Plenum Press, N. Y., 1974), p. 393.
21. G. Amsel, J. P. Nadal, E. D'Artemare, D. David, E. Girard, and J. Moulin, Nucl. Instrum. Methods 92, 481 (1971).
22. J. Lindhard, K. Dan. Vidensk. Selsk. Mat.-Fys. Medd. 34, No. 14 (1965).

23 P. P. Matyash, N. A. Skakun, and N. P. Dikii, JETP Lett. 19, 18 (1974).
24 H. D. Carstanjen, to be published.
25 L. C. Feldman, E. N. Kaufmann, J. M. Poate and W. M. Augustyniak, Ion Implantation in Semiconductors and Other Materials, ed. by B. L. Crowder (Plenum Press, N. Y., 1973), p. 491.
26 J. U. Andersen, E. Laegsgaard, and L. C. Feldman, Radiation Effects 12, 219 (1972).
27 Yu. Yu. Krychkov, N. V. Slavin, Yu. A. Timoshnikov, and I. P. Chernov, report given at the VII-th All-Union Conference on Charged-Particle Interaction with Single Crystals, Moscow 1975.
28 J. A. Davies, L. Eriksson, N. G. E. Johansson, and I. V. Mitchell, Phys. Rev. 181, 548 (1969).
29 J. A. Cairns, R. S. Nelson, and J. S. Briggs, Ion Implantation in Semiconductors, ed. by I. Ruge and J. Graul (Springer-Verlag, Berlin, 1971), p. 299.
30 J. A. Cairns, Nucl. Instrum. Methods 92, 507 (1971).
31 R. C. Der, R. J. Fortner, T. M. Kavanagh, and J. M. Khan, Phys. Rev. A4, 556 (1971).
32 B. Domeij, Nucl. Instrum. Methods 38, 207 (1965).
33 J. P. Biersack and D. Fink, Applications of Ion Beams to Metals, ed. by S. T. Picraux, E. P. EerNisse, and F. L. Vook (Plenum Press, N. Y., 1974), p. 307.
34 H. D. Carstanjen and K. Morita, to be published.
35 K. Morita and H. D. Carstanjen, Atomic Collisions in Solids, ed. by S. Datz, B. R. Appleton, and C. D. Moak (Plenum Press, N. Y., 1974), p. 825.

MULTI-STRING STATISTICAL EQUILIBRIUM CALCULATION OF
SCATTERING YIELD FOR FOREIGN ATOM LOCATION USING
CHANNELING EFFECT

K. Komaki[*]

Department of Physics, Rutgers University[†],
New Brunswick, N. J. 08903 and Bell Laboratories, Murray Hill, N. J. 07974

ABSTRACT

The continuum-model calculation using a multi-string potential, and the assumption of statistical equilibrium is applied to the analysis of foreign-atom location in the crystal lattice. Calculations are performed for various crystal types, orientations, and impurity-atom distribution conditions. Several cases in which impurity atom distributions have been proposed to explain experimental results are examined in detail.

INTRODUCTION

Since the discovery of particle channeling, this effect has been used to obtain information about the lattice location of foreign atoms in a single crystal[1]. The subsequent discovery of flux-peaking effects introduced more complexity, and at the same time gave the possibility of more precision in the analysis of non-substitutional impurity atoms[2,3]. The main analysis techniques used in these studies have been the single-string continuum calculation which is convenient but approximate, and the Monte Carlo computer simulation which is precise but expensive.

A detailed study of different analysis techniques

for nuclear lifetime measurements using the blocking effect, which corresponds to a special and well-defined impurity atom distribution, has indicated the usefulness of a multi-string continuum calculation assuming statistical equilibrium(4). This report examines the use of this technique for impurity-atom location studies.

METHOD OF CALCULATION

Lindhard's string potential(5)

$$V(r) = (E_1/2) \log[(Ca_{TF}/r)^2 + 1]$$

with $E_1 = 2Z_1Z_2e^2/d = E\psi_1^2$ and $c^2 = 3$ was used as the interaction potential between a projectile with atomic number Z_1 and a rigid string of atoms with atomic number Z_2 and spacing d. The thermal vibration of atoms was taken into account by folding a Gaussian distribution with mean square amplitude $\rho^2 = \rho_x^2 + \rho_y^2$;

$$U(r) = (1/\pi\rho^2) \int dx\, dy\, V([(r-x)^2+y^2]^{1/2}) e^{-(x^2+y^2)/\rho^2}.$$

To avoid divergence of the multi-string potential due to the asymptotic form r^{-2} of $U(r)$ at large r, $U(r)$ was multiplied by a damping factor of $\exp(-\alpha r^4)$ with $\alpha = 8 \times 10^{-5}$ Å^{-4}. The general expression of the multi-string potential is

$$U(x,y) = \sum_m \sum_n \sum_k \sum_i N_{ki}\, U_i(|\vec{r}-m\vec{a}-n\vec{b}-\vec{r}_k|)$$

where $\vec{a}$ and $\vec{b}$ are the unit vectors of the two-dimensional lattice, $\vec{r}_k$ the position of the k-th string in a two-dimensional unit cell and N_{ki} the number of the i-th atoms in a length d of the k-th string.

In the continuum theory assuming statistical equilibrium, the distribution of particles in phase space is described by means of an accessible area function defined by

$$S(E_t) = \int_A dx\, dy\, \theta(E_t - U(x,y))$$

where $\theta(x) = 1$ for $x > 0$, $\theta(x) = 0$ for $x < 0$ and A is the area of the two-dimensional unit cell. According to the conservation of transverse energy at the entrance surface, the distribution function of transverse energy inside the crystal is given by

$$F(E_t,W) = \frac{1}{A} \int_A dx\, dy\, \delta(E_t - U(x,y) - W) = \frac{1}{A} S'(E_t - W)$$

where $W = E\psi^2$ is the transverse energy of the projectile outside the crystal. In this paper changes in $F(E_t,W)$ due to random scattering during penetration are ignored. If statistical equilibrium in the transverse motion is achieved, the particle flux distribution is given by

$$P(x,y;W) = \int_{U(x,y)} dE_t\, S'(E_t - W)/S(E_t).$$

The yield of close-impact processes from impurity atoms with distribution $G(x,y)$ on the transverse plane is calculated as a function of incident direction;

$$Y(W) = \int_A dx\,dy\, G(x,y) P(x,y;W).$$

RESULTS AND COMPARISON WITH EXPERIMENTS

Calculations were performed on $Si-He^+$, $W-H^+$, $Fe-He^+$, $GaP-He^+$, and silver β-alumina-He^+ systems. Root-mean-square values of the thermal vibration amplitudes perpendicular to the axis for Si, W, and Fe crystals were chosen as 0.106, 0.071, and 0.086 Å, respectively. Vibration amplitudes of impurity atoms were assumed to be the same as those of host atoms, unless otherwise mentioned, and the angular width of the incident beam was neglected. Since height and width of the sharp peak near the center of the yield curve strongly depend on these quantities and also on the neglected multiple scattering, care must be taken when they are compared with experimental results.

Fig. 1 shows the yield curve across the <110> axis of a silicon crystal, for substitutional impurities with various vibration amplitudes and impurities displaced from the string. In the case of nearly substitutional impurities, the single-string model assuming statistical equilibrium is considered to be a good approximation. Results of the single-string calculation using a rigid string potential for the same system were reported by Picraux et al.(6). The single-string results show systematically larger dip width which is caused by the smaller value of the potential at the channel center. The dip width normalized to that of the host

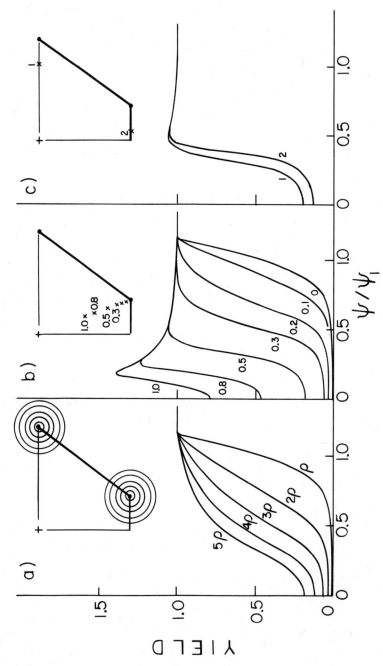

Fig. 1. Yield curve across Si <110> axis for He$^+$ probe as functions of (a) vibration amplitude around the substitutional site (ρ=0.106Å), (b) displacement distance (in Å) from the string, and (c) direction of displacement (displacement is 0.5Å). The inserts show (a) rms vibration amplitude and (b, c) position of displaced atoms.

atoms, however, shows good agreement with present results, as long as vibration amplitude and displacement are small. As seen in Fig. 1c, the dip width shows a noticeable dependence on direction of displacement, in the case of displacements larger than about 0.5Å. This indicates that the cylindrical symmetry around the string which is implicitly involved in the single string calculation breaks down beyond this distance. Yield curves from Bi atoms in Si crystal were calculated (not shown) for the <110> and <111> axes, based on proposed location by Picraux et al.(6). The results are in good agreement with their single string results and also with experimental ones.

Fig. 2 shows the yield curves from various sites in a silicon crystal across the <110>, <100>, and <111> axes. Impurity sites investigated and their coordinates are tetrahedral (1/2,1/2,1/2), hexagonal (5/8,5/8,5/8), bond-centered (1/8,1/8,1/8), split (1/8,0,0) and ytterbium sites (3/8,0,0)(2). Experimental data on Zr atoms in a Si crystal by Domeij et al.(3) show good agreement with the calculated curve for a tetrahedral site. Agreement between calculated and experimental(2) curves for Yb impurities is somewhat poor. Calculated yield curves for hexagonal sites and the experimental ones for diffusion doped Au atoms(7) show noticeable discrepancies. Further investigations might result in refinements of Yb and Au sites in Si.

Fig. 3 shows the <111>, <100>, and <110> yield curves for substitutional, octahedral and tetrahedral sites in tungsten. Experimental results on B(7) and D(8) atoms show almost perfect agreement with calculated curves for octahedral and tetrahedral sites, respectively, except for the detailed shape of peaks at $\psi \sim 0$.

The <111>, <100>, and <110> yield curves for Br atoms in an iron crystal were calculated (not shown) according to the impurity distribution analyzed by Alexander et al.(9) using a continuum-model calculation and Monte Carlo simulation. The present results show fairly good agreement with those of the single string calculation and Monte Carlo simulation, as well as with experiment.

Lattice location of B atoms in an iron crystal has been proposed by Andersen et al.(7) to be (1/2,1/6,0). Yield curves for this site across the <111>, <100>,

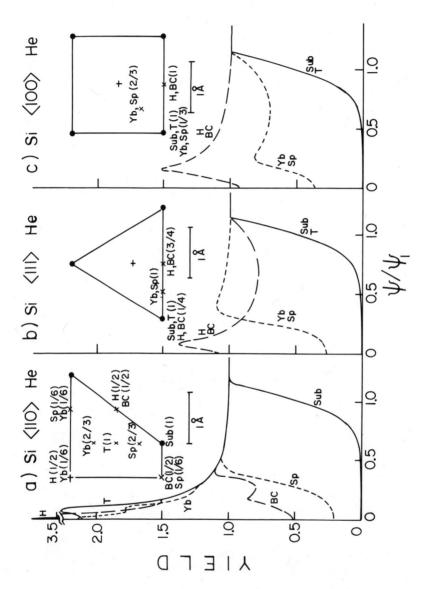

Fig. 2. (a) <110>, (b) <111>, and (c) <100> yield curves for various lattice sites in the case of He$^+$ probe on Si. Sub: substitutional, T: tetrahedral, H: hexagonal, BC: bond-centered, Yb: ytterbium, and SP: split sites. The inserts show position of each site.

<110>, <210>, and <310> axes were calculated (not shown). Agreement between calculated and experimental results is good for the <210> and <310> axes but somewhat poorer for <111>, <100>, and <110> axes. The location of B atoms might be refined by further calculations.

Merz et al.(10) have concluded that the Bi atom in GaP crystal occupy a P site from comparison of the

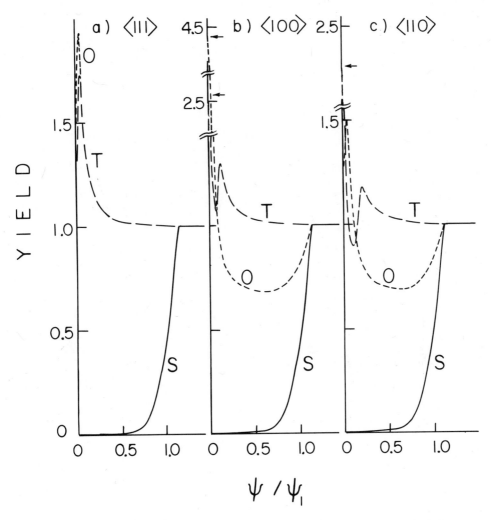

Fig. 3. (a) <111>, (b) <100>, and (c) <110> yield curves from substitutional (S), octahedral (O), and tetrahedral (T) sites in tungsten crystal for H^+ ion probe.

<110> and <111> yield curves from Ga and Bi atoms. Yield curves from impurities on Ga and P sites with various thermal vibration amplitudes were calculated for the <110> and <111> axes (not shown). Calculated values(11) of transverse vibration amplitudes were used, namely 0.145 and 0.098 Å for Ga and P atoms. Good agreement with experimental results was obtained in the case of P site impurities with a vibration amplitude 0.170 Å.

The present method can also be applied to structure analysis of polyatomic crystals. Yield curves from Al, O, and Ag atoms in silver β-alumina $Ag_{2.4}Al_{22}O_{34.2}$, across the c-axis were calculated at crystal temperatures of 300° and 110°K. Fig. 4 shows the results at 300°K. Calculations are based on the structure deter-

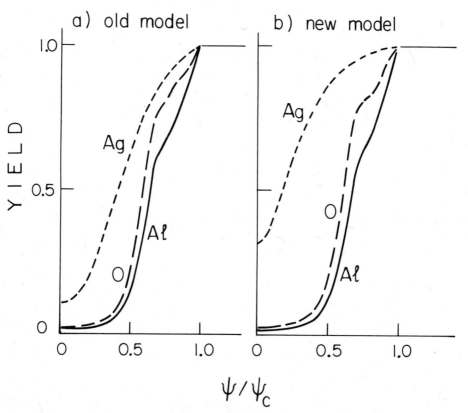

Fig. 4. The <0001> yield curves from Al, O, and Ag atoms in silver β-alumina for He$^+$ ion probe. Calculations are based on (a) old and (b) new X-ray results. $E\psi_c^2$ = 326(a), 273(b) eV.

mined by old(12) and new(13) X-ray studies. Thermal vibration amplitudes at 300°K were taken from the new X-ray data, i.e., 0.098, 0.088, and 0.41 Å for Al, O, and Ag atoms, respectively. Vibration amplitudes at 110°K were calculated from those at 300°K, assuming that each element has its own Debye temperature. According to the old model Ag atoms occupy a site of high symmetry; on the other hand, in the new model Ag atoms are distributed among three sites of lower symmetry. Experimental yield curves for Ag atoms(14) show a preference to the new model.

DISCUSSION

It is known that the assumption of statistical equilibrium is not valid at extremely shallow depths, where depth oscillation of the particle flux takes place. Comparison between results by the present method and those by trajectory calculation using an identical multi-string potential(4) has indicated that this assumption can be used as long as yields are averaged over the depth oscillation, even if statistical equilibrium itself is not achieved.

However, the present method still has some problems to be solved in the future, namely (i) minimum yields of nearly substitutional atoms are underestimated, (ii) effects of dechanneling processes are ignored, and (iii) dependence of the yield curve on azimuthal direction of the scanning plane can not be explained.

ACKNOWLEDGMENTS

The author wishes to acknowledge helpful discussions with Dr. L. C. Feldman and Dr. W. M. Gibson.

REFERENCES

[*]On leave from The University of Tokyo, Komaba, Meguro-ku, Tokyo 153, Japan.

[†]Supported in part by the National Science Foundation.

(1) J. A. Davies, J. Denhartog, L. Eriksson, and J. W. Mayer, Canad. J. Phys. 45, 4053 (1967).
(2) J. U. Andersen, O. Andreasen, J. A. Davies, and E. Uggerhøj, Radiation Effects 7, 25 (1971).

(3) B. Domeij, G. Fladda, and N. G. E. Johansson, Radiation Effects 6, 155 (1970).
(4) J. H. Barrett, Y. Hashimoto, K. Komaki, and W. M. Gibson, to be published.
(5) J. Lindhard, Kgl. Dan. Videnskab. Selskab. Mat.-Fys. Medd., 34, #12 (1965).
(6) S. T. Picraux, W. L. Brown, and W. M. Gibson, Phys. Rev. B6, 1382 (1972).
(7) J. U. Andersen, E. Lægsgaard, and L. C. Feldman, Radiation Effects 12, 219 (1972).
(8) S. T. Picraux and F. L. Vook, Phys. Rev. Lett. 33, 1216 (1974); Ion Implantation: Science and Technology, Ed. by S. Namba (Plenum Press, N. Y. 1975).
(9) R. B. Alexander, P. T. Callaghan, and J. M. Poate, Phys. Rev. B9, 3022 (1974).
(10) J. L. Merz, L. C. Feldman, D. W. Mingay, and W. M. Augustyniak, Ion Implantation in Semiconductors, Ed. by I. Ruge and J. Graul (Springer-Verlag, Berlin, 1971) p. 182.
(11) J. F. Vetelino, S. S. Mitra, and K. V. Namjoshi, Phonons, Ed. by M. A. Nusimovici (Flammarion, Paris, 1971) p. 64.
(12) C. A. Beevers and M. A. Ross, Z. Kristallogr. 97, 59 (1937).
(13) W. L. Roth, J. Solid State Chem. 4, 60 (1972).
(14) L. C. Feldman, W. M. Augustyniak, J. P. Remeika, and P. J. Silverman, this conference.

DISCUSSION

Q: (R. Behrisch) How much does the form of your axial dips and peaks depend on the plane in which you scan through the dips ?

A: (A. Komaki) Nothing at all. Dependence of yield on the azimuthal angle of scanning plane cannot be explained by the present method.

Q: (E. Rimini) You may distinguish between small displacements from the row and a high value of the thermal vibration amplitude by measurements at different crystal temperatures.

A: (A. Komaki) Yes.

Q: (H.D. Carstanjen) In the case of Bi implanted into GaP, would you see any possibility to distinguish between enlarged thermal vibrational amplitudes and a small displacement of the Bi atoms ?

A: (A. Komaki) Certainly, but I have not tried to.

ION CHANNELING STUDIES OF THE LATTICE LOCATION OF INTERSTITIAL

IMPURITIES: HYDROGEN IN METALS*

S.T. Picraux[+]

Sandia Laboratories, Albuquerque, New Mexico 87115

ABSTRACT

The ability to detect distortions of interstitial impurities away from high symmetry sites by the channeling technique has been examined for axial channeling. Multi-row continuum calculations of the angular distributions as a function of displacement from the center of the channel and increased impurity vibrational amplitude suggest localization of interstitials can be made to $\lesssim 0.4$ Å from flux peaking structure and $\lesssim 0.2$ Å when projections can be found which give appreciable dips ($\gtrsim 20\%$). Experimental examples are given for deuterium interpreted from channeling measurements to be located near the tetrahedral interstitial site in W and in distorted near-octahedral interstitial sites in Cr and Mo. Calculated and measured $\langle 100 \rangle$ angular distributions for these sites show good agreement.

I. INTRODUCTION

The lattice location technique for determining the position of foreign atoms in crystals is an important application of the channeling effect [1,2]. Impurities are located by comparing the signals from the interaction of the channeled beam with the impurity and the host atoms. The ability to determine locations is due to the spatial variation of the particle flux density across the channel.

---

* This work was supported by the United States Energy Research and Development Administration, ERDA.

+ Short term visitor at Institute of Physics, University of Aarhus.

Interstitial impurites can control important properties of solids. Hydrogen in metals is a system of great technological interest, which at the same time provides a good test of the channeling technique because of the diversity of interstitial behaviour which can arise due to dilute hydrogen and to hydrogen-defect, hydrogen-impurity and hydrogen-hydrogen interactions. In order to study detailed interstitial behavior it is valuable to fully develop the lattice location technique for the interstitial case. Good insight into the technique can be gained by analytic calculation, but comparisons with experiment are needed to develop a full understanding.

Two features of the channeling angular distribution make interstitial location possible. For a particular channeling direction, interstitials which are located near lattice rows or planes will give a channeling dip, and interstitials near the center of channels will give an enhanced yield (flux peak). In addition, the dips becomes narrower with increasing displacement from a row, and the flux peaks become broader with increasing displacement from the center of the channel.

In this study we examine how distortions away from high symmetry sites, such as at the center of a channel, influence channeling angular distributions. Such distortions can be equilibrium positions at lower symmetry sites or a loss of localization due to large amplitude thermal motion. The object of these investigations is to obtain a better understanding of how precisely an interstitial atom can be located, and of the degree of confidence in the assignment by channeling of a given interstitial site. In practice, of course, both axial and planar channeling are important for locating an interstitial impurity. In this short paper interstitial angular distributions are discussed only for the axial channeling case. Continuum analytic calculations are used to examine the sensititvity to position and vibrational amplitude of interstitials. Analytic examples are given for ⟨100⟩ channeling in the bcc lattice and are compared with experimental measurements of deuterium lattice locations in Cr, Mo and W.

II. ANALYTIC METHOD

A multi-row continuum calculation has been used to determine the spatial distribution of the particle flux density in the channel as a function of the angle between the incident beam and the lattice rows. This has been shown previously to give a reasonable description of the near-surface, average flux density in the channel [3,4]. The flux density, f, as a function of incident angle, ψ, and position in the channel, $\vec{r}$, is given by

$$f(\vec{r}, \psi) = \int \frac{dA(\vec{r}_{in})}{A(E_\perp)}, \qquad (1)$$

$$E_\perp = E\psi^2 + U(\vec{r}_{in}) \geq U(\vec{r})$$

where E is the beam energy, U the continuum potential, $E_\perp$ the transverse energy of a particle entering the channel at position $\vec{r}_{in}$ with angle ψ, and A the area available to a particle entering at $\vec{r}_{in}$, i.e. the area inside the equipotential contour corresponding to $E_\perp = E\psi^2 + U(\vec{r}_{in})$. Equation 1 has been solved numerically and Fig. 1 gives an example of the continuum potential contours for He incident along the $\langle 100 \rangle$ direction in bcc Cr and the flux density along the line EF for $\psi = 0$ and along AC as a function of ψ. The flux density is finite at the center since a minimum value for $A = \pi a^2$ has been imposed, where a is the Thomas Fermi screening distance. The Moliere potential with a static lattice has been used.

In this simple model statistical equilibrium is assumed and depth effects, due to lattice atom thermal vibrations and multiple scattering by electrons, are not taken into account [3,5]. However, measurements typically are near the surface and average over depths ~ 1000 Å which will average out most oscillation effects. We can easily take into account the influence of the impurity vibration on the angular distribution by using the impurity spatial probability

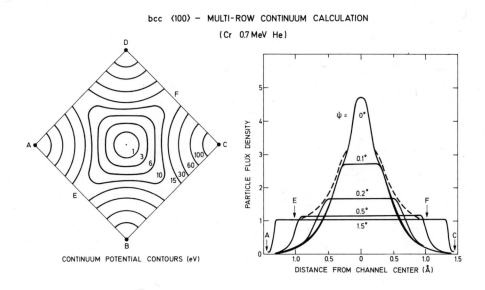

Fig. 1. Calculated $\langle 100 \rangle$ continuum potential contours and flux density curves for 0.7 MeV He in bcc Cr with lattice rows at ABCD and distance AC = 2.88 Å. The flux density along direction AC is given as a function of incident angle ψ relative to $\langle 100 \rangle$ direction (solid lines) and along direction EF is given for $\psi = 0$ (dashed line).

distribution, dP, in determining the yield from the flux distributions. For a gaussian distribution in cylindrical coordinates

$$dP = e^{-(\frac{r'}{\rho})^2} 2r'dr'/\rho^2 \quad , \quad (2)$$

where ρ is the rms vibrational amplitude transverse to the row. This is equivalent to allowing the emitting atom to vibrate in blocking calculations [6].

III. ANALYTIC EXAMPLES

As seen from the flux density curves in Fig. 1 the yield from an impurity near the center of the channel will decrease rapidly with increasing angle. In Fig. 2 the calculated angular distributions are shown for an impurity located in the center of the channel as a function of its rms vibrational amplitude transverse to the row (ρ), and for zero vibrational amplitude as a function of displacement (δ) from the center toward a row. The dip for the substitutional atoms is also shown, as calculated with room temperature lattice atom vibrations and single row continuum potential (Eq. 8 of Ref. 6). The effect of increased vibration is only to reduce the intensity of the central flux peak and thus effectively increase the peak width at half maximum, whereas for displacements side band structure develops for $\delta \gtrsim 0.4$ Å. As the impurity position approaches sufficiently close to the row, this evolves into a narrowed dip.

The influence of vibration and equilibrium displacement on the flux peaking structure can be compared to the corresponding influence on dips for impurities on lattice rows. Previous single-row continuum calculations are given in Fig. 3 for 1 MeV He incident along ⟨110⟩ direction in Si[7]. In this case a vibration of $\rho = 0.106$ Å for the displaced atoms is included in the calculation. Comparison of Figs. 2 and 3 indicate that the dips are much more sensitive for localization of impurities than the flux peaks. This is simply a reflection of the fact that the continuum potential changes rapidly near the rows and is relatively flat, near the channel center.

The flux peak is an important and distinct feature, but its magnitude, as yet, has not been very useful for interpretation of experimental results. This is because its magnitude can oscillate strongly with depth and, due to the narrowness of the peak, can be sensitive to such experimental parameters as surface oxide layers, electron multiple scattering, beam divergence, crystal mosaic spread, depth, and lattice thermal vibrations. Thus as seen in Figs. 2 and 3, localization of interstitials by the flux peaking structure can only be made to ~ 0.3 - 0.4 Å, whereas if appreciable components ($\gtrsim 20\%$) can be found with projections near lattice row, the localization can sometimes be improved to an accuracy of ~ 0.1 - 0.2 Å.

LATTICE LOCATION OF INTERSTITIAL IMPURITIES 531

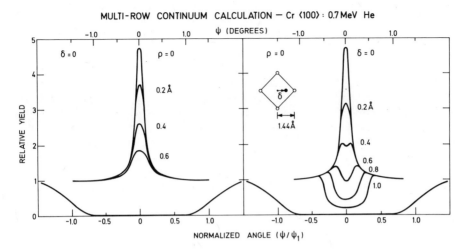

Fig. 2. Calculated ⟨100⟩ angular distributions for 0.7 MeV He in Cr for an impurity located in the center of the channel ($\delta = 0$) as a function of transverse vibrational amplitude (ρ) and for $\rho = 0$ as a function of displacement toward a row (δ). The beam angle is normalized by the characteristic angle for channeling

$$\psi_1 = \left[\frac{2Z_1 Z_2 e^2}{Ed} \right]^{1/2} = 1.5°$$

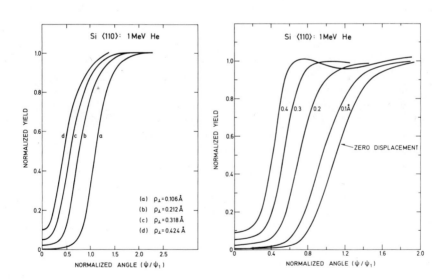

Fig. 3. Si ⟨110⟩ angular distributions for 1 MeV He as a function of impurity vibrational amplitude and equilibrium displacement(right) calculated by single-row continuum model. From Ref. 7.

In general the symmetry of an interstitial site is such that several projected positions will result for a given channeling direction. Thus the angular distributions will have several contributions which can reduce the magnitude of characteristic features useful for locating impurities. An example of the calculated $\langle 100 \rangle$ angular distribution is shown in Fig. 4 for the octahedral (O-site) tetrahedral (T-site) and an intermediate interstitial site located 1/2 way between the O- and T-site. The $\approx$ 1/3 dip of width similar to that for the lattice rows is due to 1/3 of the O-sites being contained within the rows for any given $\langle 100 \rangle$ direction. This is the feature which in practice most clearly distinguishes it from the T-site. Since the magnitude of the flux peak at $\psi = 0$ is difficult to relate quantitatively to experiment the only feature which can distinguish the T from the intermediate site is the small enhancement and reduction in yields near $\psi = 0.4^\circ$. However this region at

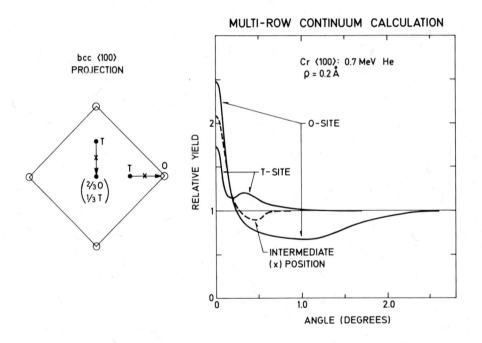

Fig. 4. Calculated $\langle 100 \rangle$ angular distributions for 0.7 MeV He in Cr for an impurity with $\rho = 0.2$ Å located in: the octahedral interstitial site (O); the tetrahedral interstitial site (T); and a position 1/2 way between these two sites (x). The projection shows equivalent positions with 2/3 O, 1/3 T and 1/3 intermediate sites in center and the remainder in other positions shown where $\delta/\delta_o = 1/4$, 1/2, 3/4 and 1, and $\delta_o = 1.44$ Å is the distance from the channel center to the row.

intermediate angles is subject to additional experimental difficulties, since for a given scan planar structure may give rise to weak intensity variations. Thus, from the flux-peaking structure the T- and intermediate sites, which are separated by 0.36 Å, would be only marginally distinguishable along this particular direction, whereas the O- and intermediate sites are separated by the same distance and would be easily distinguishable.

IV. COMPARISON WITH EXPERIMENTAL EXAMPLES

Experimental studies have been carried out for D implanted into the bcc crystals Cr, Mo and W as a function of implantation and anneal temperature 8,9). The nuclear reaction D (^{3}He,p)^{4}He was used to detect D and backscattering was used to monitor the lattice signal. Three examples of ⟨100⟩ angular distributions for interstitial and distorted interstitial positions are given and compared with calculations in Fig. 5 and 6.

Figure 5 shows a ⟨100⟩ scan from Ref. 10 for 1×10^{15} D/cm^2 in W, implanted at 15 keV at room temperature and measured at room temperature with a 750 keV ^{3}He beam. The D has previously been interpreted to

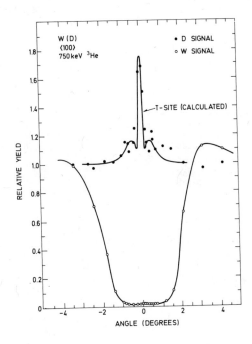

Fig. 5. Comparison of measured ⟨100⟩ angular distribution using 0.75 MeV ^{3}He for D in W with the calculated curve for T-site.

INTERSTITIAL SITES
bcc LATTICE

Fig. 6. Comparison of measured ⟨100⟩ angular distributions using 0.7 MeV ³He for D in Cr and Mo with calculated O-site and distorted O sites I and II where δ = 0.36 and 0.2 Å, respectively, along the ⟨100⟩ directions shown.

○ LATTICE ATOMS
● OCTAHEDRAL SITE
▲ TETRAHEDRAL SITE
✕ DISTORTED O-SITES (I and II)

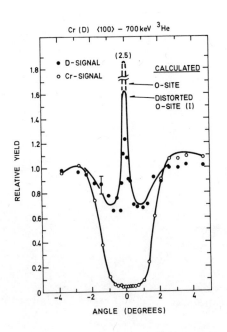
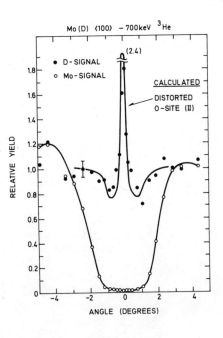

LATTICE LOCATION OF INTERSTITIAL IMPURITIES

be near the T-site and the curve labled "T-site" is the calculated angular distribution for the D located in the tetrahedral interstitial position with a transverse vibrational amplitude of $\rho = 0.2$ Å (corresponds to a 3-dimensional $\rho = 0.24$ Å). This fairly large value of ρ is consistent with neutron scattering measurements of ρ for H in β-Pd [11].

The good agreement in the region of the maximum of the flux peak may be fortuitous, although the W surface is known to have very little oxide to contribute to scattering of the channeled beam. The enhancement in yield in the side-band region ($\psi \approx 0.5°$) is also evident in the data. However, the experimental secondary maxima appear to be at larger angles and the central region is somewhat broader than given by the calculation. This could indicate the D is distorted a few tenths of an Å off the T-site, but this is within the limits of the ability to localize the impurity from only flux peaks (approximately ± 0.3 Å here).

Figure 6 shows angular distributions for 1×10^{15} and 6×10^{14} per cm^2, 15 keV D implanted at 87°K in Cr and Mo, respectively, and annealed to 295°K. The measurements were made at 87°K. The lines for the D signal are calculated whereas for the lattice dips lines are drawn through the data. Prior to annealing an appreciable fraction of the D is found to be near the tetrahedral site in both Cr and Mo [8], whereas upon annealing it moves to well-defined positions near the O-site which we refer to as distorted O-sites. The changes arises presumably from D-D or D-defect interactions and will be discussed separately [8].

Based on the $\langle 100 \rangle$ axial scans in Fig. 6 and on $\{100\}$ and $\{110\}$ planar scans (not shown) the D has been interpreted to be in the distorted O-sites I and II in Cr and Mo, respectively. The calculated curves are shown in Fig. 6 for D located in the O-site and distorted O-sites I and II, again with $\rho = 0.2$ Å. The distorted O-site I has a [001] displacement $\delta = 0.36$ Å toward a nearest neighbor lattice atom and the distorted O-site II has a [010] displacement $\delta = 0.2$ Å in the plane as shown in Fig. 6. These two distortions are <u>not</u> equivalent for $\langle 100 \rangle$ channeling due to the symmetry of the bcc O-site.

Distortion I will not affect the 1/3 dip since this is the component along the row, whereas for distortion II that component is moved off of the row. This is reflected in the appreciable narrowing of the envelope of the dip for O-site II and this feature is in good agreement with the data in Fig. 6. Also the relative displacement directions mean the flux peak is reduced more for distortion I than II and this is consistent with the relative magnitude of the experimental flux peaks. The choice between the O-site and distorted O-site I for D in Cr cannot be made from the $\langle 100 \rangle$ scans. However a 2/3's dip observed for the $\{100\}$ plane implies a displacement > 0.3 Å and the 1/3 dip for the $\langle 100 \rangle$ implies this must be predominately in the direction given by distortion I. Also, from the flux peak in Cr, displacement I cannnot be appreciably greater then $\delta = 0.4$ or the structure and reduction in the flux peak will become apparent (as in the right-hand side of Fig. 2). For the D site in Mo the projected displacement of $\delta \approx 0.2$ Å is rather well defined (± 0.1 Å by the narrowing of the dip), although the direction of the displacement within the shaded plane in Fig. 6 cannot be determined from these data.

In conclusion the channeling lattice location technique allows localization of interstitial impurities to $\lesssim 0.4$ Å from the structure of flux peaks and to $\lesssim 0.2$ Å from the structure of dips. While it is usually difficult to eliminate all ambiguity in site interpretation, in favorable cases rather detailed information can be obtained on distorted interstitial sites. This has been demonstrated for axial scans by comparison of continuum model calculations with experimental data for deuterium in the tetrahedral site in W (± 0.3 Å) and in distorted octahedral sites in Cr and Mo (± 0.1 Å).

Acknowledgements

Discussions with J.U. Andersen, L.C. Feldman, and F.L. Vook are gratefully acknowledged.

REFERENCES

1. J.A. Davies, Channeling ed. by D.V. Morgan (John Wiley, 1973) p. 391

2. S.T. Picraux, New Uses of Low-Energy Accelration, ed. by J.F. Ziegler (Plenum Press, 1975)

3. D. van Vliet, Radiation Effects, 10, 137 (1971)

4. R.B. Alexander, P.T. Callaghan and J.M. Poate, Phys.Rev. B9, 3022 (1974)

5. J. Lindhard, K.Dan.Vid.Selsk.Mat.Fys.Medd. 34, no. 14 (1965)

6. J.U. Andersen, K.Dan.Vid.Selsk.Mat.Fys.Medd. 36, no. 7 (1967)

7. S.T. Picraux, W.L. Brown, and W.M. Gibson, Phys.Rev. B6, 1382 (1972)

8. S.T. Picraux (submitted to Phys.Rev.)

9. S.T. Picraux and F.L. Vook, Phys.Rev.Lett. 33, 1216 (1974)

10. S.T. Picraux and F.L. Vook, Proc.International, Conference on Ion Implantation (Osaka, 1974) p. 355

11. J. Bergsma and J.A. Goedkoop, Physica 26, 744 (1960)

DISCUSSION

Q: (P. Müller) Have you made a comparison with other methods like Huang-scattering with neutrons or X-rays or diffuse scattering ?

A: (S.T. Picraux) No. But I might comment that these methods, when applicable, tend to give complimentary information.

DEPTH DISTRIBUTION OF DAMAGE OBTAINED BY RUTHERFORD BACKSCATTERING COMBINED WITH CHANNELING

Rainer Behrisch and Joachim Roth

Max-Planck-Institut für Plasmaphysik, EURATOM-Association, D-8046 Garching b.München, Fed.Rep.of Germany

ABSTRACT

The different approaches to determine depth distributions of damage in solids by Rutherford backscattering combined with channeling are reviewed. These methods are best applicable for damage introduced by ion bombardment. Most investigations up to now have been done at semiconductors where the ion damage seems to be more suited for analysis by this method than the ion damage in metals. The quantity used for getting depth profiles is mostly the increase in minimum yields in single alignment Rutherford backscattering, while only few measurements have been done at double alignment and at slight misalignment, i.e. the sides of the channeling dips.

INTRODUCTION

Channeling, i.e. the steered motion of energetic ions moving at an angle ψ smaller than a critical angle ψ_c relative to the close packed lattice rows in a crystal and having no close encounter with a lattice atom / 1 to 5 / depends largely on the perfection of these lattice rows. In an undamaged crystal channeled ions, i.e. ions guided by the lattice rows, become deflected to angles larger than the critical angle ψ_c i.e. dechanneled, only due to collisions with the electrons and the thermally vibrating row atoms. If damage is introduced this dechanneling is increased. This has been found experimentally already in the first more detailed measurements of the channeling effect /6,7,8/ as demonstrated in Fig.1 and Fig.2.

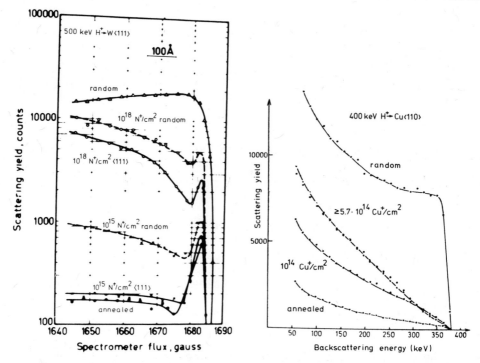

Fig.1 (left). Backscattering yields for 500 keV protons from a tungsten single crystal at ⟨111⟩ incidence. As most of the incident ions are channeled the backscattering is largely reduced for the annealed crystal. After damaging the crystal with 200 keV N^+ ions we get a considerably increased backscattering from the damaged region (~100 Å) but also an increased backscattering from depth larger than the damaged layer (dechanneling) /7/.

Fig.2 (right). Backscattering yields for 400 keV protons bombarding a Cu single crystal in a ⟨110⟩ direction. Increased backscattering from the whole depth probed by protons is found after damaging with 250 keV Cu ions of different doses. /8/

The increase in dechanneling has since been widely used to especially determine the damage distribution caused by ion bombardment of solids /5,9 to 30 /. Most investigations have been done up to now at semiconductors (Si, Ge) because of their technical interest and as the form of damage created, mostly amorphisation, could be analysed more straightforward. In metals the primarily created vacancies and interstitials generally agglomerate already at temperatures much below room temperature to large dislocations and the injected ions may cluster to bubbles or percipitations. Thus strain is generated which may extend further than the range of the damaging ions. This kind of damage is more difficult to measure by channeling methods as it occurs mostly as small displacements of the atoms of the lattice rows.

The methods of the extraction of damage distributions from channeling measurements are still not yet fully developed. The information obtainable will presumably always be incomplete and has to be supplemented by other methods as transmission electron microscopy. However channeling measurements are fast and easy to perform and thus they are a useful technique in getting sometimes sufficient information about the amount, the kind and the distribution of damage especially in ion bombardment investigations.

In the following, the theoretical basis and the different approximations used for the determination of damage and damage distribution from single alignment measurements and double alignment measurements with improved sensitivity are outlined. Other channeling informations and their possible sensitivity to damage are discussed. Examples of results from measurements and from computer simulations are introduced at several places. A systematic investigation and comparison of the results of the different approximations used up to now, has still not yet been performed.

BASIC FORMULAE

The basic analytical formulae describing channeling phenomena have first been published by J.Lindhard /2/ and have been extended by several groups /3-5,31-36 /. The crystal lattice is regarded as being made up of close packed lattice rows. Energetic particles moving at a small angle ψ relative to these close packed lattice row experience a repulsive force. This can be described in a first approximation by a continuum potential U (r) around the lattice rows depending only on the distance r. This picture holds as long as the angle ψ is smaller than a critical angle ψ_c which is of the order of the characteristic angle ψ_1 given by /2/

$$\psi_1 = \left(\frac{2 Z_1 Z_2 e^2}{E d} \right)^{1/2} \qquad (1)$$

where Z_1 and Z_2 are the atomic numbers of the moving ion and the lattice atom, d is the distance of the atoms in the close packed lattice rows and E is the energy of the moving ions. Generally

$\psi_c \simeq \sqrt{2}\,\psi_1$. The approximation of the potential of the lattice rows by a continuum potential implies the conservation of the transverse energy $E_\perp$ of the ions relative to the lattice rows. A better approximation is, that the transverse energy is conserved only at the positions half way between the row atoms /2,33/.

If an ion beam will enter into a lattice parallel to a close packed direction, depending on the point of entry r in the unit cell and the corresponding potential $U(r)$, the different ions will achieve different transverse energies. This is demonstrated in Fig.3. The transverse energy $E_\perp$ of an ion inside the crystal is then given by

$$E_\perp = U(r) \qquad (2)$$

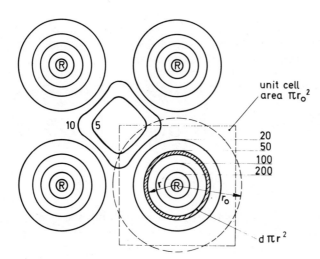

Fig.3. Contour line plots of equal potential for protons in Cu in a <100> direction. R = lattice rows, the numbers at the lines are in eV. The square unit cell is generally replaced in the calculations by a circle of radius r_o with equal area. Ref./37/.

For an angle of incidence ψ relative to close packed lattice rows this is given by: $E_\perp = U(r) + E\psi^2$. Only particles with $E_\perp < E\psi_c^2$ will be channeled, i.e. confined by the continuum potential to an area $\pi r^2 \leq \pi \hat{r}^2$ with $E_\perp = U(\hat{r})$. The other particles with larger transverse energies will see the lattice as built up of nearly randomly distributed atoms.

After having passed through the crystal surface, i.e. at depth z=o, the parallel beam has achieved a distribution in transverse energy[+] $g(E_\perp,0)$ which can be calculated using Fig.3.

[+] J.H.Barret named $g(E_\perp,z)$ the density of trajectories /34/

$$g(E_\perp, 0)\, dE_\perp = \frac{d(\pi r^2)}{\pi r_0^2} \quad (3)$$

$E_\perp$ is generally not conserved during the motion through a real lattice. Due to scattering at electrons, at the thermally vibrating lattice atoms and at defects in the crystal the distribution in transverse energy changes with depth z to $g(E_\perp, z)$ having higher probabilities for larger $E_\perp$. In a first approximation the transverse energies $E_\perp$ of the ions increase uniformly with depth. This increase depends on $E_\perp$ and is the faster, the higher $E_\perp$ /2,35,38/. For a more accurate calculation a diffusion equation for $g(E_\perp, z)$ has to be solved where the diffusion coefficient depends on $E_\perp$ /39-41/.

The probability of a close encounter of the ions with a lattice atom $X(z)$ at depth z relative to this probability in a random lattice named the single aligned yield relative to the normal (or random) yield is given by /2/

$$X(z) = \int_0^\infty g(E_\perp, z)\, \pi(E_\perp)\, dE_\perp \quad (4)$$

Here $\pi(E_\perp)$ is the relative probability that an ion with $E_\perp$ will hit a lattice atom. The introduction of relative yields, i.e. normalising the yields to the random yield largely simplifies the equation as cross sections and solid angles are omitted. This means however that the electronic energy loss for those ions at aligned incidence which react with the lattice atoms have been assumed to have the same electronic energy loss as the ions in an amorphous material. This is not generally fulfilled, however, the errors introduced in heavily damaged layers are small. Attempts have also been made to correct for this error /15/.

The relative probability $\pi(E_\perp)$ will depend on the distribution of elongations $P(r)$ of the lattice atoms from the lattice rows by thermal vibrations or damage. We may write:

$$\pi(E_\perp) = \int_0^{r_0} P(r)\, W(r, E_\perp)\, dr \quad (5)$$

$W(r, E_\perp)$ is the probability that an ion with the transverse energy $E_\perp$ will hit an atom at a distance r from a lattice row relative to this probability in a random lattice.

Most channeling measurements have been done in Rutherford ion backscattering, where the reaction yield is the Rutherford backscattering yield. In this case we obtain the depth z of the reaction from the geometry and the energy loss of the backscattered particles /9,14,42/. In backscattering the detector for observation may be also aligned with a close packed lattice row. This double aligned yield $\chi'(z)$ can be calculated from the equation (4) and (5) assuming reversibility of the particle trajectories and we obtain /2,9,38/.

$$\chi'(z) = \int_0^\infty P(R)dR \iint g(E_{\perp 1}, z_1) W(r_1, E_{\perp 1}) g(E_{\perp 2}, z_2) W(r_2, E_{\perp 2}) dE_{\perp 1} dE_{\perp 2} \quad (6)$$

where the index 1 refers to the ingoing ions and 2 to the outgoing ions and P(R) is the 3 dimensional distribution of elongations of the lattice atoms.

The relation of equation (4) for the backscattering yield in an undamaged crystal is shown schematically in Fig.4 taken from ref. /34/. The upper curve shows the distribution in transverse energies $g(E_\perp, z)$ of 3 MeV protons in a $\langle 111 \rangle$ channel in tungsten at a mean depth of z = 800 Å as calculated by a computer simulation /34/. The middle curve shows the quantity $\pi(E_\perp)$ assuming a thermally vibrating lattice at 1200°C without damage. The product $g(E_\perp,z)\pi(E_\perp)$ is shown in the bottom curve. In order to obtain the reaction yield we have to integrate this curve. It is seen that the reaction yield for $E_\perp \simeq E\psi_c^2$ is well above 1, i.e. larger than in a random lattice.

The exact form of $\pi(E_\perp)$ in an undamaged crystal as seen in Fig.4 is generally replaced by single step function which is also introduced in the figure.

$$\pi(E_\perp) = \begin{cases} 0 \text{ for } E_\perp < E\psi_c^2 \\ 1 \text{ for } E_\perp \geq E\psi_c^2 \end{cases} \quad (7)$$

This approximation is especially useful in calculating damage distributions. It means that we may now divide the particles moving in the crystal into two parts: The random fraction and the aligned fraction of the beam. The random fraction is equal to the relative reaction yield in a channeling experiment. Thus the same symbol $\chi(z)$ is generally taken for both. The relative reaction yield in single alignment is thus given by

$$\chi_1(z) = \int_{E\psi_c^2}^\infty g(E_\perp, z) dE_\perp \quad (8)$$

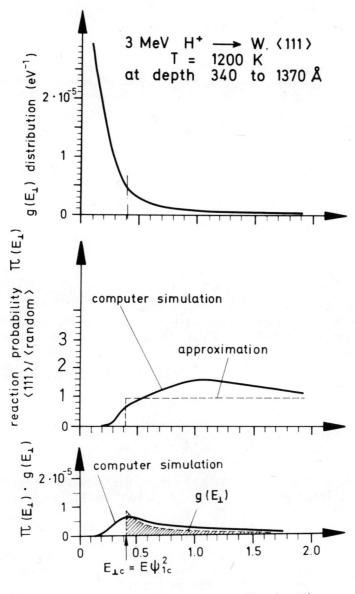

Fig.4 (top). Distribution of transverse energy of 3 MeV protons incident in a ⟨111⟩ direction on a tungsten crystal at 1200°C in a depth of 340 to 1370 Å. (middle). Normalized reaction probability of 3 MeV protons in the ⟨111⟩ channel with the vibrating lattice atoms as a function of transverse energy. The dashed line shows the approximation generally used. (bottom). Product of the probability of a certain transverse energy times the normalised reaction probability /34/.

which is equal to 1 in a random lattice. In double alignment the relative reaction yield is given by /43,38/

$$\chi'_1(z) = \int P(R)dR \int_{E\psi^2_{C_1}}^{\infty} g(E_{\perp 1}, z_1)dE_{\perp 1} \int_{E\psi^2_{C_2}}^{\infty} g(E_{\perp 2}, z_2)dE_{\perp 2} \qquad (9)$$

For symmetric trajectories in and out this is equal to /43/

$$\chi'_1(z) = h(\Theta) \chi_1^2(z) \qquad (10)$$

where $h(\Theta) = 2 - \frac{1}{2}\sin^2(\pi - \Theta)$ is a geometrical factor caused by the redistribution in $E_\perp$ after backscattering.

The aligned fraction of the beam is given by

$$1 - \chi_R(z) = \int_0^{E\psi_C^2} g(E_\perp, z)dE_\perp \qquad (11)$$

These ions have no interaction with the atoms of the lattice rows in an undamaged crystal. They may however interact with atoms displaced by damage.

Equation (4) and (8) may as well be used to define the critical angle ψ_c by:

$$\int_{E\psi_C^2}^{\infty} g(E_\perp, z) dE_\perp = \int_0^{\infty} g(E_\perp, z) \Pi(E_\perp)dE_\perp \qquad (12)$$

From Fig.4 it is obvious that the crude approximation made in (7) may be give large errors especially if many particles have transverse energies slightly above $E\psi_C^2$.

REACTION YIELDS IN DAMAGED CRYSTALS

With the approximations introduced in equation (7) for an undamaged crystal, the relative reaction yield in a damaged crystal can be split up into two parts:

The yield obtained from the crystal, if the atoms displaced by damage at depth z are removed, i.e. the random fraction of the beam $\chi_R(z)$ at depth z (yield due to dechanneled ions).

The yield from the atoms displaced by damage at depth z (yield due to direct backscattering).

If we take a concentration $N_D(z)/N$ of randomly distributed displaced lattice atoms we get for the relative reaction $\chi_2(z)$

for single alignment /9/:

$$\chi_2(z) = \chi_R(z) + \left[1 - \chi_R(z)\right] \frac{N_D(z)}{N} \qquad (13)^{+)}$$

The random fraction $\chi_R(z)$ of the beam at depth z in a damaged crystal generally differs from the random fraction $\chi_1(z)$ in an undamaged crystal due to dechanneling at the damage sites at depths smaller than z.

Similarly, to the relative single alignment reactions yield we can calculate the relative double alignment reaction yield $\chi'_2(z)$ i.e. the relative reaction yield for ion incidence and observation in close packed directions at an angle Θ to each other /19,38,43-45/

$$\chi'_2(z) = h(\Theta) \chi_R^2(z) + \left[1 - \chi_R(z)\right]^2 \frac{N_D(z)}{N} \qquad (14)$$

where $h(\Theta)$ had been introduced in equation (10).

The basic problem is how to get $\chi_R(z)$, if we know $\chi_1(z)$ and have some information about the amount of damage at depth smaller than z. Several models have been used:

a) THE SINGLE SCATTERING MODEL /9,11-15/

Disregarding the details of the change of the distribution in transverse energies $g(E_\perp,z)$ with depth z, it is assumed that those ions of the aligned beam which are scattered in one single collision at a displaced lattice atom by an angle larger than ψ_c become part of the random beam. Scattering angles smaller than ψ_c are disregarded, which consequently gives too small dechanneling at large depth. This single scattering model gives an exponential decrease of the aligned beam:

$$\left[1 - \chi_R(z)\right] = \left[1 - \chi_1(z)\right] P_s \qquad (15)$$

+) If the atoms displaced by damage are not randomly distributed $N_D(z)$ is a mean value given by /13/

$$N_D(z) = \int_0^{r_0} \int_0^{E\psi_c^2} g(E_\perp,z) W(E_\perp,r) N_D(r,z) \, dr \, dE_\perp \qquad (13a)$$

Here $N_D(r,z)$ is the distribution of elongations r of the lattice atoms perpendicular to the ions. Further the notations of the equations (4) and (5) are used.

where P_s is obtained by integrating the Rutherford cross section between ψ_c and $\pi/2$.

$$P_s = e^{-\frac{\pi}{4}\frac{\psi_1^4}{\psi_c^2} N d^3 \int_0^{z/d} \frac{N_D(x)}{N} dx} \qquad (16)$$

Equation (15) may be rearranged to:

$$\chi_R(z) = \chi_1(z) + \left[1 - \chi_1(z)\right]\left[1 - P_s\right] \qquad (17)$$

b) THE LINEAR INCREASE MODEL /15,44,45/

For small depth and small damage concentrations the term $1-P_s$ may be expanded to:

$$1 - P_s = \frac{\pi}{4}\frac{\psi_1^4}{\psi_c^2} N d^3 \int_0^{z/d} \frac{N_D(x)}{N} dx + \ldots \qquad (18)$$

If we further take into account that $\chi_1(z) \ll 1$ we may approximate equation (17) by

$$\chi_R(z) = \chi_1(z) + \beta \int_0^{z/d} \frac{N_D(x)}{N} dx \qquad (19)$$

where β can be regarded as an adjustable parameter of the order of $\beta \simeq (\pi/4)(\psi_1^4/\psi_c^2)d^3 N$. This linear increase has the advantage of great simplicity.

c) MULTIPLE AND PLURAL SCATTERING MODEL /9,12-15,28,46-48/

A parallel ion beam having passed an amorphous layer has attained an angular spread which is described by plural or multiple scattering models depending on the thickness of the amorphous layer. The form of such distributions has been calculated by several authors /49-53/. The passage of the aligned beam through the damaged layer in a crystal can also be treated as a plural or multiple scattering problem. The increase in random beam is obtained in this model by integrating the emerging angular distribution for angles larger than ψ_c which may give a value

P_m, and we obtain

$$\chi_R(z) = \chi_1(z) + \left[1 - \chi_1(z)\right] P_m \qquad (20)$$

Using the multiple scattering formula derived by J. Lindhard for Coulomb potentials /2/ we get

$$P_m = e^{-\frac{\psi_c^2}{\Omega^2(z)}} \qquad (21)$$

with

$$\Omega^2(z) = \frac{\pi}{2} \psi_1^4 \, Nd^3 \, \ln(1.23\,\varepsilon) \int_0^{z/d} \frac{N_D(x)}{N} \, dx$$

$$\varepsilon = \frac{aE}{Z_1 Z_2 e^2} \cdot \frac{M_2}{M_1 + M_2}$$

$$a = \frac{0.8853 \, a_0}{(Z_1^{2/3} + Z_2^{2/3})^{1/2}}$$

with Z_1, Z_2, M_1, M_2 the atomic numbers and masses of the ions and target atoms respectively and a_0 the Bohr radius. Several other forms have been used for P_m /49,50/. However, these multiple scattering formulae are only correct for small angles and do not include the single scattering distributions for large deflection angles. Thus the random fraction calculated is underestimated for small depth in the work published in the past. A correct multiple scattering fomula including large angle single scattering /53/ should give for small depth z the same results for $\chi_R(z)$ as the single scattering model.

d) THE DIFFUSION MODEL /27,28,41,54/

According to equation (8) the random fraction of the beam is approximated by integrating the distribution of transverse energies $g(E_\perp,z)$ for transverse energies larger than $E\psi_c^2$. Thus the more correct procedure to obtain $\chi_R(z)$ is first to calculate the change of $g(E_\perp,z)$ with depth similarly as it was done for thermal dechanneling in undamaged crystals using a diffusion equation for $g(E_\perp,z)$ /39-41,54/.

$$\frac{\partial g(E_\perp,z)}{\partial z} = \frac{\partial}{\partial E_\perp}\left[A(E_\perp)D(E_\perp)\frac{\partial}{\partial E_\perp}\frac{g(E_\perp,z)}{A(E_\perp)}\right] \quad (22)$$

where $A(E_\perp)$ is the region $r < \hat{r}$ where the particle with $E_\perp$ is confined to and $D(E_\perp)$ is the diffusion coefficient. $D(E_\perp)$ consists of 3 contributions due to the scattering at electrons $D_e(E_\perp)$, the scattering at the thermally vibrating lattice atoms $D_n(E_\perp,T)$ where T is the temperature and due to the scattering at lattice atoms displaced by damage $D_d(E_\perp)$ /47,54/

$$D(E_\perp) = D_e(E_\perp) + D_n(E_\perp,T) + c(z) D_d(E_\perp) \quad (23)$$

where $c(z)$ is the concentration of the damage. $D_d(E_\perp)$ further depends largely on the kind of damage, i.e. the mean displacement from the lattice rows.

Though this approach to calculate $\chi_R(z)$ by solving a diffusion equation still contains approximations as equal distribution in the transverse phase space several conclusions can be drawn.

1) The increase of $\chi_R(z)$ with depth z is not additive for dechanneling by damage and by the thermal vibrations as confirmed experimentally /55/.

2) The damage at a depth smaller than z may still cause increased dechanneling at depth larger than z due to a large distortion of $g(E_\perp,z)$. This effect named "delayed dechanneling" has been found in experiments /23/ and in computer simulations /56/.

3) As in a diffusion $E_\perp$ may also be decreased we may find a decrease of χ_R for large χ_R and large depth /57,41,58/ (rechanneling).

The different approaches to calculate χ_R are summarized in Fig.5. It is seen that the single scattering model underestimates χ_R at large depth while the multiple scattering model gives too low values at small depth due to not using the approriate multiple scattering formula. The diffusion model gives slightly larger values for χ_R than the single and the multiple scattering model. The results from the linear increase model should coincide with the single scattering model. Fig.6 shows an example for $g(E_\perp,z)$ in an undamaged and a damaged crystal.

DETERMINATION OF DAMAGE DISTRIBUTIONS

Among the different methods used to calculate damage distributions from Rutherford backscattering spectra, mostly the step by step procedure was applied /5,12-15/. Here we start with the aligned backscattering spectra of the undamaged crystal, of the

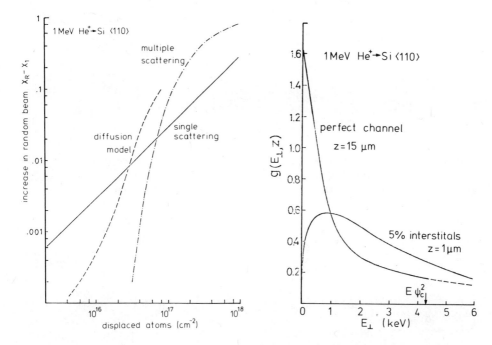

Fig.5 (left). Increase in random fraction for 1 MeV He$^+$ ions in Si $\langle 110 \rangle$ as calculated by the single scattering, the multiple scattering and the diffusion model after transmission through a surface layer of 10^{15} to 10^{18} displaced atoms per cm^2. For the diffusion model the defect concentration has been taken to $3 \cdot 10^{-3}$; from ref. /41/.

Fig.6 (right). The distribution of transverse energies $g(E_\perp, z)$ for 1 MeV He$^+$ ions in Si $\langle 110 \rangle$ at a depth of 15 µm in an undamaged crystal and at a depth of 1 µm in a damaged crystal with 5 % interstitials obtained by solving a diffusion equation from ref. /58/.

damaged crystal and with a random spectrum. From this we get $\chi_1(z)$ and $\chi_2(z)$ or $\chi_1'(z)$ and $\chi_2'(z)$ respectively. For getting the depth scale the usual Rutherford backscattering analysis is used /9,14,15,42/.

At a depth z_0 just below the surface peak the random fraction of the beam $\chi_R(z_0)$ in a damaged crystal is approximated by the random fraction $\chi_1(z)$ in the undamaged crystal. From equation (13) we then get for the damage at $z_0 + \Delta z = z_1$

$$\frac{N_D(z_1)}{N} = \frac{\chi_2(z_1) - \chi_1(z_0)}{1 - \chi_1(z_0)} \tag{24}$$

The next step is to calculate $\chi_R(z_1)$ from $N_D(z_1)$ from equation (15), (19), (20) or (8) with (22) and (23) depending on the model used. Knowing $\chi_R(z_1)$ we can calculate $N_D(z_2)$, the damage of the next depth interval

$$\frac{N_D(z_2)}{N} = \frac{\chi_2(z_2) - \chi_R(z_1)}{1 - \chi_R(z_1)} \tag{24a}$$

By this procedure we get the whole distribution $N_D(z)$, which should go to zero at depth larger than the damage. For double alignment, and symmetry between the in and outgoing ions, equation (24) has to be replaced by /44,45,59/

$$\frac{N_D(z_1)}{N} = \frac{\chi_2'(z_1) - \chi_1'(z_0)}{\left[1 - \left(\frac{\chi_1'(z_0)}{h(\theta)}\right)^{\frac{1}{2}}\right]^2} \tag{25}$$

and (24 a) by

$$\frac{N_D(z_2)}{N} = \frac{\chi_2'(z_2) - h(\theta)\chi_R^2(z_1)}{[1 - \chi_R(z_1)]^2} \tag{25a}$$

Here the relation (10) has been used which is also necessary to calculate $\chi_R(z)$. Double alignment is more sensitive to N_D as the contributions of the random fraction of the beam to backscattering is largely reduced.

In the equations (24) and (25) it is assumed that at each depth interval the atoms displaced by damage contribute some "direct backscattering". This is valid at least for not too large depth. From equation (13) together with (18) we can estimate the relative amount of direct backscattering and dechanneling and get

$$\chi_2(z) = \chi_1(z) + \left[1 - \chi_1(z)\right] \left[\frac{\pi}{4} \frac{\psi_1^4}{\psi_c^4} Nd^3 \int_0^{\frac{z}{d}} \frac{N_D(x)}{N} dx + \frac{N_D(z)}{N} \right] \quad (26)$$

Here the first term in the last square braket represents the contribution due to the dechanneled beam and the second represents the direct backscattering. For the close packed lattice rows Nd^3 is of the order of 1, while ψ_1 and ψ_c are about 1 to 2 degrees. This gives $(\pi/4)Nd^3(\psi_1^4/\psi_c^4) \simeq 10^{-3}$ to 10^{-4}. For uniformly and randomly distributed lattice atoms this means that only after about 10^3 to 10^4 atomic layers the dechanneling contribution becomes as large as the direct backscattering contribution. Such a backscattering distribution with a so called "damage peak" was shown already in Fig.1 and is obtained in most damage measurements at semiconductors. However, as we will see in the next paragraph damage configurations predominant in metals, as dislocations are much more effective in dechanneling than calculated in the single scattering model used in deriving equation (26). This was shown already in Fig.2. At small depth direct backscattering will presumably prevail even at these damage configurations.

Equation (26) further shows that for increasing energies of the analysing beam the dechanneling decreases as the critical angle ψ_c gets smaller. Thus if the alignment and the divergence of the beam are good enough a direct backscattering peak should always be obtainable if the analysing energy is large enough.

Finally, some other methods for extracting damage distributions should be mentioned.

E. Bøgh et al. /11/ used the single scattering model and an iterating procedure instead of a step by step calculation. First they calculated $N_D(z)$ from (13) putting $\chi_R(z) = 0$ and for large z $N_D(z) = 0$. Then they calculated $\chi_R(z)$ from (17) and the result is used in turn to get a better approximation for $N_D(z)$ from (13). This procedure converges already after ~ 3 iterations. It works however only if it is well known at which depth the damage is zero and if delayed dechanneling is negligible.

A different scheme was used by Matsunami and Itoh /58/ who analysed data of Eisen /12/. They assumed that the damage increases proportional to $1 - \exp(-\alpha W)$ where W is the energy deposited per unit volume by the damaging ion beam and α is a constant. This allowed to get $\chi_R(z)$ and $N_D(z)$ from the difference of backscattering spectra from a sample damaged by different doses.

Pabst /19/ and Schulze /59/ have used the difference in single and double alignment spectra to obtain damage distributions. However, the formulae given by Pabst do not fulfill reversibility of the particle paths and Schulze made very crude approximations. Single and double alignment backscattering spectra from the same probe allow to eliminate $\chi_R(z)$ from the equations (13) and (14) and we obtain directly:

$$\frac{N_D(z)}{N} = -A + \left\{ A^2 + \left[\chi_2'(z) - h(\theta)\chi_2^2(z)\right]\left[h(\theta) - \chi_2'(z)\right]^{-1} \right\}^{\frac{1}{2}} \quad (27)$$

with $\quad A = \dfrac{\left[1 - \chi_2(z)\right]^2 - 2\left[\chi_2'(z) - h(\theta)\chi_2(z)\right]}{2\left[h(\theta) + \chi_2'(z)\right]}$

The results of this equation have not yet been tested with experimentally measured backscattering spectra.

Pronko /60/ has neglected direct backscattering in formula (13) which gives $\chi_2(z) = \chi_R(z)$. Then $N_D(z)$ can be calculated from (15) and (16) or (19) or (20) and (21). This gives for the single scattering model:

$$N_D(z) = -\frac{4}{\pi} \frac{\psi_c^2}{\psi_1^4} \frac{1}{d^2} \frac{d}{dz} \ln \frac{1 - \chi_2(z)}{1 - \chi_1(z)} \quad (28)$$

For the linear increase model we get

$$N_D(z) = \frac{Nd}{\beta} \frac{d}{dz}(\chi_2(z) - \chi_1(z)) \quad (29)$$

and for the multiple scattering model using Lindhards approximation we obtain

$$N_D(z) = -\frac{4}{\pi} \frac{\psi_c^2}{\psi_1^4} \frac{1}{d^2} \frac{1}{2\ln(1.23\varepsilon)} \frac{d}{dz} \ln \frac{1}{\frac{\chi_2(z) - \chi_1(z)}{1 - \chi_1(z)}} \quad (30)$$

Equation (30) has to modified if other multiple or plural scattering formulae are used.

Finally Ziegler /15/ and recently Baeri et al. /48/ analysed the damage in Si by measuring the backscattering for different small angles of incidence relative to a close packed direction. Thus the ion beam probed effectively different areas of the unit cell of the channel and Baeri et al. could deduce the concentration of atoms displaced by damage at different distances relative to the lattice rows.

RESULTS OF DAMAGE MEASUREMENTS

The damage introduced in Si and Ge due to the implantation of 50 to 400 keV Sb ions has been analysed by Bøgh et al./11/ by 500 keV He-ions as well as H-ions using the iteration procedure. The results are shown in Fig.7. They found reasonable agreement for the distributions measured by different ions and it is reported that the measured distributions agree also well with the calculated distributions.

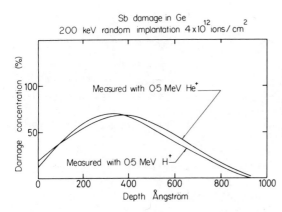

Fig.7. Damage distribution in Ge implanted by 200 keV $4 \cdot 10^{12}$ Sb ions/cm^2 analysed by 0.5 MeV He$^+$ as well as H$^+$ ions /11/.

Feldmann and Rodgers /13/ tried to evaluate the damage profile introduced in Si by implantation with 250 keV Ne ions by the calculation of $\chi_R(z)$ with the single and multiple scattering model. They found that the dechanneling is overestimated by the multiple scattering mechanism leading to negative damage and that the single scattering mechanism underestimated the dechanneling so that at no depth zero damage was reached.

For the case of boron implanted Si Westmoreland et al. /12/ analysed the defect profile using the plural scattering mechanism. At damage densities below $2 \cdot 10^{22} cm^{-3}$ displaced atoms the distribution obtained by this procedure returned to zero behind the disorder peak. Above $2 \cdot 10^{22} cm^{-3}$ the plural scattering treatment begins to overestimate the dechanneling similar to the multiple scattering mechanism. Fig.8 a and b show the results for $\chi_R(z)$ and for $N_D(z)$. In order to overcome these difficulties the critical angle is generally adjusted until the damage distribution smoothly

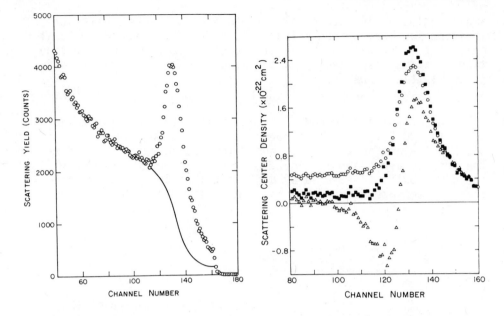

Fig.8 a(left). Backscattering spectrum for analysing Si, implanted with $5 \cdot 10^{14}$ 200 keV B ions/cm^2, with 1.8 MeV He ions. The solid line shows the random fraction as calculated by plural scattering /12/.

Fig.8b (right). Damage distribution (scattering center densities) calculated from the spectrum of Fig.8 using the plural scattering (■), the single scattering (o), and the multiple scattering model (△) /12/.

approaches zero damage at deep layers. The uncertainties of the analytic calculations with the single, multiple and plural scattering models have led to the proposal of the linear increase model by Ziegler /15/ where the free parameter β is used to be varied until the best return to zero damage is reached. The results of this model agree within the uncertainties with the results of the multiple scattering model /15/ where ψ_c was adjusted to get zero damage at large depths.

Ziegler/15/ has further studied the influence of different electronic stopping powers for random and channeled particle trajectories on the damage distributions. Some of his results are shown in Fig.9.

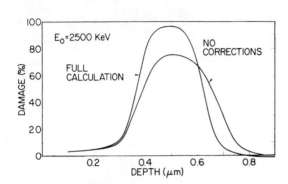

Fig.9. Damage distribution of Si implanted with $8 \cdot 10^{15}$, 175 keV B-ions/cm^2 analysed with 2.5 MeV He ions using <110> channeling. The spectra have been evaluated assuming the same electronic stopping for channeled and randomly moving ions (no correction) as well as for correcting for different energy losses (full calculation). In this case the electronic stopping power for the channeled ions has been further assumed to depend on the damage concentration.

Only few experiments have been carried out on the depth profiling of radiation damage in metals by Rutherford backscattering combined with channeling. In metals primary defects produced during ion bombardment as interstitials and vacancies are highly mobile at room temperature and cluster mostly to large agglomerates as dislocation loops and stacking faults producing finally a dislocation network. These defects generally only slightly disturb a large number of lattice planes and only very few atoms are displaced more than the screening radius a (Fig.10). Such defect clusters produce predominantly dechanneling of channeled particles as has been shown by Quéré /61,62/ and was also found in the computer calculations of Morgan and van Vliet /63/. The mean dechanneling cross section per unit length of a dislocation line averaged over all possible directions in respect to the channel axis has been calculated by Quéré /61,62/ to

$$\sigma = \frac{\pi}{2} \frac{ba}{x} \frac{1}{\psi_c^2} \qquad (31)$$

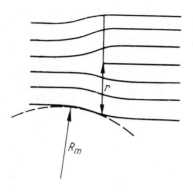

Fig.10. Lattice distortion around an edge dislocation. R_m is the minimum radius of curvature of the lattice rows for a given distance r from the dislocation.

with b the burgers vector of the dislocation, a the screening length and x a number depending on the nature of the dislocation. For screw dislocations $x = 12.5$ and $x = 4.5$ for an edge dislocation was obtained /61,62/. This dechanneling cross section is several orders of magnitude larger than the Rutherford cross section for direct backscattering. Due to this high dechanneling probability in a layer with dislocations the aligned backscattering spectra do generally not show a well defined damage peak from direct backscattering as in semiconductors /8,20,25,60,64/. Further the strain introduced in the lattice can reach much larger depths than the range of the implanted ions. Additionally it depends on the damaging ion dose so that the necessary boundary condition of zero damage at large depth is difficult to establish. This fact discouraged most authors to try to extract depth profiles from their backscattering spectra at metals /20, 25,64/.

As an approximation in this case of high dechanneling Pronko /60/ neglected the contribution of direct backscattering in equ. (13) completely and calculated damage distributions using equation (28). Fig.11 shows his results for the damage profile in Au after bombardment with $3 \cdot 10^{12}$ Au^{++} ions of 540 keV.

Another type of defects introduced in materials after ion implantation are precipitations of the implanted atoms, as for example gas bubbles after implantation of inert gas ions. Additional to the dechanneling at the dislocations and the lattice strain produced also direct backscattering can occur at the inner surfaces of the gas bubbles. By choosing the energy of the analysing particles high enough the direct backscattering can be increased over the dechanneling background so that the damage peak in the aligned backscattering spectra can be observed. This can further be improved by using the double aligned technique.

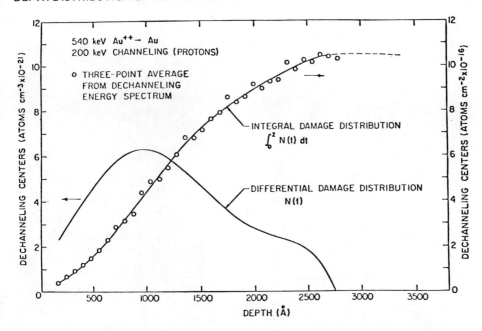

Fig.11. Damage distribution, i.e. dechanneling center distribution, in Au after bombardment with $3 \cdot 10^{12}$ 540 keV Au^{++} ions/cm^2 analysed with 200 keV protons. The integrated damage profile was calculated from the aligned yields before and after damaging. This was fitted by a 6th degree polynomal curve and analytically differentiated /60/.

Fig.12 shows a set of such backscattering spectra taken in <100>/ <111> double alignment with 150 keV H^+ ions from a Nb single crystal damaged with different doses of 4 keV He^+ ions/45,66/. For the annealed crystal the surface peak is well seen while the backscattering from large depth is reduced by $\sim 10^3$. After damaging by 4 keV He^+ ions backscattering from the larger depth increases. At a damaging dose of $5.1 \cdot 10^{16}$ He^+/cm^2 a damage peak shows up at a depth of 450 Å. As this peak is not seen at lower damaging doses, it may be caused by the formation of inner surfaces of the gas filled bubbles. At about $1.4 \cdot 10^{17}$ He^+/cm^2 the backscattering spectrum changed again. A surface layer of ~ 450 Å with nearly random backscattering occurs. Observation of the crystal in the scanning electron microscope shows that after this very dose blisters can be seen on the surface as shown in Fig.13. Part of the surface layer is bent up so it is no longer aligned with the incident beam. Measurements at higher energies, where the deckeldicke (thickness of the blister scin) could be measured also optically, confirmed that the thickness

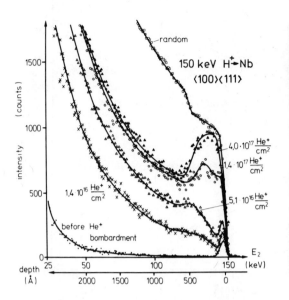
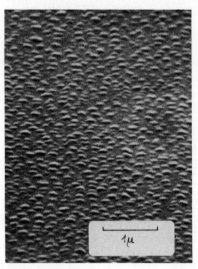

Fig.12 (left). Double aligned backscattering spectra of 150 keV H^+-ions from a Nb single crystal damaged by different doses of 4 keV He^+ ions /66/.

Fig.13 (right). Surface of the Nb single crystal bombarded with $4 \cdot 10^{17}$ 4 keV He^+ ions/cm^2 as observed in the scanning electron microscope. /66/.

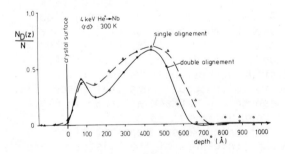

Fig.14. Damage distribution in a Nb single crystal after bombardment with $5.1 \cdot 10^{16}$ He^+ ions/cm^2 measured with 150 keV H^+-ions in 100 111 double alignment and 100 single alignment. The Rutherford backscattering spectra have been analysed using the linear increase model /44/.

of the misaligned layer as determined by these Rutherford backscattering spectra is equal to the deckeldicke.

The results of these measurements in double as well as in single alignment have been used to determine damage distributions before blistering occurs taking the linear increase model /44/. The results are shown in Fig.14. Two damage peaks are well resolved. The peak close to the surface corresponds to the range and/or damage distribution for 4 keV He^+ in Nb as calculated from theory /67/ while the second large peak could not yet be clearly interpreted. It may be due to cavities formed in the region of the highest stress at the interface between the implanted layer and the bulk material. More investigations especially also with a transmission electron microscope are necessary. It is interesting to note, that the same number was taken for the adjustable parameter ß in evaluating both the single and the double aligned spectra from the same probe.

CONCLUSIONS

We have tried to briefly summarize the status of damage distribution measurements in crystals by channeling methods. Best results have been obtained for the semiconductors Si and Ge and brittle materials as W, while for most metals at room temperature, due to clustering of the damage and the built up of stress, less information can be obtained up to now. However, the method had been very successfull for the investigation of bubble and blister formation in metals. The sensitivity of the measurements can be increased by double alignment and by cooling the crystal to $\leq 30°K$ /70/.

In the formulae used to evaluate the measured spectra the simple model of the aligned and the random beam in channeling experiments has been used. Further energy loss of the probing ions has been generally neglected, except for getting information about the depth of the backscattering. Thus the results are only correct for small depth compared to the total depth probed by the beam. More systematic investigations seem to be necessary especially about the dependence of the method on the energy of the analysing beam on the divergence and the angle of incidence of the beam and about double alignment compared with single alignment. Especially a detailed comparison between the results of channeling measurements with those of other methods as transmission electron microscope is necessary in order to get better correlation with the observed dechanneling and the damage configurations /68,69/. More well defined measurements on different metals should be done.

REFERENCES

1) M.T.Robinson, D.K.Holmes, O.S.Oen, Appl.Phys.Lett. $\underline{2}$, 30 (1963) and Phys.Rev. $\underline{132}$,2385 (1963)

2) J.Lindhard, Mat.Fys.Medd. $\underline{34}$ no 14 (1965)

3) S.Datz, C.Erginsoy, G.Leibfried, H.O.Lutz, Ann.Rev.Nucl. Sci. $\underline{17}$, 129 (1967)

4) D.S.Gemmell,Rev.Mod.Phys.$\underline{46}$, 129 (1974)

5) Morgan (Editor) "Channeling", John Wiley & Sons (1973)

6) B.W.Farmery, R.S.Nelson, R.Sizmann, M.W.Thompson, Nucl. Instr. Meth. $\underline{38}$,231 (1965)

7) E.Bøgh, Cairo Solid State Conf., Interaction of radiation with solids, ed. by A.Bishay (Plenum Press) 1966

8) J.U.Andersen. M.W.Gibson, E.Uggerhoj, Proc.Int.Conf. on Application of Ion Beams to Semiconductor Technology Grenoble (1972) Ed.E.Glotin

9) E.Bøgh, Can.J. of Phys. $\underline{46}$,653 (1967)

10) J.A.Davies, J.Denhartog, L.Eriksen, J.W.Mayer, Can.J.Phys. $\underline{45}$, 4053 (1967)

11) E.Bøgh, P.Høgild, I.Steensgaard, "Radiation Damage in Reactor Materials", Vol. 1, p.77 (1969), IAEA, Vienna

12) F.H.Eisen, B.Welch, J.E.Westmoreland, J.W.Mayer, "Atomic Collisions Phenomena in Solids", Proc.Int.Conf. Univ. Sussex p111, Sept.1969. North Holland (1970).J.E.Westmoreland, J.W.Mayer, F.H.Eisen, B.Welch, Rad.Eff. $\underline{6}$, p.161 (1970)

13) L.C.Feldmann, J.W.Rodgers, J.Appl.Phys. 41, 3776 (1970)

14) J.W.Mayer, L.Eriksen, J.A.Davies "Ion Implantation in Semi-conductors" Abroad Press (1970)

15) J.F.Ziegler, J.appl.Phys.$\underline{43}$, 2973 (1972)

16) J.W.Mayer, Proc.Int.Conf.Rad.Eff.in Semiconductors, Albany (1970) and Rad.Eff. $\underline{8}$, 269 (1971).

17) "Ion implantation in Semiconductors", ed. J.Ruge and J. Graul, Springer Verlag (1971)

18) J.Hirvonen, W.L.Brown, P.M.Glotin, see ref. (17)

19) H.J.Pabst, D.W.Palmer, Proc.Int.Conf.Defects in Semicond. Reading, July 1972

20) G.Linker, M.Gettings, O.Meyer, Int.Conf.Ion Implantation New York, Dez.1972

21) H.J.Hollis, Phys.Rev. 8B, 931 (1973)
22) Y.Akasaka, K.Horie, K.Yoneda, T.Sakurai, H.Nishi, S.Kawabe, A.Pohi, J.Appl.Phys. 44, 220 (1973)
23) S.U.Campisano, G.Foti, F.Grasso, E.Rimini, "Proc.Int.Conf. on Atomic Collisions in Solids", Sept.1973, Plenum Publ. Corp. p.905 (1974)
24) P.P. Pronko, J.B.Mitchell, J.Shewchun, J.A.Davies, Rad. Eff. 20, 257 (1973)
25) P.P.Pronko, J.Bøttiger, J.A.Davies, J.B.Mitchell, Rad.Eff. 21, 25 (1974)
26) F.H.Eisen, J.Bøttiger, Appl.Phys.Lett. 24, 3 (1974)
27) H.Kudo, T.Oshiyama, M.Mannami, J.Phys.Soc.of Japan 36, 214 (1974)
28) N.Matsunami, N.Itoh, Proc.Int.Conf.Ion Impl.in Semicond. Osaka, Aug.(1974)
29) G.Götz, K.Hehl, F.Schwabe, E.Glaser, Rad.Eff. 25, 27 (1975)
30) P.Baeri, S.U.Campisano, G.Foti, E.Rimini, Appl.Phys.Lett. 26, 422 (1975)
31) C.Erginsoy, Phys.Rev.Lett. 15, 360 (1965)
32) A.F.Tulinov, Dokl.Akad.Nauk.SSSR 162, 546 (1965)
33) J.U. Andersen, L.C.Feldmann, Phys.Rev.B 1 2063 (1970)
34) J.Barrett, Phys.Rev. B 3 1527 (1971)
35) Proc.Int.Conf.Atomic Coll. in Solids IV, ed.by S.Andersen, V.Björkqvist, N.G.E.Johansson, Gausdahl, Sept.1971, Gordon and Breach (1972)
36) Proc.Int.Conf. Atomic Coll.in Solids V, ed.by S.Datz, B.R.Appleton, C.D.Moak, Gatlinburg, Sept.1973, Plenum Publ.Corp.(1974).
37) D.V.Morgan, D. vanVliet ref./ 35/
38) R.Behrisch, B.M.U.Scherzer, H.Schulze, see ref./35/ and Rad.eff. 13, 33 (1972)
39) E.Bonderup, H.Esbensen, J.U.Andersen, H.E.Schiøtt, see ref./35/ and Rad.Effects 13 (1972)
40) H.E.Schiøtt, E.Bonderup, J.U.Andersen, H.Esbensen, see ref /36/.
41) N.Matsunami, N.Itoh, see ref /36/, p.175
42) Proc.Int.Conf.on Ion beam Surface layer analysis, Yorktown Heights (1973) ed.J.W.Mayer, J.Ziegler and Thin Solid films 19 (1973).

43) B.R.Appleton, L.C.Feldmann, Appl.Phys.Lett.15, 305 (1969) and Proc.Int.Conf. on Atomic Collisions in Solids, Univ.Sussex p.417, Sept. 1969, North Holland (1970)

44) J.Roth, R.Behrisch, B.M.U.Scherzer, F.Pohl, Proc.8th Symp. on Fusion Technology (SOFT) Nordwijkerhout 1974.

45) J.Roth, Thesis Technical University of Munich (1974) and Report IPP 9/17, Max-Planck-Institut für Plasmaphysik, Garching, 1974

46) F.Fujimoto, K.Komaki, H.Nakayama, phys.stat.sol.(a) $\underline{5}$ 725 (1971)

47) F.Fujimoto, K.Komaki, N.Nakayama, M.Ishi, see ref. /35/ and Rad.Effects $\underline{13}$, 43 (1972)

48) P.Baeri S.U.Campisano, G.Ciavola, G.Foti, E.Rimini, Appl.Phys.Lett. to be publ.(1975)

49) G.Molière, Z.Naturforschg. $\underline{3a}$ 78 (1948)

50) E.Keil, E.Zeitler, W.Zinn, Z.Naturforschung $\underline{15\ a}$, 1031 (1960)

51) W.T.Scott, Rev.Mod.Phys. $\underline{35}$ 231 (1963)

52) L.Meyer, Phys.Stat.Sol (b) $\underline{44}$, 253 (1971)

53) P.Sigmund, K.B.Winterbon, Nucl.Instr.Meth.$\underline{119}$ 541 (1974)

54) N.Matsunami, J.phys.Soc.Japan (1974)

55) M.L.Swanson, L.M.Howe, A.F.Quennville, Rad.Eff. $\underline{25}$, 61 (1975) and Rad.Eff. to be publ. (1965)

56) K.Morita, R.Sizmann, Rad.Eff. $\underline{24}$, 281 (1975)

57) K.Morita, N.Itoh, J.Phys.Soc.Japan, 30, 1430 (1971)

58) N.Matsunami, N.Itoh, this conference proceedings (1975)

59) H.Schulze, Thesis Technical University of Munich (1971) and IPP Report 7/3, Max-Planck-Institut für Plasmaphysik, (1970)

60) P.P.Pronko, preprint ANL, to be publ. Proc. of the Int. Conf. atomic collisions in solids, Amsterdam, Sept.1975

61) Y.Quéré, Phys.Stat.Sol. $\underline{30}$, 713 (1968)

62) Y.Quéré, Ann.Phys. $\underline{5}$, 105, (1970)

63) D.V.Morgan, D.van Vliet, Proc.Int.Conf. on Atomic Coll. Phenomena in Solids, Univ. of Sussex, p.476, Sept.(1969) North Holland (1970)

64) D.K.Sood, G.Dearnaley, J.Vac.Sci.Techn.$\underline{12}$, 463 (1975)

65) J.Roth, R.Behrisch, B.M.U.Scherzer, J.Nucl.Mat. $\underline{53}$, 147 (1974)

66) J.Roth, R.Behrisch, B.M.U.Scherzer, Proc.Int.Conf.Appl.of Ion Beams to Metals, Albuquerque, p.573, Sept.1973

67) H.Schiøtt, Mat.Fys. Med. $\underline{35}$, no 9,(1966)

68) K.L.Merkle,P.P.Pronko, D.S.Gemmell, R.C.Michelson, J.R.Wrobel, Phys.Rev. $\underline{B8}$, 1002 (1973)

69) P.P.Pronko, K.L.Merkle, Proc.Int.Conf. Appl.Ion Beams to Metals Albuquerque p 481, Sept.1973, Plenum Publ.Corp.(1974)

70) J.Davies, private communications.

ACKNOWLEDGEMENTS

It is our great pleasure to thank J.Davies, N.Itoh, P.Pronko and E.Rimini for kindly sending us their informations prior to publication and for several helpful remarks in preparing this manuscript. B.M.U.Scherzer contributed in many discussions on dechanneling and damage distributions. Appreciations also to Otto Meyer for patiently waiting for the delayed manuscript.

DISCUSSION

Q: (S. Das) In your Rutherford backscattering spectra you obtain the thickness of blister skin by measuring the width of the misaligned peak. Do you not think there will be some contribution to this misaligned peak from the bottom of the blister because of the very high gas pressure. This may result in an overestimation of the thickness of blister cover (deckeldicke).

A: (R. Behrisch) The region below the deckeldicke may well contain strain in the lattice. However, as pointed out in this talk, this will cause predominantly dechanneling while from the deckeldicke direct backscattering dominates. Further we have also measured the deckeldicke in the scanning electron microscope at He implantation energies above 9 keV. There is good agreement with the results of the Rutherford backscattering measurements.

Comment: (F. Fujimoto) Dr. K. Komaki and myself investigated the dechanneling due to lattice defects and a method to analyze the defects distribution. This one is one kind of the diffusion model. This work was reported in phys. stat. sol. (1970), in the conference of atomic collision in solids in Gausdal (1971) and in US-Japan-Seminar in Kyoto (1971) (JSPS press) for ion implantation into semiconductors. The results are similar to those of Itoh and Matsunami.

HELIUM TRAPPING IN ALUMINUM AND SINTERED ALUMINUM POWDERS*

S. K. Das, M. Kaminsky and T. Rossing[†]

Argonne National Laboratory

Argonne, Illinois   60439

ABSTRACT

The surface erosion of annealed aluminum and of sintered aluminum powder (SAP) due to blistering from implantation of 100-KeV ^{4}He$^+$ ions at room temperature has been investigated. A substantial reduction in the blistering erosion rate in SAP was observed from that in pure annealed aluminum. In order to determine whether the observed reduction in blistering is due to enhanced helium trapping or due to helium released, the implanted helium profiles in annealed aluminum and in SAP have been studied by Rutherford backscattering. The results show that more helium is trapped in SAP than in aluminum for identical irradiation conditions. The observed reduction in erosion from helium blistering in SAP is more likely due to the dispersion of trapped helium at the large Al-Al$_2$O$_3$ interfaces and at the large grain boundaries in SAP than to helium release.

INTRODUCTION

During the operation of thermonuclear reactors the surfaces of components exposed to bombardment by energetic particles from the plasma region can be seriously eroded by radiation blistering.[1,2] In this process, particles from the plasma strike exposed surfaces with sufficient energy to be implanted in the lattice.[2,3] Near the end of their range, where damage is intense and the gas concentration is sufficiently high, gas bubbles form. Such bubbles near the surface region can grow and deform the surface skin to form blisters. Helium blistering, caused by implantation of helium projectiles, can result in serious surface erosion due to peeling of the blister skin upon rupture.

One possible way to reduce surface erosion due to helium blistering is to maintain the surfaces at a high temperature (e.g., > 900°C for Nb and V) at which some of the implanted helium is released without forming large bubbles.[3-5] However, the operating temperatures of various fusion reactor components may be limited by other design criteria. A more desirable solution would be to choose a material with a microstructure which minimizes the formation of blisters. A promising material appeared[6] to be sintered aluminum powder (SAP), which consists of an aluminum matrix strengthened by a dispersion of fine Al_2O_3 particles in concentrations ranging from 3 to 15%. It also has a very small grain size in comparison to pure annealed aluminum. SAP maintains its tensile strength and creep resistance at temperatures approaching the melting point of aluminum. A comparison of the surface erosion due to helium blistering in aluminum and sintered aluminum powder (SAP) had been made for irradiation at room temperature with 100-KeV He^+ ions to a dose of 1.0 C/cm^2. The results showed a reduction in erosion rate in SAP by more than three orders of magnitude from the erosion rate in pure aluminum.[6]

In order to determine if the observed reduction in blistering is due to enhanced helium trapping (e.g., at the trapping sites along the large $Al - Al_2O_3$ interface) or due to helium release, Rutherford backscattering studies of the implanted helium profile in annealed aluminum and in SAP have been conducted.

EXPERIMENTAL TECHNIQUES

The SAP used in this study was prepared at the Holifield National Laboratory and contained a nominal 10.5% Al_2O_3 (SAP895).[6] Thin discs of SAP of $\sim$ 60 μm thickness were cut from the billet by spark cutting. The surfaces of SAP were optically polished and then cleaned in ultrasonic baths of trichloroelthylene, acetone, distilled water and methanol, in that order. For Rutherford backscattering measurements the discs were thinned by argon-ion sputtering to a thickness of $\sim$ 5 μg. The polycrystalline aluminum foils used in this study were of high purity ($\sim$ 99.99%). The samples were mechanically polished, degreased in the same four ultrasonic baths used for SAP, annealed for 2 hours at 300°C in a vacuum of $1 - 3 \times 10^{-8}$ Torr, and finally electropolished in a solution containing 617-ml H_3PO_4, 134-ml H_2SO_4, 240-ml H_2O and 156-g CrO_3 at 70°C. For Rutherford backscattering measurements, thin foils of aluminum were prepared first by electropolishing to a thickness of $\sim$ 20 μm and then, thinning by argon-ion sputtering to a final thickness of $\sim$ 3-4 μm. The targets were irradiated at room temperature with 100-keV He^+ ions from a 2-MeV Van de Graaff accelerator to a dose of 0.05 C/cm^2. The vacuum in the target chamber was maintained at $\sim$ 5 x 10^{-8} Torr by ion pumping during irradiation. After helium-ion irradiation Rutherford-backscattering measurements were made with a 1.5-MeV H$^+$ beam. The backscattered ions were energy-analyzed using solid-state surface-barrier detectors; the scattering angle was 90°. Some thick (> 75 μm) SAP and Al targets were irradiated to higher doses with 100-keV ^{4}He$^+$ ions and were examined in a Cambridge Stereoscan S4 - 10 scanning electron microscope. Examination of polished SAP surfaces, before irradiation, at high magnifications (resolution < 200 Å) did not reveal any porosity of the surface.

RESULTS

The significant reduction of helium blistering in SAP as compared to annealed aluminum for identical irradiation conditions (100-keV He$^+$ ions to a dose of 1.0 C/cm^2 at room temperature) is illustrated in Figure 1. Figure 1 (a) shows an enlarged view of a portion of the irradiated area of aluminum. One can see that four exfoliated layers have been removed [layers marked 1 to 4 in Figure 1 (a)]. Figure 1 (b) shows a typical example of blisters formed on SAP. In contrast to the aluminum, where large exfoliation is observed, only a few blisters were ruptured in the case of SAP. The erosion rates estimated from the ruptured and lost skins for aluminum and SAP are 1.75 $\pm$ 0.25 and 0.001 atoms per helium ion, respectively, as reported earlier.[6]

The blister-skin thicknesses in the annealed aluminum and SAP were measured from the scanning electron micrographs to be 0.61 $(^{+0.05}_{-0.04})$ μm and 0.66 $\pm$ 0.02 μm, respectively. The value for aluminum, based on a number of recent observations,[7] is slightly larger than the one reported earlier,[6] but falls within the error limits quoted. The corresponding calculated projected range for 100 keV ^{4}He$^+$ in aluminum is 0.70 μm,[8] a value which is considered to be accurate within 12%.[9] The calculated peak in the distribution of energy deposited into damage in aluminum for 100-keV ^{4}He$^+$ ions according to Brice[8] is 0.59 μm, with an estimated accuracy of $\sim$ 12%. It may be noted that the measured skin-thickness values for aluminum and SAP fall between the calculated projected range and the peak in the distribution of energy deposited into damage.

Rutherford backscattering studies of implanted helium profiles in annealed aluminum and in SAP were conducted in order to determine whether the observed reduction in blistering in SAP is due to enhanced helium trapping or due to helium release. This technique has been described earlier.[10] Figures 2 (a) and 2 (b) show Rutherford-backscattering data for 1.5-MeV protons incident on 100-keV helium-ion-implanted aluminum and SAP, respectively. The

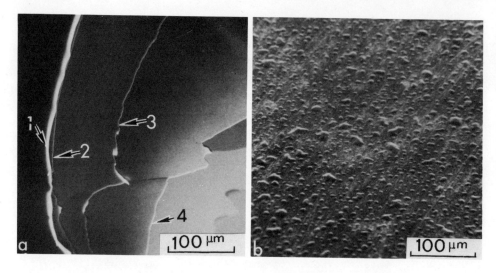

Figure 1. Scanning electron micrographs of surfaces of (a) aluminum and (b) SAP after irradiation at room temperature with 100-keV ^{4}He$^+$ ions to a total dose of 1.0 C/cm^2.

helium-ion dose for both targets was kept at 3.15 x 10^{17} ions/cm^2, low enough to prevent helium release by blister rupture, and the implantations were done at room temperature. An examination of the irradiated surfaces by scanning electron microscopy revealed no visible blisters in either material. The backscattering spectra of 1.5 MeV protons on both aluminum [Figure 2 (a)] and SAP [Figure 2 (b)] reveal the appearance of aluminum, oxygen and helium peaks at onset energies which agree with the calculated onset energies (Figure 2) within the experimental error. One readily notices that the oxygen peak in SAP is significantly stronger than for the aluminum sample as expected. A comparison of the area under the helium peaks [see insert in Figures 2 (a) and 2 (b)] for aluminum and SAP samples reveals that a larger fraction of the implanted helium is trapped in SAP than in aluminum by about a factor of two. The distance between the onset energy and the maximum energy to the helium peak can be converted to a depth scale using known stopping cross-sections for protons in aluminum[8] as shown in the insert of Figures 2 (a) and 2 (b). For SAP the peak of the helium distribution lies at 0.5 ± 0.2 μm [Figure 2 (b)], a value which agrees more closely with the calculated maximum in the damage energy distribution ($\sim$ 0.58 μm) than with the maximum in the projected range distribution (0.70 μm).

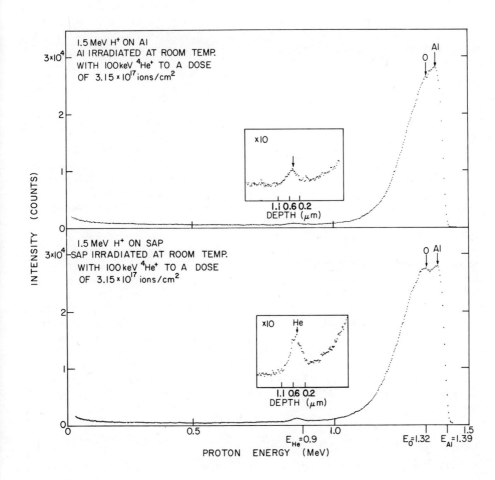

Figure 2. Multichannel analyzer out/put exhibiting the number of protons backscattered from (a) an aluminum foil (b) a SAP foil (both foils implanted with 100 keV ^{4}He$^+$ at room temperature to a dose of 3.15×10^{17} ions/cm^2) as a function of energy. The incident proton energy is 1.5 MeV. The insets in (a) and (b) show a 10 x magnification of the counts observed for helium peaks. The onset energies for aluminum, oxygen and helium are indicated as E_{Al}, E_O and E_{He}, respectively.

Another set of backscattering data was obtained for a higher dose of 4.4×10^{17} helium ions/cm^2. At this dose no blisters were visible. The observed increase in the area under the helium peaks for both SAP and aluminum corresponded with $\pm$ 5% to the increase in dose. Again, at this dose the trapped helium concentration in SAP was higher by a factor of two than in aluminum.

DISCUSSION

The results presented in this paper indicate that more helium is trapped in SAP than in aluminum for identical irradiation conditions given above. In contrast to aluminum the erosion due to helium blistering in SAP is very low. These results imply that the reduction in erosion due to blistering in SAP at room temperature is not due to an enhanced helium release through the grain-boundaries (a typical average grain size in SAP is $\sim$ 0.5 µm as compared to $\sim$ 300 µm in annealed aluminum[6]). The observed reduction is more likely related to helium trapping at the Al-Al$_2$O$_3$ interfaces. In addition the large grainboundary area in SAP (which is more than 5 orders of magnitude larger then for the annealed aluminum used here) provides very effective trapping sites for helium. A dispersion of trapped helium will help to prevent helium-bubble coalescence to larger diameters and subsequent blister rupture and exfoliation. There exists other evidence for helium bubble nucleation at the Al-Al$_2$O$_3$ interface in aluminum alloys containing dispersed Al$_2$O$_3$ particles.[11] Furthermore, the fact that the yield strength of SAP ($\sim$ 35,600 psi) is much higher than for annealed aluminum ($\sim$ 1,700 psi) will help to reduce the blister rupture and exfoliation in SAP from that observed in aluminum.

REFERENCES

[†]Work performed under the auspices of the U. S. Atomic Energy Commission and the Energy Research and Development Administration.

[1] M. Kaminsky, IEEE Trans. Nucl. Sci. NS18, 208 (1971).
[2] S. K. Das and M. Kaminsky, J. Appl. Phys. 44, 25 (1973).
[3] S. K. Das and M. Kaminsky, J. Nucl. Mat. 53, 115 (1974).
[4] M. Kaminsky and S. K. Das, Nucl. Tech. 22, 373 (1974).
[5] W. Bauer and G. Thomas, J. Nucl. Mat. 53, 127 (1974) and also 53, 134 (1974).
[6] S. K. Das, M. Kaminsky and T. D. Rossing, Appl. Phys. Lett. 27, 197 (1975).
[7] S. K. Das, M. Kaminsky and G. Fenske, Paper presented at the International Conference on Applications of Ion Beams to Materials, University of Warwick, Coventry, England, Sept. 8-12, 1975 (to appear in the proceedings).

[8] D. K. Brice, *Ion Implantation Range and Energy Deposition Distributions*, (IFI/Plenum Data Company, New York, 1975) p. 27.
[9] D. K. Brice, private communication (1975).
[10] R. S. Blewer, *Applications of Ion Beams to Metals*, Eds. S. T. Picraux, E. P. EerNisse and F. L. Vook (Plenum Press, New York, 1974) p. 557.
[11] E. Ruedl, *Irradiation Effects on Structural Alloys for Nuclear Reactor Applications*, ASTM STP 484 (American Society for Testing Materials, Philadelphia, 1970) p. 300.

DISCUSSION

Q: (W.K. Chu) In your backscattering spectra have you integrated the area of He peak and calculated the fraction of He gas traped to the total implantation dose? This information will be important for estimation of emission and diffusion.

A: (S.K. Das) We did not estimate the absolute helium concentration in the two targets Al and SAP. We were mainly interested in relative helium contents in order to see whether more helium is released from SAP as compared to Al or not. A comparison of the area under the helium peak in Al and SAP showed that helium concentration in SAP was about a factor of 2 higher than in annealed Al; from which we concluded that reduction in blister exfoliation in SAP is not due to enhanced helium release as compared to Al.

Q: (B.R. Appleton) If SAP should prove to avoid the blistering erosion problem, could one fabricate a fuse wall from SAP?

A: (S.K. Das) It is certainly not easy to fabricate a huge structure like first wall out of SAP. As I mentioned in my talk we chose Al and SAP as a test case to study microstructural effects on blistering.

Q: (R. Behrisch) As I understood, you propose that the reason for not seeing blisters on SAP is due to the longer retention of the implanted helium in SAP compared to Al. Does this mean that blistering and flaking may as well occur on SAP, only at higher doses and a real comparison can only be done for much higher doses than used in your work?

A: (S.K. Das) First let me correct one point in your question. I did not say that we do not see blisters in SAP, what I tried to show is that for same irradiation conditions (irradiation with 100 keV He$^+$ at room temperature to a dose of 1.0 C/cm^2) there is no severe exfoliation of blisters in SAP as compared to Al. It may be possible that at very high doses finely dispersed helium bubbles may coalesce and exfoliation of blister skin layers may occur. However, irradiating annealed Al to the same high dose will cause many more exfoliated blister skin layers and thus the erosion rate of Al will be still higher than SAP. Thus I think our comparison at this dose of 1.0 C/cm^2 gives a good idea of the relative erosion rates due helium blistering in SAP and Al.

HELIUM RE-EMISSION AND SURFACE DEFORMATION IN NIOBIUM DURING MULTIPLE TEMPERATURE HELIUM IMPLANTATION*

W. Bauer and G. J. Thomas

Sandia Laboratories
Livermore, CA 94550

INTRODUCTION

The first wall of controlled thermonuclear reactor (CTR) devices will be exposed to large fluences of He atoms at a variety of temperatures. It has been shown by a number of authors[1] that this type of bombardment results in considerable surface deformation such as blistering and exfoliation. It has also been shown[2] that the surface deformation is due to the agglomeration of He atoms into bubbles. The buildup of He bubbles is a result of incomplete He re-emission during bombardment. It is very desirable for CTR device first wall applications to achieve a surface state which results in complete He re-emission and little or no surface deformation. Furthermore, it is desirable to achieve this state on materials that are structurally suitable and vacuum compatible.

In previous work[2] on He bombarded Nb, the He re-emission reached 100% and a reasonably stable form of surface deformation was achieved at temperatures of 1200°C. In contrast, at temperatures of 400-600°C, highly undesirable flaking occurred with erratic He re-emission. This type of surface deformation results in continuing plasma contamination and wall erosion. Therefore, because of these He bombardment effects, the first wall application of Nb would be restricted to high temperatures. It was suggested[3] that pre-treatment of Nb at 1200°C might result in

---

*This work is supported by the U. S. Energy Research and Development Administration.

considerably less surface deformation at lower temperatures because shorter He diffusion paths are created. In this paper we describe a series of experiments in which the effects of high temperature He pre-treatment of Nb on lower temperature He implantations were investigated.

EXPERIMENTAL RESULTS

The implantations, re-emission measurements, and sample preparation followed a procedure described in detail elsewhere.[2] For each of the implantations discussed in this paper, the samples were pre-implanted at 1200°C to a dose of 2×10^{18} He atoms/cm^2. This treatment results in a typical surface structure shown in the scanning electron micrograph of Fig. 1. This structure is not only typical of this pretreatment but is not substantially changed by subsequent low temperature He implantations. The holes on the surface are due to the intersection of very large faceted He bubbles which have grown from the region where the implanted helium comes to rest. An example of this type of He agglomeration is shown in the transmission electron micrograph of Fig. 2. In this case, the sample has received 4×10^{18} He atoms/cm^2 at 400°C in addition to the 2×10^{18} He atoms/cm^2 at 1200°C. The effect of the lower temperature implantation on the bulk microstructure will be discussed later.

We now discuss the effects of subsequent low temperature He implantation illustrated in the scanning electron micrograph of Fig. 3. The Nb sample was pre-implanted at 1200°C to 2×10^{18} He atoms/cm^2 and then implanted at 400°C to a dose of 4×10^{18} He atoms/cm^2. The lower right-hand portion of the figure shows the unimplanted portion of the sample. More toward the center of the sample there is a narrow strip of almost pure flaking typical of <u>only</u> 400°C He implantation. This occurred over a portion of the sample due to a small ($\sim$ 200 μm) thermally induced difference in sample position relative to the He beam between the 1200°C and 400°C implantations. The remainder of the sample, with the exception of a few small grains, was devoid of flaking, as can be seen in part in the upper left portion of Fig. 3. The major part of the sample received the low and high temperature He implantations and shows surface deformation characteristic of only the pretreatment as shown in Fig. 1. A higher magnification scanning electron micrograph of the implanted region near the edge of another sample is shown in Fig. 4. Since holes are seen below the blister, the skin

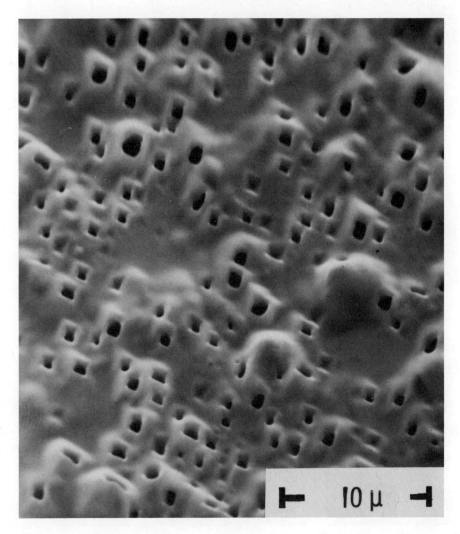

Figure 1. Scanning Electron Micrograph of Nb surface after 2×10^{18} 300 keV He atoms/cm^2 at 1200°C. This surface structure appears to be very stable.

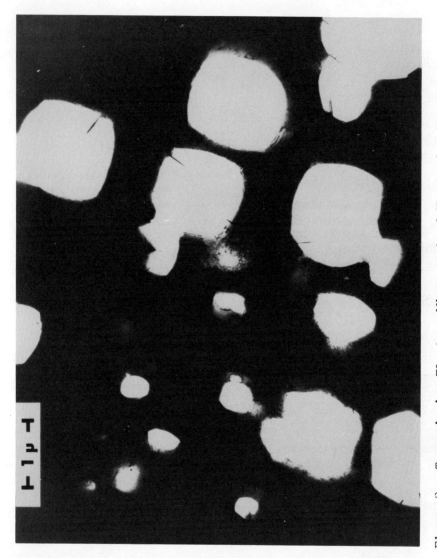

Figure 2. Transmission Electron Micrograph of Nb showing large faceted He bubbles after 2×10^{18} He atoms/cm^2 at 1200°C and 4×10^{18} He atoms/cm^2 at 400°C implantation. 300 keV

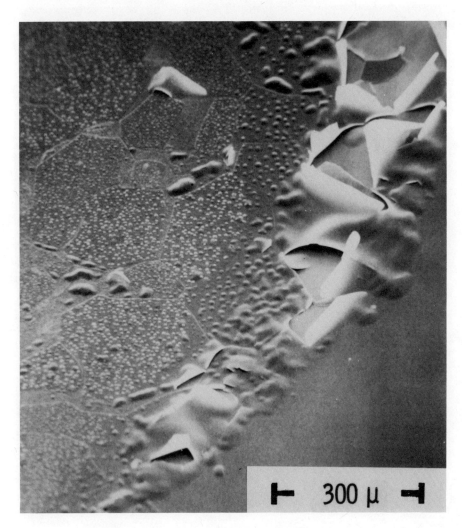

Figure 3. Scanning Electron Micrograph of Nb surface after 2×10^{18} 300 keV He atoms/cm^2 at 1200°C and 4×10^{18} He atoms/cm^2 at 400°C implantation.

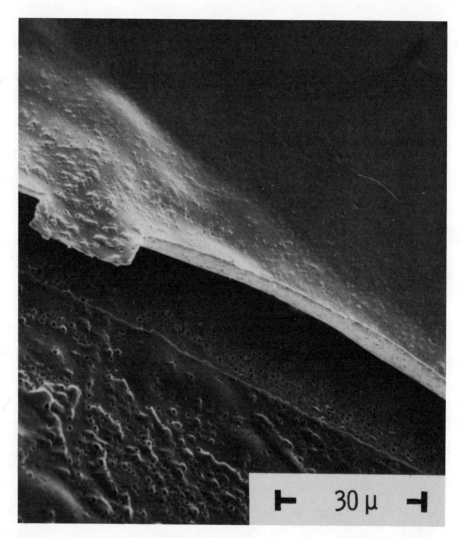

Figure 4. Scanning Electron Micrograph of Nb surface after 2×10^{18} He atoms/cm^2 at 1200°C and 4×10^{18} He atoms/cm^2 at 600°C implantation. 300 keV.

thickness in this region is less than the width of the large bubble-containing layer generated at 1200°C.

The He re-emission during implantation at 400°C of a number of samples is shown in Fig. 5. The re-emission data shown with the open squares is characteristic of a sample without pre-treatment. In contrast, the re-emission at 400°C of the two pre-implanted samples rises immediately and then continues to increase relatively smoothly. Furthermore, the re-emission of the sample implanted at 150 keV (triangle data) is larger than that of the sample implanted with 300 keV. At the lower energy implantation the He atoms come to rest at a range of 0.59 μm with a distribution of FWHM 0.37 μm, which is completely within the large bubble structure which extends from a range of ~1μ m (300 keV) to the surface. Therefore, there is a much higher probability of He re-emission in the lower energy implantation case due to a shorter diffusion path of the He atom to a surface. It should be pointed out that the re-emission at 400°C without pretreatment is the same for 150 and 300 keV implantations. These re-emission data are in general agreement with the understanding developed from the scanning electron microscopy data.

However, some He agglomeration and subsequent release, as indicated by the more sensitive re-emission measurements, occurs even in the pretreated samples. Further evidence for this He agglomeration can be found by examining the transmission electron micrograph shown in Fig. 6. This is a higher magnification micrograph of one of the large bubbles shown in Fig. 2. There exists a distribution of small bubbles which are not found in samples implanted only at 1200°C. The bubble size and density is similar to that observed earlier for He implantation at 400°C to a dose of 1×10^{18} He atoms/cm^2 (35 Å bubbles at a density of 7×10^{16}/cm^3). Therefore their existence must be attributed to the 400°C implantations.

SUMMARY

It has been shown that the creation of a porous surface state by He implantation at 1200°C in Nb prevents further surface deformation up to a He dose of ~4×10^{18} He/cm^2 at 400°C - 600°C where flaking normally occurs. Some residual He trapping still takes place at 400°C as is evidenced by He re-emission and transmission electron microscopy. It appears that a porous structure consisting of micron size holes with a spacing of

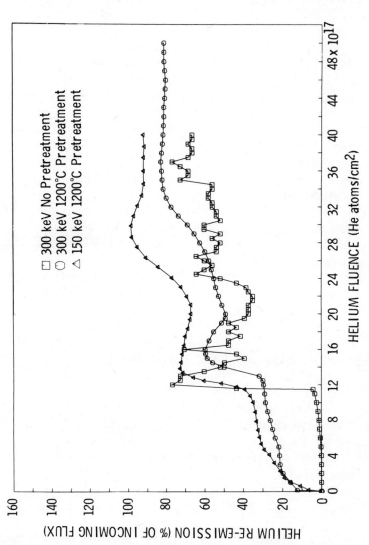

Fig. 5. He re-emission during 400°C implantation of Nb for two identically pretreated (2x10^{18} He atoms/cm^2 at 1200°C) samples and one annealed sample. One of the two pretreated samples was implanted with 150 keV, the other with 300 keV He atoms.

HELIUM RE-EMISSION AND SURFACE DEFORMATION IN NIOBIUM

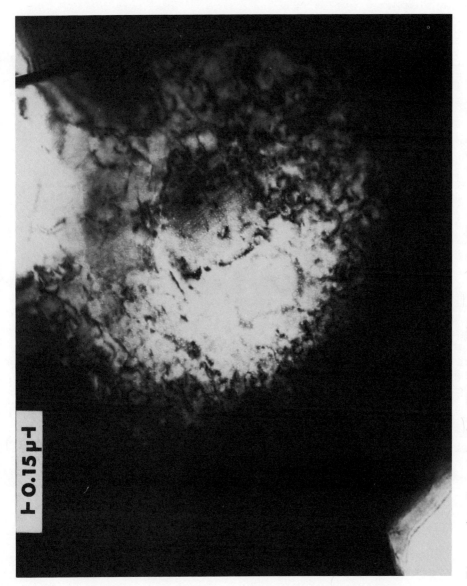

Figure 6. Transmission Electron Micrograph to Higher magnification of a region of the sample shown in Fig. 2.

several microns prevents most of the deleterious effects of He implantation.

ACKNOWLEDGEMENTS

The authors gratefully acknowledge the invaluable assistance of D. H. Morse and L. A. Brown.

REFERENCES

1. See for example

    G. J. Thomas and W. Bauer, J. Nucl. Mat. 53, 134 (1974).
    S. K. Das and M. Kaminsky, J. Nucl. Mat. 53, 115 (1974).

2. W. Bauer and G. J. Thomas, Nucl. Metallurgy 18, 255 (1973).

3. R. L. Hirsch, DCTR/ERDA - private communication.

DISCUSSION

Comment: (J. Biersack) J. Roth and I got such surface structures with 6 keV He$^+$ on Nb; however, after increasing the dose by 2 orders of magnitude (compared to your present dose) the holes opened up completely until finally just the rims remained between closely adjacent craters.

A: (W. Bauer) At lower implantation energies (such as 6 KeV) sputtering plays an important role at large doses. This probably accounts for the microstructure observed in your work. At our higher energies (300 keV) even large dose irradiations $10^{19} - 10^{20}$ He/cm^2 does not appear to alter the hole structure.

STUDIES ON COMPOUND THIN FILM SEMICONDUCTORS BY ION BEAM AND

ELECTRON MICROSCOPY TECHNIQUES

S.U. Campisano[1], G. Foti[1], E. Rimini[1], G. Vitali[2], and C. Corsi[3]

[1]Instituto di Struttura della Materia, Università di Catania, Corso Italia, 57 - I95129 Catania, Italy
[2]Instituto di Fisica-Facoltà di Ingegneria, Università di Roma
[3]Laboratoria di Elettronica dello Stato Solido C.N.R., Roma

Multilayer heteroepitaxial growth of $Pb_xSn_{1-x}Te$ films obtained by radio frequency sputtering on BaF_2 and Ge substrates has been analyzed by ion beam backscattering techniques. Channeling effect measurements have been used to investigate both the heteroepitaxial relationship between substrate and deposited layer also in the presence of an interposed gold film, and the lattice perfection of the compound semiconductor. Reflection high energy electron diffraction (RHEED), electron surface replicas and scanning electron microscopy have been adopted to provide complementary information. Single crystal and well oriented structures of $Pb_xSn_{1-x}Te$ films have been obtained both on single crystal of germanium and on gold underlying films in spite of the high mismatch in the lattice parameters between the grown film and the substrate.

INTRODUCTION

High quality $Pb_xSn_{1-x}Te$ large area single crystal films have been recently obtained by radio frequency sputtering on alkali halides and semiconductor substrates(1,2). These ternary chalcogenides have shown interesting photoconductive properties and are used in the manufacturing process of infrared detector arrays. In addition metallic layer are involved either as free-of-noise ohmic contacts or as active Schottky barrier heterojunctions(2,3) in integrated electronic structures. The role played by the interposed metal film in determining the heteroepitaxial growth structure of the thin film compound semiconductor is then of relevance and must be detailed for an understanding of the mechanisms involved.

Information about the multilayer thin film structure and the underlaying bulk substrate material are generally required in a correlated way to characterize thin film integrated optoelectronics. As diagnostic tools X-rays or electron diffraction and electron microscopy have been mainly used to investigate the epitaxial growth and the structure of the deposited film (4).

Ion backscattering techniques (5,6,7), widely adopted in these last years to investigate implanted semiconductors and metals (8,9), interdiffusion in thin films (10), silicide (11) and compound (12) formation, is now used in analyzing thin film compound semiconductors. Thickness, atomic composition, profile-concentration uniformity, heteroepitaxial growth relationship, lattice perfection and possible interface reactions between sputtered film and single crystal substrate are investigated by ion beam analysis as a function of the most important experimental parameters involved in the deposition process. The obtained results are compared to those supplied by reflection high energy electron diffraction (RHEED) analysis on the crystalline order of the outermost atomic layers. Scanning electron microscopy and surface replicas are also used to examine surface morphology and topography and to provide lateral resolution.

EXPERIMENTAL

$Pb_xSn_{1-x}Te$ and PbTe films have been deposited by radio frequency sputtering technique which provides a mixed Ar-Te gaseous plasma (1). Cleaved BaF_2 and optically polished and etched Ge substrates have been maintained at fixed temperature ($\pm$ 5°C) in the range 250°-450°C. Gold layers have been deposited at a substrate temperature of about 200°C by normal vacuum technique. The depositions have been made on masked samples so that in the same sample there was an uncovered substrate portion.

Analyses have been performed by MeV He^+ beam backscattering using usual set-up with an overall energy resolution of 20 keV. Channeling effect measurements have been carried out by means of a goniometric stage which allowed sample orientation in any space direction with a sensitivity of 0.05°, comparable with the beam angular spreading. Samples have been initially aligned along a major axis in the uncovered part and then the incident beam has been translated to various portions of the sample to allow a direct comparison of the yields.

Reflection high energy electron diffraction patterns have been obtained by an A.E.I. EMG6 electron microscope at 60 kV. A two stage carbon replica shadowed with Ni-Cr has been used to study surface morphology of the film. Scanning electron microscope analyses have been performed to investigate surface topography.

RESULTS AND DISCUSSION

Typical energy spectra of 2.0 MeV He^+ particles backscattered from PbTe and $Pb_xSn_{1-x}Te$ films sputtered on Ge substrates are reported in Fig.1a and 1b respectively. The energies corresponding to He ions backscattered from Pb, Sn and Te surface atoms, are marked in the upper part of the figure with the Pb signal separated from those of the other elements. The atomic composition of the PbTe

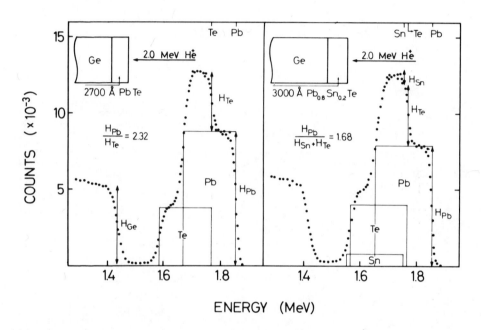

Fig.1 - Energy spectra of 2.0 MeV He^+ beam scattered from a 2700 Å thick PbTe (left) and from 3000 Å thick $Pb_xSn_{1-x}Te$ (right) layers r.f. sputtered on Ge substrates. The surface energy positions of Te, Sn and Pb are indicated by arrows. The contributions of the different elements to the resulting spectra are shown separately in a schematic way.

layer can be obtained straightforwardly by the signal height ratio Pb/Te. The experimental value 2.32 is in good agreement with the calculated 2.45 figure for the stochiometric composition PbTe. The film thickness, given in Å, has been determined from the energy width of the signals, through the knowledge of the stopping cross section (13) and assuming their additivity (Bragg's rule) and bulk density values.

The Te and Sn atom signals cannot be separated because the

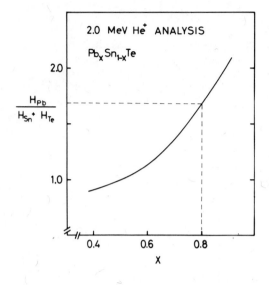

Fig.2 - Ratio between the Pb and the Te plus Sn signal heights as a function of the x value in the formula $Pb_x Sn_{1-x} Te$.

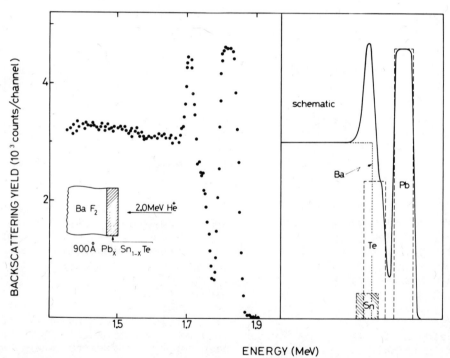

Fig.3 - Energy spectrum of 2.0 MeV He^+ beam scattered from 900 Å thick $Pb_{0.8}Sn_{0.2}Te$ layer sputtered on BaF_2 substrate. The schematic spectrum as due to the overlap of the several contributions is reported in the right hand side.

backscattering energy difference (15 keV) for 2.0 MeV He^+ is within
the energy resolution of the system. Assuming a composition
$Pb_xSn_{1-x}Te$, the x value can be determined with a very good accuracy
from the ratio $H_{Pb}/(H_{Sn}+H_{Te})$ as argued from Fig.2 in which it is
plotted versus x. In the x value range of interest in the present
investigation the high slope of the curve implies a small error for
the experimental determination of the composition, e.g. an error of
$\pm 10\%$ in the experimental height ratio corresponds to an uncertain--
ty of $\pm 3 \%$ in the evaluation of the x composition value. For the
case analized in Fig.1b the height ratio value is 1.68 and it corresponds to $x=0.8\pm0.02$, i.e. $Pb_{0.8}Sn_{0.2}Te$. A fair good agreement has
been found between this x value and that computed from the ratio of
the Pb to Ge or Ba substrate signal height.

As another example of application of the ion beam technique,
the backscattering energy spectrum of a 900 Å thick $Pb_{0.8}Sn_{0.2}Te$
film r.f. sputtered on BaF_2 is reported in Fig.3, together with its
reconstruction shown in the right hand part. The absence of any
tail in the low energy edge of the Pb or Sn-Te signals and in the
high energy edge of Ba indicate that if any reaction of the sputtered film with the substrate occurs it extends over a thickness less
than 200 Å, i.e. less than the depth resolution of MeV He^+ backscattering.

Channeling effect measurements have been used to provide information on the crystal structure, on the lattice perfection and on
the heteroepitaxial relationship between film and substrate. Backscattering energy spectra obtained for 2.0 MeV He^+ beam impinging

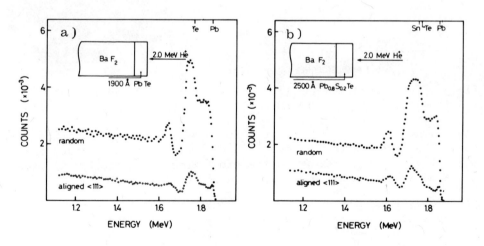

Fig.4 - Random (●) and aligned (o) energy spectra of 2.0 MeV He^+ beam
scattered from a 1900 Å thick PbTe (left) and from a 2500 Å thick
$Pb_{0.8}Sn_{0.2}Te$ layers sputtered on cleaved (111) BaF_2 substrates.

along a random direction and parallel to the <111> axis of the BaF_2 substrate are reported in Fig.4a and 4b for a 2500 Å thick $Pb_{0.8}Sn_{0.2}Te$ and for a 1900 Å thick PbTe layer respectively. The simultaneous decrease of the yields due to backscattering from the substrate and from the deposited film atoms indicate that the layer grows epitaxially on the BaF_2 crystal. The minimum yield of the Pb signal is about 12% for both compositions pointing out that a good crystal structure with a low content of defects has been obtained. The parallelism of the <111> substrate planes with those of the deposited film is also evidenced by the overlap of the stereographic maps recorded by Rutherford backscattering on the film and on an uncovered portion of the substrate. A further check of the heteroepitaxial growth has been obtained through the angular yield profiles of Pb and of Ba covered substrate signals shown in Fig.5a. The minimum Ba and Pb yield occurs at the same angular position indicating that no misalignement occurs within 0.05°. The high aligned

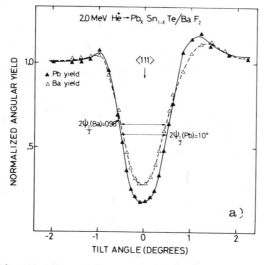

Fig.5 - Normalized angular yield profiles for 2.0 MeV $He^{\pm}$ beam scattered from a 900 Å thick $Pb_{0.8}Sn_{0.2}Te$ layer suputtered on (111) cleaved BaF_2 substrate. The Ba signal (Δ) in the covered portion reaches the minimum value at the same angular position as the Pb signal (▲) does. The RHEED pattern of 500 Å thick $Pb_xSn_{1-x}Te$ layer taken along the <11$\bar{2}$> azimuth is shown in the right hand upper part, while the electron surface replica is reported in the lower part.

yield (∼ 30%) of the covered Ba substrate signal is due to the angular spreading experienced by the beam particles in traversing the overlaying film; in the uncovered substrate region the Ba aligned yield is instead 4%. The single crystal nature of the deposited film has also been tested by RHEED whose pattern is reported in Fig.5b. The appearance of spots clearly indicates that we deal with a single crystal, at least as it concerns the outermost atomic layers of the film analyzed by electron diffraction. The Ni-Cr shadowed surface replica shown in Fig.5c evidences the structure of the growing film.

A parallel investigation carried out using germanium single crystals as substrate showed that well oriented heteroepitaxial $Pb_xSn_{1-x}Te$ films can be obtained by r.f. sputtering. This result is apparently conflicting with the assumption that strictly cristallographic relationship between the substrate and the condensing material are required for heteroepitaxial single crystal growth. The matching of the lattice parameters and thermal expansion coefficients are generally considered of relevance in assisting hetero-

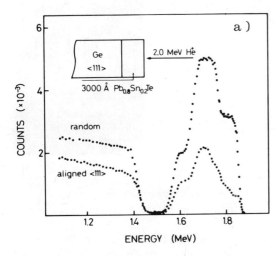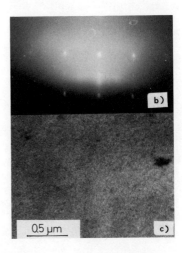

Fig.6 - Energy spectra of 2.0 MeV He^+ particles impinging on the <111> substrate direction (o) and on a non-channeled direction of a Ge single crystal covered with a 3000 Å thick $Pb_{0.8}Sn_{0.2}Te$ sputtered film. The RHEED pattern taken along the <11$\bar{2}$> azimuth and the electron scanning micrograph are shown in the upper and in the lower right-hand part respectively.

epitaxial single crystal growth. The lattice parameters of Ge and $Pb_{0.8}Sn_{0.2}Te$ are 5.675 and $\sim 6.4_1$ Å respectively, while the thermal expansion coefficient $\Delta l/l\Delta T(°K^{-1})$ are $5.8 \ 10^{-6}$ and $18.8 \ 10^{-6}$ at 300°K for Ge and $Pb_{0.8}Sn_{0.2}Te$ respectively. Fig.6 reports random and <111> aligned energy spectra for 2.0 MeV $He^{\pm}$ beam scattered from a 3000 Å thick $Pb_{0.8}Sn_{0.2}Te$ layer r.f. sputtered on Ge <111> substrate at 280°C. The minimum yield ranges from 20% to 45% on going from the surface to the interface, these values become 14% and 40% respectively for 1.0 MeV $He^{\pm}$ analyzing beam. The RHEED pattern of the film is shown in Fig.6b revealing that the outermost atomic layers are well oriented. However the high minimum yield and dechanneling rate point out the presence of defects due to the relief of the strains induced during the growth to fit the lattice of the Ge substrate.

The role played by a gold layer interposed between film and substrate in determining the structure and morphology of the film has been also investigated for the relevance of metallic contact

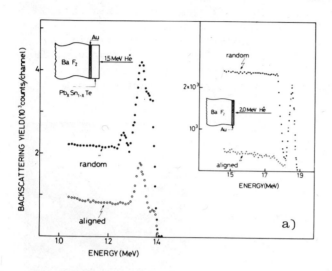
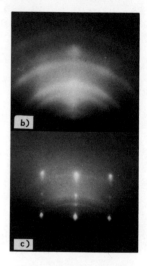

Fig.7 - Backscattering energy spectra of 2.0 MeV He^+ beam impinging on a random (o) and along the <111> axial direction of a BaF_2 single crystal covered with a 300 Å thick gold layer and 900 Å thick $Pb_{0.8}Sn_{0.2}Te$ sputtered film. The insert reports the spectra of the Au covered portion. RHEED patterns of the Au layer and of the superimposed $Pb_{0.8}Sn_{0.2}Te$ film are shown in the upper and in the lower right hand side respectively.

problems in solid state devices. $Pb_xSn_{1-x}Te$ layers grow epitaxially on gold single crystal films (4) deposited on alkali halides substrates.

A 900 Å thick $Pb_xSn_{1-x}Te$ layer deposited on BaF_2 substrate with an interposed 300 Å thick gold film has been analyzed and the results are summarized in Fig.7. The backscattering energy spectra of 2.0 MeV He^+ particles impinging on a random and on the <111> BaF_2 substrate direction are reported in Fig.7a. The insert shows the energy spectra for the portion of the sample covered only by gold; the polycrystalline nature of the metal film does not allow reduction in the gold signal as the Ba signal does. A simultaneous reduction has been observed instead for the yield of the overlaying $Pb_xSn_{1-x}Te$ film and of the Ba substrate for an aligned incident beam direction. The $Pb_xSn_{1-x}Te$ film grows parallel to the (111) plane of BaF_2, although the interposed gold film shows a polycristalline disordered structure. These results are confirmed by the RHEED patterns (see Fig.7b and Fig.7c).

The structure of a gold layer deposited onto Ge <111> substrate has been also evaluated. Channeling effect measurements reported in Fig.8 indicate that a 2300 Å thick Au layer grows uniformly and epitaxially. Angular yield profiles around the <111> direction for an uncovered portion of Ge and for the Au covered portion are reported in Fig.8b. The half-width at half minimum ($\psi_\frac{1}{2}$) of the Ge yield follows the normal $E^{-\frac{1}{2}}$ dependence on the beam energy, while only small changes are measured in the Au critical angle. The Au minimum

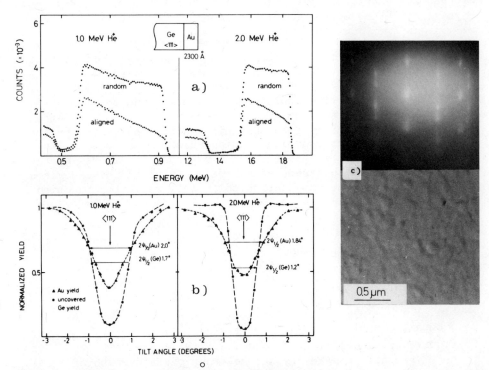

Fig.8 - Analysis of a 2300 Å thick Au film deposited on Ge <111> substrate: a) backscattering energy spectra of 1.0 and 2.0 MeV He^+ beam on the <111> and on a non channeled direction; b) angular yield profiles of Au (▲) and of an uncovered portion of Ge substrate (o); c) RHEED pattern along the <11$\bar{2}$> azimuth and Ni-Cr shadowed replica of the Au layer.

yield changes from 25% to 39% on going from 1.0 to 2.0 MeV He^+. This energy dependence of both minimum yield and critical angle is to be attributed (14) to the presence of defects inbedded in a single crystal structure characterized by a mosaic spread of width comparable with the Au critical angle ($\sim 1°$) or to large single crystals slightly misoriented to each other. This last hypothesis agrees with RHEED analysis, shown in Fig.8c, where a double pattern indicates the presence of large single crystals ($\sim 1\ mm^2$) comparable with the electron beam area.

CONCLUSIONS AND SUMMARY

Ion beam and electron microscopy techniques have been used to characterize the growth of compound semiconductors on different substrates also in the presence of an interposed metallic layer. The ion beam analysis has provided in depth information, i.e. thickness, composition, uniformity, possible interface reaction, lattice defect and epitaxial relationship between film and substrate. Electron microscope techniques gave information on surface topography and crystal structure of the outermost probed atomic layers.

In spite of the high lattice parameter mismatch between $Pb_xSn_{1-x}Te$ growing film and Ge or Au substrate, well oriented heteroepitaxial films have been obtained. Defects are present in these layers because of the relief of strains induced during the growth to fit the lattices. The substrate temperature and the deposition rate influence both the crystal structure and the lattice defect content. The overcoming of the matching lattice parameter problem is the first step for reducing considerable limitations in the choice of substrates suitable to obtain properly tailored energy gap values in film-substrate structure as in solid state devices (e.g. photovoltaic heterojunction solar cell and heterojunction infrared detectors).

The authors are indebted to Mrs M.Compagnino for her care in typing the manuscript. Thanks are due to Mr V.Scuderi and Mr.V.Piparo for their technical assistance.

REFERENCES

1) C.Corsi, J.Appl.Phys. $\underline{45}$, 3466 (1974)
2) C.Corsi, E.Fainelli, G.Petrocco, G.Vitali, S.U.Campisano, G.Foti and E.Rimini, "Single crystal heteroepitaxial growth of $Pb_xSn_{1-x}Te$ films on germanium substrates by r.f. sputtering" - submitted to Thin Solid Films.
3) G.Y.Robinson, Solid State Electronics, $\underline{18}$, 331, (1975).
4) C.Corsi and G.Vitali, Thin Solid Films, $\underline{25}$, 511 (1975)
5) W.K.Chu,J.W.Mayer,M.A.Nicolet,T.M.Buck,G.Amsel and F.Eisen, Thin Solid Films, $\underline{17}$; 1 (1973).
6) E.Rimini, in "Physics of Thin Films, International College on Applied Physics (Capri August 1974), Edited by N.A.Mancini and I.F.Quercia.
7) W.K.Chu,J.W.Mayer,M.A.Nicolet,S.U.Campisano and E.Rimini, in "Backscattering" Catania Working Data, J.W.Mayer and E.Rimini Editors,Catania 1975.
8) J.W.Mayer,L.Eriksson and J.A.Davis "Ion Implantation In Semiconductors" (Plenum Press, N.Y. 1970).

9) See the proceedings of the "International Conference on Ion Beam Surface Layer Analysis", published in Thin Solid Films 19 (1973), and the proceedings of the Conference on "Application of Ion Beams to Metals", Ed.s S.T.Picraux, E.P.EerNisse and F.L.Vook, (Plenum Press, N.Y.1974).
10) Proceedings of the Conference on "Low Temperature Diffusion and Applications to Thin Films" published in Thin Solid Films 25 (1975).
11) J.W.Mayer and K.N.Tu, J.Vac.Sci Technol. 11, 86, (1974).
12) S.U.Campisano, G.Foti, E.Rimini, S.S.Lau and J.W.Mayer, Phil.Mag. 31, 903 (1975).
13) W.K.Chu and J.F.Ziegler, Atomic Data and Nuclear Data Tables, 13, 463, (1974).
14) D.Sigurd, R.W.Bower, W.F.vanderWeg and J.W.Mayer, Thin Solid Films 19, 319, (1973).

DISCUSSION

Q: (R. Behrisch) How do you explain the increase in the aligned yield from the near surface region for the Au film in increasing the analysing energy from 1 MeV to 2 MeV ?

A: (E. Rimini) The increase can be due to a mosaic spread structure of the epilayer. With increasing energy the critical angle value decreases and then the layer fraction of He probed area (corresponding to grains misoriented at angle $\simeq \psi_{crit}$) produces a random signal.

ENERGY DEPENDENCE OF CHANNELING ANALYSIS IN

IMPLANTATION DAMAGED Al*

E. Rimini, S. U. Campisano, G. Foti and P. Baeri

Instituto di Struttura della Materia dell'Universita

Corso Italia, 57 - 195129 CATANIA

S. T. Picraux, Sandia Laboratories

Albuquerque, New Mexico 87115

Ion channeling and backscattering measurements along the $\langle 110 \rangle$ axis of implantation disorder in Al single crystal are reported as a function of the He analyzing beam energy (0.5-2.0 MeV). For 100 keV N^+ implants (8×10^{16} ions/cm^2) appreciable dechanneling is observed near the end of the projected ion range (2400 Å) and the minimum yield increases with decreasing analyzing beam energy. A different behavior has been observed for 150 keV Zn^+ implantation into Al with fluences of 6×10^{15} and 1.2×10^{16} ions/cm^2 respectively. The minimum yield in the Zn implanted region (790Å) shows a $\sqrt{E}$ dependence on the energy E of the analysis beam. These channeling measurements show that the nature of the disorder for light and heavy mass implants in Al is different. The energy dependence of the minimum yield points out that the defects for the N implant are consistent with random scattering centers while for Zn implants extended defects with associated strains, like dislocations, are produced. The energy dependence of the critical angle can also be related to the defect structure.

---

*This work has been supported in part by Centro Siciliano di Fisica Nucleare e di Struttura della Materia and by U. S. Energy Research and Development Administration (ERDA).

INTRODUCTION

Channeling effect measurements have been widely adopted (1) as a diagnostic tool to investigate defect distributions introduced by ion implantation in semiconductors. A good analytical approach (2) has been developed to obtain the depth profile of displaced atoms from the relative reduction in yield of aligned backscattering spectra. In the extension of such studies to damaged metals important differences have been observed in the shape of aligned yields. In semiconductors the direct backscattering of channeled ions by displaced atoms is usually comparable to the dechanneling yield due to small angle scattering. In this case a disorder peak is observed in the channeled spectrum. In metals, however, the direct backscattering (3) is often almost negligible with respect to the dechanneling contribution, so that only a monotonically increasing aligned yield with depth is observed. The two characteristic shapes point out a difference in the nature of the introduced damage. For instance heavy ions in semiconductors give rise to stable displaced atom configurations and amorphous regions. Where as for metals, more extended defects, such as dislocations, and the associated strains appear to be of greater importance (1,5). In addition the damage region sometimes extends to depths larger than the damaging ion projected range. Due to these differences between metals and semiconductors a different approach is needed to analyze the disorder in metals. An analysis of channeling measurements in implanted metals has been developed assuming the additivity of the dechanneling rate in the unimplanted crystal with the dechanneling rate due to defects (3). The dechanneling cross section due to the complex defect structures in metals, are not generally known, in contrast to semiconductors where displaced atoms act as individual scattering centers.

Apart from quantitative estimates of the amount of disorder, it is relevant to test the capabilities of the ion channeling/backscattering technique to distinguish different types of defects. Isolated amorphous regions are characterized by a dechanneling cross section which decreases with increasing analyzing beam energy while an opposite behavior has been predicted for extended defects (6). In this paper the energy dependence of the dechanneling of 0.5-2.0 MeV He ions is investigated for N and Zn implantation disorder in Al. The structure of defects introduced during irradiation would be expected to depend on energy densities of the atomic collision cascades. Thus the present implantations of 100 keV N^+ and 150 keV Zn^+ in Al single crystals are characterized by energy densities of about 5×10^{-5} eV/Å^3 and 5×10^{-3} eV/Å^3 respectively (7).

EXPERIMENTAL TECHNIQUE

Electropolished Al ⟨110⟩ oriented single crystals obtained from Metals Research were used. Implantations were performed along a random orientation at room temperature with 100 keV N^+ and 150 keV Zn^+

ions to fluences up to 8×10^{16} and 1.2×10^{16} ions/cm² respectively. The ion fluxes were $\sim 10^{14}$ N⁺ ions/cm² sec and $\sim 10^{13}$ Zn⁺ ions/cm²-sec. The analyses were performed with a He beam in the energy range 0.5-2.0 MeV by channeling and backscattering techniques. The particles backscattered at 165° were energy analyzed by a surface barrier detector and the usual electronics were used to handle the pulses. The overall energy resolution was about 20 keV. The crystal was oriented by means of a goniometric stage within 0.05°, comparable with the angular spread of the He beam.

RESULTS AND DISCUSSION

a) <u>N Implant</u>

Aluminum single crystals implanted with 8×10^{16} ions/cm² of 100 keV N⁺ has been analyzed by 1.0 and 2.0 MeV He beam energy respectively. The aligned and random backscattering spectra for the unimplanted (a) and implanted (b) region are reported in Fig. 1.

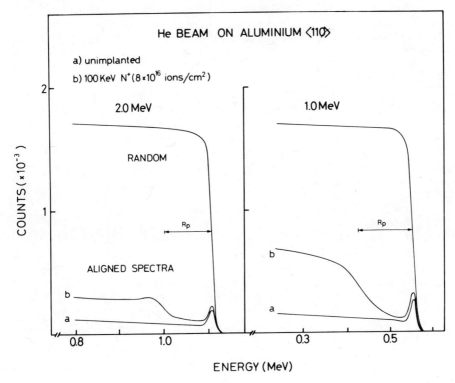

Fig. 1 - Energy spectra of 2.0 and 1.0 MeV He beam impinging along the ⟨110⟩ axis and a nonchanneled direction of an Al crystal (a) implanted and (b) damaged with 8×10^{16} ions/cm² of 100 keV N⁺. The implant and the analysis have been performed at R. T. The energy width of projected range R_p is reported.

The area of the surface peak in the aligned spectrum of the unimplanted crystal corresponds to an oxide layer of about 60 Å; the value (3.5%) of the minimum yield just behind the surface and the low dechanneling rate indicate the presence of a single crystal with a relatively low density of defects.

The 2.0 MeV He$^+$ aligned $\langle 110 \rangle$ yield after implantation exhibits relatively little dechanneling in a depth region which extends from behind the surface peak to about 70% of projected range ($R_p \simeq 2400$ Å) (2) of the implanted 100 keV N$^+$ ions. The aligned yield then increases rapidly and after a small bump it becomes practically flat and parallel to the unimplanted crystal yield. A similar behavior is shown by the spectra obtained for the 1.0 MeV He$^+$ analysis, however the bump disappears and the normalized aligned yield is about a factor of two higher than that corresponding to 2.0 MeV analysis and for depths $\geqslant 1.5$ times the projected range. The observed increases of the normalized aligned yield with decreasing analyzing beam energy agrees with an energy dependence of the dechanneling cross section $\sigma_D \propto E^{-1}$, consistent with a random distribution of scattering centers. This dechanneling behavior may be due to intrinsic Al defects or due to possible influence of the N such as possible nitride precipitation.

b) <u>Zn Implants</u>

A quite different trend of the aligned yield has been observed in the channeling analysis of Al crystals implanted with 150 keV Zn$^+$. Backscattering energy spectra for 2.0, 1.0 and 0.5 MeV He$^+$ analyzing beam energies are shown in Fig. 2 for unimplanted (a) and implanted crystals with 6×10^{15} (b) and 1.2×10^{16} (c) ions/cm^2 of 150 keV Zn$^+$. With decreasing analyzing beam energy the normalized aligned yield decreases for both the fluences, thus indicating an energy dependence of the dechanneling cross section opposite to that found for the N$^+$ implants.

The energy dependence of the dechanneling cross section can be estimated through the difference $\Delta\chi$ of the dechanneled fraction in the damaged and undamaged crystal evaluated at a depth z comparable with the 150 keV Zn$^+$ in Al projected range ($R_p = 750$ Å, $\Delta R_p = 275$ Å(7)). The $\Delta\chi$ values are reported in Fig. 3 as a function of the square root of the beam energy. A linear increase with $(E)^{\frac{1}{2}}$ is clearly shown for both fluences. For comparison the same quantity $\Delta\chi$ is reported in the insert for the N$^+$ implant and it follows an opposite trend.

The energy dependence of $\Delta\chi$ for the Zn implanted Al crystal suggests that the dechanneling of the aligned beam is due mainly to strains (6). This suggests that the predominant crystal imperfections giving rise to a channeling signal are extended defects in host lattice, e.g., dislocation and vacancy loops. In a transmission study of the dechanneling caused by such defects, Queré (6) proposed a model to compute the dechanneling cross section of dislocations.

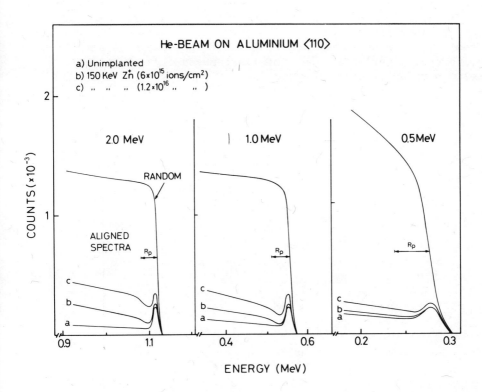

Fig. 2 - Energy spectra of 2.0, 1.0 and 0.5 MeV He$^+$ beam impinging along the $\langle 110 \rangle$ axis and a nonchanneled direction of an Al crystal (a) unimplanted, (b) damaged with 6.0×10^{15} ions/cm^2, and (c) 1.2×10^{16} ions/cm^2 of 150 keV Zn$^+$. The implants and the analyses have been performed at room temperature.

Apart from their exact values, these cross sections depend on energy as $\sigma_Q \propto (E)^{\frac{1}{2}}$, this dependence is related to the local curvature of the lattice near a dislocation. The observed experimental energy dependence is thus in agreement with strain causing dechanneling, although the evaluation of defect density with confidence would require complementary analysis to evaluate σ_Q.

For a further understanding of the nature of damage, channeling angular yields have been measured in Zn implanted Al and the results are summarized in Fig. 4. Here the angular widths are plotted vs. $E^{-\frac{1}{2}}$ for an unimplanted and for a crystal damaged with 1.2×10^{16} ions/cm^2. The critical angle of the undamaged crystal follows the theoretical expected energy dependence and the extrapolated value reaches a null value for E → ∞ as expected from theory. A different trend is observed for the damaged crystal. At low energy (0.5 MeV) the critical angle of the damaged crystal is narrower than the unimplanted one and this decrease can be interpreted in terms of a defect distribution characterized by small displacement from the atomic row. With increasing beam energy the two straight lines cross and the critical angle of the damaged crystal has an extrapolated value of about 0.2°.

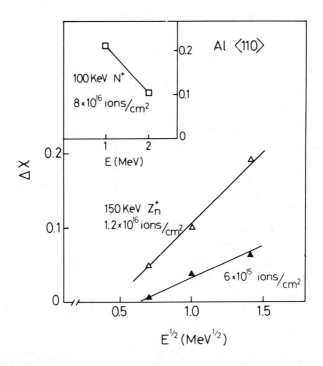

Fig. 3 - Difference $\Delta\chi$ between the normalized aligned yields for Zn damaged and undamaged Al crystals as a function of the square root of the analyzing beam energy. The insert shows the same difference $\Delta\chi$ for the N^+ implant. All the differences have been evaluated at a depth of about R_p from the surface.

In metals, arrays of dislocations produced by irradiation may give rise to small angle grain boundaries or distortions of lattice rows which would result in slightly misoriented regions. At very high energy the channeling technique is sensitive to these small misalignments due to the low values of the critical angle (11).

Thus the energy dependence of the dechanneling and of the angular width are believed to be consistent and to give further information on the nature of the implantation disorder in metals.

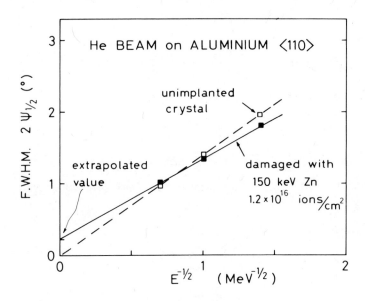

Fig. 4 - The full width at half minimum ($2\Psi_{\frac{1}{2}}$) of the angular yield profiles around the $\langle 110 \rangle$ axis as a function of $E^{-\frac{1}{2}}$ for Zn^+ implant in Al and for a virgin sample.

The authors are indebted to J. A. Davies for clarifying discussions.

REFERENCES

1. J. W. Mayer, L. Eriksson and J. A. Davies, Ion Implantation in Semiconductors (Academic Press, New York, 1970) Chap. 4.

2. E. Bøgh, Can. J. Phys. 46, 653 (1968).
   L. C. Feldman and J. W. Rodgers, J. Appl. Phys. 41, 766 (1970).
   J. F. Ziegler, J. Appl. Phys. 43, 2973 (1972).
   S. U. Campisano, G. Foti, F. Grasso and E. Rimini in Atomic Collisions in Solids, Eds. S. Datz, B. R. Appleton and C. D. Moak (Plenum Press, New York, 1975), p. 905.

Ref. 2 (continued)

J. E. Westmoreland, J. W. Mayer, F. H. Eisen, and B. Welch, Rad. Eff. 6, 51 (1970).

3. K. L. Merkle, P. P. Pronko, D. S. Gemmell, R. C. Mikkelson and J. R. Wrobel, Phys. Rev. B 8, 1002 (1973).

4. M. O. Rualt, B. Jouffrey and P. Joyes, Phil. Mag. 25, 833 (1972).

5. C. Gómez-Giraldes, B. Hertel, M. Rühle and M. Wilkens in Applications of Ion Beams to Metals, Eds. S. T. Picraux, E. P. EerNisse and F. L. Vook (Plenum Press, New York, 1974), p. 469.

6. Y. Quére, Ann. Phys. (N. Y.) 5, 105 (1970).

7. The values of R_p, ΔR_p and ΔR have been provided by D. K. Brice.

8. L. Meyer, Phys. Stat. Solidi B44, 253 (1971).

9. P. Sigmund and K. Winterbon, Nucl. Inst. Methods 119, 541 (1974).

10. J. F. Ziegler and W. K. Chu, Atomic Data and Nuclear Tables 13, 463 (1974).

11. J. A. Davies, private communications.

DISCUSSION

Q: (R. Behrisch) Is it possible that the energy of maximum 2 MeV used for analyzing the damage produced by the 150 keV Zn implantation is still too low in order to get a direct backscattering peak?

A: (G. Foti) No! Maybe too high because by decreasing energy it will decrease the dechanneling on dislocation or strains while the Rutherford cross section will increase thus producing a "direct backscattering peak".

Comment:(G. Dearnaley) With reference to the point raised by Dr. Behrisch we have measured ^{4}He backscattering from ion-bombarded Al crystals after similar doses of heavy ions, using a 3.5 MeV analysing beam. There is still no near-surface peak but the same linear dechanneling described by Dr. Foti.

Comment: (N. Itoh) I wonder that the difference in your energy dependence of $\Delta\chi$ for N^+ and Zn^+ implantations may be due to the difference in the damage concentration rather than the difference in the kind of implanted ions. In view of the diffusion calculation I think that the relative magnitude of the diffusion constant for defect scattering to that for the lattice and electronic scatterings may be an important factor for determining the flux distribution of the channeled particles and may effect the energy dependence of $\Delta\chi$.

Comment: (J. Davies) The lattice location of the two implanted ions (N vs Zn) is probably quite different and this may be responsible for the difference in the type of defect involved. Zinc is almost certainly substitutional in Al, whereas nitrogen may well be in some interstitial or precipitation site.

A: (G. Foti) The atomic concentration of nitrogen and zinc in aluminum are about 10 % and 2 %, respectively, then possible aluminum nitrate precipitation cannot be ruled out.

CHARACTERIZATION OF REORDERED (001) Au SURFACES BY THE COMBINED TECHNIQUES OF POSITIVE ION CHANNELING SPECTROSCOPY (PICS), LEED-AES, AND COMPUTER SIMULATION*

B. R. Appleton, D. M. Zehner, T. S. Noggle, J. W. Miller,
O. E. Schow III, L. H. Jenkins, and J. H. Barrett

Solid State Division, Oak Ridge National Laboratory

Oak Ridge, Tennessee 37830

ABSTRACT

The applicability of positive ion channeling techniques for studying the structure of single crystal surfaces was investigated. Characterization of the (001) surface of epitaxially grown Au single crystals by LEED and AES showed that when the (001) surface was sputter-cleaned and annealed in ultra high vacuum, the normal (1x1) symmetry of the (001) surface reordered into a structure which gave a complex (5x20) LEED pattern as previously reported. The yield and energy distributions of 1 MeV He ions scattered from the Au surfaces were used to determine the effective number of monolayers contributing to the normal and reordered surfaces. Computer simulation calculations were used to correlate the ion scattering results with model constructions for the normal and reordered surfaces. The combined techniques of PICS, LEED-AES, and computer simulation enabled the nature of the restructured surface to be characterized as one monolayer of hexagonal symmetry superimposed on the (001) substrate.

---

*Research sponsored by the U. S. Energy Research and Development Administration under contract with Union Carbide Corporation.

INTRODUCTION

1. Reordered Au Surface

It is known from low energy electron diffraction (LEED) studies[1-3] that the outermost atoms of a clean (001) Au surface are not in the atomic configuration of the cube planes in the bulk. The cleaned Au surfaces exhibit a complex diffraction pattern frequently designated (5 x 20) instead of the primitive (1 x 1) symmetry expected of a normal (001) surface. Although suggested atomic arrangements which might produce this complex pattern include ordered arrays of vacancies,[1] an impurity stabilized hexagonal topmost layer[2] and the presence of volume twins which terminate at the surface,[4] recent LEED[5,6] and RHEED[7] results support the original suggestion by Palmberg and Rhodin[3] that a reordered hexagonal array of surface atoms with an interatomic spacing 5% smaller than that found in bulk gold is responsible. Nevertheless, the exact interatomic spacing and suggested (5 x 20) unit cell are still in question, and the number of atom layers contributing to the reordered surface structure is uncertain. A three layer upper limit can be imposed on the reordered surface by realizing that a greater depth would eliminate the (001) contribution which is observed in the LEED patterns due to the limited escape depth of the electrons.[8]

2. Positive Ion Channeling Spectroscopy (PICS)

In this paper we seek a new approach to characterizing this reordered Au surface by utilizing the phenomenon of positive ion channeling. Channeling has been extensively studied and is well understood in most instances.[9] It is the orderly arrangement of atoms comprising the single crystal lattice that is responsible for the channeling phenomenon and any deviation from this order can be expected to cause a corresponding perturbation of the channeling effect. In some instances it is possible to deduce from measurements of these perturbations a model for the change that has occurred in the crystal lattice. This method of positive ion channeling spectroscopy (PICS) is not new and has been successfully utilized since the very early channeling studies using the phenomena of energy loss, ion scattering x-ray excitation, nuclear reactions and others as sensors of the channeling effect.[9-22] It is the surface sensitivity of the channeling effect which we propose to exploit to investigate the nature of the reordered Au surface. A well collimated beam of ions incident parallel to an axial direction in a single crystal interacts normally with the exposed surface atoms but has a greatly reduced probability for interacting with those atoms in the bulk which are shadowed by the surface atoms. Although this surface sensitivity has been used to investigate, for example, surface disorder,[9,14] thermal vibrations,[9,15,16] oxide formation,[9,16,17,18,19] and numerous other surface sensitive phenomena[9-19] it has not been utilized to study clean well-characterized surfaces until recently.[20,21,22]

The investigations reported here combine the techniques of LEED and AES to characterize the Au surfaces and establish a model for the surface order, PICS to determine the surface atom densities of the normal and reordered surfaces, and computer simulation calculations to aid in extracting the nature of the reordered surfaces from the PICS measurements. Particular emphasis is given in this paper to the limitations and capabilities of the PICS technique for such investigations.

EXPERIMENTAL

1. Sample Preparation

The Au samples were self-supporting single crystals of a (001) orientation ranging in thickness from 3000 to 7000 Å. The method of preparing and characterizing these samples has been described in previous publications.[23] Briefly, Ag was evaporated onto cleaved surfaces of NaCl, this substrate was annealed, and Au was subsequently evaporated to the desired thickness. A collodion membrane was used both to support the films while NaCl and Ag were dissolved and then to transfer the films from the substrate to 12.5 mm diameter, 0.5 mm thick Au discs containing $\simeq$ 3 mm diameter apertures. The collodion was burned away during annealing in air at 600°C, leaving self-supporting films over the apertures. X-ray rocking curves[23] made after this anneal gave FWHM of 6-9'.

2. Sample Characterization, LEED and AES

The Au sample surfaces were characterized in a three grid LEED-AES ultrahigh vacuum system, which has been described previously.[24] Following the surface reordering process and characterization in the LEED-AES system, the cleaned Au samples were transported to the ion scattering apparatus in a dry nitrogen environment. After the ion scattering measurements, the samples were transferred in a dry nitrogen environment back to the LEED-AES system to verify that the previously cleaned surfaces were still reordered. Such a transfer process was possible because of the extremely low sticking coefficients of common gases on the (001) Au surface. The LEED patterns produced by the differing atomic configurations on the (001) Au surfaces at these various stages are shown in Fig. 1. Figure 1a is representative of the Au samples in the "as-prepared" condition, that is, before any cleaning or vacuum annealing treatments. The primitive

Au(001)

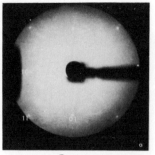

84 V

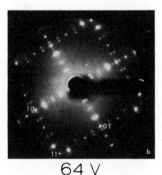

64 V

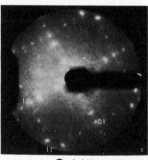

64 V

Fig. 1. LEED patterns from (001) Au surfaces: (a) Normal surface contaminated with S and C (as prepared), (b) After cleaning by Ar ion bombardment and annealing (reordered surface), (c) After exposure to a dry nitrogen environment for $\simeq$ 30 min. (following transfer to and from PICS apparatus.)

(1 x 1) diffraction pattern indicated that the atoms were arranged in the proper square array characteristic of a normal (001) surface. Auger analysis indicated that both S and C impurities were present. Although the high background intensity was due to the presence of disordered impurities, one can estimate from the intensity of the LEED pattern and AES signal that less than $\simeq$ 3 monolayers of impurities were present. The distinctive pattern shown in Fig. 1b was taken after the Au surface had been sputter cleaned with 50-100 eV Ar ions and annealed at 450°C for 30 minutes.[6] Auger analysis of this reordered surface revealed no detectable impurities. In addition to the diffraction spots from the underlying (001) structure, extra spots and splittings are seen which are consistent with the proposed hexagonal (5 x 20) structure. That the reordered region is highly unstable with respect to impurities is apparent by noting that deposition of one monolayer of Ag, Pd or Cu[25] and as little as 0.15 and 0.05 monolayers of Pb[26] and S,[27] respectively, result in the conversion of the (5 x 20) atomic configuration to the normal (1 x 1) array. Figure 1c was obtained from a surface which had been reordered, transferred to the ion scattering system where the measurements reported in Fig. 7 were made, and returned to the LEED-AES system. Auger analysis indicated the presence of S and C apparently picked up in the transfer process. The (5 x 20) pattern characteristic of the reordered surface is still visible but the spots are diffuse and the splittings less well

resolved. This is possibly due to some decreases in domain size of the (5 x 20) structure resulting from impurities.

3. Ion Scattering Measurements, PICS

The method of the PICS measurements made on Au single crystals with surfaces in the normal and reordered conditions are summarized in this section. A beam of 1 MeV ^{4}He ions from the Solid State Division's 2.5 Mv Van de Graaff accelerator was collimated to an angular divergence $\leq 0.02°$ and a beam size at the sample of 1 mm in diameter. This beam impinged on the Au single crystals which were held in a three axis goniometer for orientation purposes. The vacuum in the scattering chamber was 5×10^{-9} Torr. The yield vs. energy distributions of ions Rutherford scattered 94° from the Au crystals were recorded using a solid state detector with an acceptance angle of 7°, cooled to -30°C. The data were acquired, stored and analyzed using a Tennecomp System PDP 11/05 computer based pulse height analysis system and standard pulse height analysis techniques. A beam monitoring system was used for quantitative yield measurements.

Figure 2 shows two spectra which illustrate how the phenomenon of Rutherford scattering is altered by the channeling effect, a consequence which makes it suitable for PICS. Both spectra were taken on an "as-prepared" (001) Au single crystal. The upper curve was obtained with the ion beam incident along a misoriented direction in the crystal, far from any low index direction, so that the ions encountered the atoms approximately at random. This spectrum is characteristic of the scattering distribution one would obtain from an amorphous or polycrystalline Au sample. The onset of scattered ions just below 0.96 MeV is due to 1 MeV ^{4}He ions scattered $\simeq 94°$ from Au atoms at the surface of the sample, and the yield of ions at decreasing energies is the result of the Rutherford scattering process occurring at increasing depths into the sample. The lower curve in Fig. 2 was obtained from the (001) Au crystal tilted 45° so that the beam was incident parallel to the [$\bar{1}$01] direction. This channeling spectrum was taken for the same total incident beam as the misoriented or random spectrum, and the overall decrease in scattering yield is a result of the channeling effect.[9] The pronounced peak near 0.95 MeV results from ions which scattered from all Au atoms on or near the surface which were "visible" to the incident ion beam. The width of this surface peak, in our case, is entirely due to instrumental, kinematic and energy loss effects so that the area under the surface peak is directly proportional to the density of exposed atoms. The FWHM of this peak corresponds to a depth resolution of about 55 Å. This estimate required using the stopping power for randomly directed ions since the appropriate value for channeled ions is not known. Since there is some uncertainty in the random value itself, this should be regarded as an approximate value. Moreover, the sharp decrease in the yield of scattered ions directly behind the surface peak

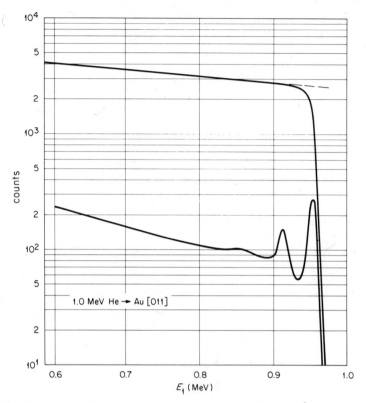

Fig. 2: Rutherford scattering spectra for 1 MeV ^{4}He ions incident in a misoriented direction (upper curve) and incident parallel to the [$\bar{1}01$] direction of a (001) Au sample with normal atomic surface arrangement.

(so-called minimum yield region) is also a result of the ion channeling phenomenon. Because the [$\bar{1}01$] rows of atoms in the Au crystals were aligned with the incident ion beam, the surface atoms shadowed all other atoms in the lattice. Those ions which did not hit the surface atoms directly were steered by a series of correlated glancing collisions with the atoms in the rows into oscillatory trajectories, and Rutherford scattering was suppressed to about 0.02 of that observed in a randomly oriented single crystal. This is the meaning of the normalized ordinate used in succeeding figures where the scattering yields have been normalized relative to the random yield which is taken as 1.0. The oscillatory structure behind the surface peak is also a channeling effect and has been discussed elsewhere.[16,28] Ions scattered from light impurities such as C or S on the surface of the Au crystals would have appeared as peaks at 0.476 MeV and 0.764 MeV, respectively, had they been present in detectable concentrations. Light impurities in concentrations such as those encountered on the normal surfaces will have negligible effect on the PICS measurements.

COMPUTER SIMULATION

The program used for doing computer simulation calculations in connection with the present experiments has been described previously.[16] It follows trajectories of individual ions through a lattice with thermal displacements of the lattice atoms. It uses the binary collision approximation and computes the deflection at each collision by the impulse approximation using Moliére's[29] approximation to the Thomas-Fermi potential. The effects of beam divergence and mosaic spread may be included, and this has been done using values from the experiments for all results to be presented below.

The first use of computer simulations herein is to normalize the measurements to those for a random direction in the solid. The difficulty of selecting a truly random direction is well recognized from both experimental and theoretical work. One method of performing this normalization was suggested several years ago by means of a series of simulation calculations.[30] The method consists in averaging the yield over all the directions lieing on a cone whose axis is a major row direction in the solid and whose half angle is several times the channeling half angle of the row. Care should be taken that the cone chosen is not near any other important row. The resulting normalization to random should be accurate to within 1 or 2%. This same method has been tested experimentally[31] and has been used frequently. However, in the present case, because of the ready availability of the computer simulation program, a different approach has been taken. The present method consists of measuring the yield over a convenient range of incident directions and then normalizing the results to a computer simulation for the same range. The results are shown in Fig. 3. The normalization was performed over the range of ψ from 1 to 2 degrees, where the yield is varying slowly. The overall agreement is quite good. The only significant point of disagreement is the width of the (200) channel; the reason for this is not understood. The combined errors of the measurements and simulation probably amount to a few percent.

The second use of computer simulations was to understand the yield from the unrestructured surface. The results of the simulation are shown in Fig. 4. For the present purposes, the main feature of interest is the area under the peak at the surface. The most convenient way of expressing this area is in terms of L, effective number of surface atoms per row,[32] which is defined to be the area divided by d, the distance between atoms in the row. The effective number of surface atoms per row from the simulation is 2.14 ± 0.02; it is greater than unity because the thermal vibrations keep the atoms at the ends of the rows from completely shielding all of the other atoms in the row. Of secondary interest is the yield at depths just beyond the sub-surface oscillations, which is 0.023 ± 0.001. The sub-surface oscillations themselves are not of interest here, but as noted

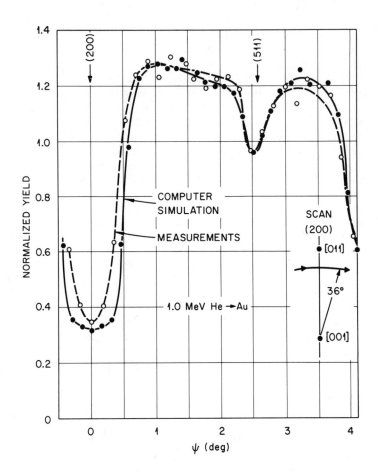

Fig. 3. Comparison over a range of orientations of measured yields with yields from computer simulations. The range of orientations is shown in the insert in the lower right hand corner; ψ is the angle from the (200) channel. The measured values have been normalized to those from computer simulation in the manner explained in the text.

earlier, their origin is understood.[16,28] In addition to the simulation results, Fig. 4 shows these results convoluted with an experimental depth resolution of 70 Å FWHM. The depth resolution is shown convoluted with the entire simulation results and also separately with the surface peak and with the sub-surface structure. While the surface peak and sub-surface structure are separate in the unconvoluted results, they overlap after convolution. The best procedure for obtaining the area under the surface peak for the

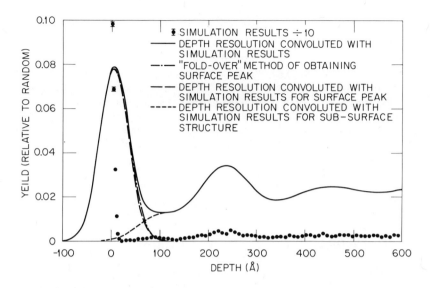

Fig. 4. Backscattering yield from 1 MeV He ions directed along [$\bar{1}$01] in Au at room temperature.

experimental results seems to be a "fold-over" method in which the high energy side of the surface peak is folded across the energy of the maximum and the area under the resulting curve is taken. The equivalent construction for the simulation results is shown in Fig. 4. For the simulation, it is possible to assess what effect this procedure has. In Fig. 4, it adds 0.07 effective atoms per row to the area under the surface peak, but this value is sensitive to the value of depth resolution used in the convolution. When the normalization procedure described above is applied to the measured channeled result in Fig. 2, the height of the surface peak is 0.12 rather than 0.08 as in Fig. 4. Since the shape of the peak is determined almost entirely by the shape of the depth resolution function and the area under the peak is very insensitive to this shape, this comparison between Figs. 2 and 4 suggests that a FWHM value of 70 Å is too large. Agreement for the peak heights is obtained with a FWHM value of about 50 Å which is in reasonable agreement with the estimated experimental depth resolution of 55 Å. Such a value is probably within the uncertainties of the estimate. For this smaller width, the correction due to overlap of the sub-surface structure with the surface peak is only 0.01. In view of the uncertainty in the FWHM value, a correction of 0.03 ± 0.02 will be used for comparison with experiment. With this correction value, the area under the surface peak from the simulation calculations becomes 2.17 ± 0.03.

Simulation was not applied to the restructured surface. Almost all of the atoms in the restructured region lie sufficiently far from the rows to make a contribution to the surface peak independent of that of the rows. On the basis of general channeling results, 0.2 Å was selected as a distance from rows which would divide those surface atoms that act as part of the row from those that act independently. The exact value of this distance is not critical since it is small. If increased accuracy of future measurements justify the need, computer simulations could be used to refine this distance.

One further point should be considered; this is the relationship between monolayers and effective number of surface atoms per row. This relationship is illustrated in Fig. 5, which shows the appearance of the (001) surface when viewed along two different directions. Viewed along [001], the ends of half of the rows are in the first monolayer and the ends of the other half are in the second monolayer; in this orientation one atom per row constitutes two monolayers on the surface. Viewed along [$\bar{1}$01], the ends of all of the rows are in the first monolayer so that in this orientation, which is the one actually used in the experiments, one atom per row constitutes one monolayer. The term monolayer as used throughout the paper will refer to the (001) surface which has an atomic density of 1.21×10^{15} atoms/cm^2. Other experimental arrangements may correspond to different relationships between atoms per row and monolayers, but the relationship is always simple.

ANALYSIS AND CONCLUSIONS

1. Normal Surface

Following the philosophy of the PICS technique, the measurements reported in connection with Fig. 2 should serve as a source function characteristic of the channeling effect for a normal (001) Au surface. We can check the validity of this assertion by comparing these measurements to the computer simulation calculations for a normal surface reported in the previous section. It is clear from a comparison of the measured (Fig. 2) and calculated (Fig. 4) scattered yield distributions that the general features are reproduced. The area contained within the surface peak is the comparison of importance. Since this is a quantity which must be measured, some discussion of the experimental factors involved are appropriate.

We determined the effective number of atoms per row contributing to the surface peak from an accurate measurement of the surface peak

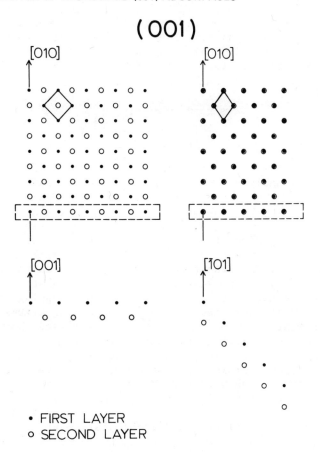

Fig. 5. Schematic representation of a (001) surface viewed along the [001] direction (top lefthand illustration) and along the [1̄01] direction (top righthand illustration). The solid circles represent the top most layer and the open circles the second layer. The lower illustrations on both sides of the figure show a side view of the arrangement of the two top most layers enclosed within the dotted rectangles.

yield relative to the scattering yield from a "randomly" oriented Au sample using the expression:

$$L = \frac{(A_{sp})(\delta x)}{(N_R')(C)(d)} \cdot \qquad (1)$$

In equation (1), A_{sp} = the area (in counts) under the surface peak in the aligned spectrum obtained by the "fold-over" method discussed in connection with the computer calculations; N_R' = counts per δE or δx obtained from the randomly oriented yield measurement; C = a random correction factor discussed below; d = distance between atoms along the aligned direction, in our case [$\bar{1}01$]; and

$$\delta x = \frac{\delta E}{\left.\frac{dE}{dx}\right|_R \left(K^2 + \frac{\cos\theta_i}{\cos\theta_o}\right)} \quad (2)$$

where δE is the analyzer channel width in energy units, $\left.\frac{dE}{dx}\right|_R$ the stopping power, K^2 the kinematic scattering factor and θ_i and θ_o the incident and emergent angles of the ions respectively. Determination of the effective number of atoms per row, L, requires a knowledge of the random stopping power, an accurate relative yield measurement, and some assurance that the random direction chosen for the measurement gives a truly random scattering yield. Accurate relative yield measurements are not difficult and the random stopping power was selected from a standard tabulation.[33] Obtaining a truly random yield measurement is more complicated, as discussed above, and quite important since it serves as the normalizing factor for the surface peak yield measurement. The random correction factor, C, in Eq. (1) was selected by normalizing the measurements to the calculations as discussed in connection with Fig. 3. The advantage of this normalization technique over previous methods[30,31] is that it requires only a limited number of measurements at a known orientation; the disadvantage is that it requires a model calculation of the random yield at these orientations.

The calculated values for the effective number of atoms per row contributing to the surface peak, L, and the minimum scattering yield, χ_{min}, for the normal Au surface are compared to the measured values in Table I. These expected values are in excellent agreement with the measurements. Thus, our measurements appear to be a good representation of a normal (001) Au surface and we can proceed to PICS measurements on the reordered surface.

2. Reordered Surfaces

The LEED and AES measurements served two essential functions. First, they characterized the surfaces at the various stages of investigation verifying, in particular, that the PICS measurements were made on a reordered surface. Second, they provided the necessary information for a model construction of the reordered surface. A model for the atomic configuration of the reordered Au surface which is consistent with the complex LEED patterns in Fig. 1 was suggested by Palmberg and Rhodin[3] and is illustrated for the conditions of our

TABLE I

Comparison of Measured and Calculated Surface Peaks and Minimum Yields

Au Surface	Date Source	L	X_{min}
Normal or As Prepared	Computer Simulations	2.17 ± 0.03	0.023 ± 0.001
	Measurement	2.2 ± 0.1	0.021
Reordered	Model Predictions		
	i) Reordered Top Layer	1.2	
	ii) 001 Substrate	2.2 / 3.4	
	Measurement	3.2 ± 0.1	0.024

experiment in Fig. 6. The upper illustration shows a normal (001) surface (solid circles) and a superimposed hexagonal layer (open circles) with a 5% contraction in the Au interatomic spacing which represents the reordered layer. Such a configuration can be positioned on a (001) surface so that its unit cell has dimensions (5 x 20) with respect to the underlying (1 x 1) array[2] and has an atomic density 1.26 of that for the (001) Au surface. Viewing this arrangement along the [$\bar{1}$01] direction as in the lower illustration of Fig. 6 and assuming that only atoms in the reordered layer lying more than 0.2 Å from a row make an additional contribution to the surface peak, additional scattering equivalent to 1.20 monolayers is expected from the reordered layer. Here it is assumed that the interplanar spacing between the reordered and normal layers is identical to the interplanar spacing along the [001] direction in the bulk crystal, although this assumption is not critical. If the reordered region were two layers in depth with AB type stacking as occurs along the [111] direction in fcc crystals, then no overlap occurs between the first and second layers and our model construction predicts scattering from 2.40 additional monolayers. Even if the second layer overlapped the first for some reason, one would still expect ≃ 2 monolayers to contribute to the surface peak yield because of thermal vibrations.

The results of measurements on the reordered Au surface are compared in Fig. 7 to those already discussed for the normal (001) surfaces. Clearly, the reordered surface has altered the channeling effect in such a way as to cause an increase in the surface peak yield. From analysis such as that already discussed, the reordered surface peak yield corresponded to scattering from 3.2 ± 0.1 monolayers. The results of the PICS measurements are summarized in

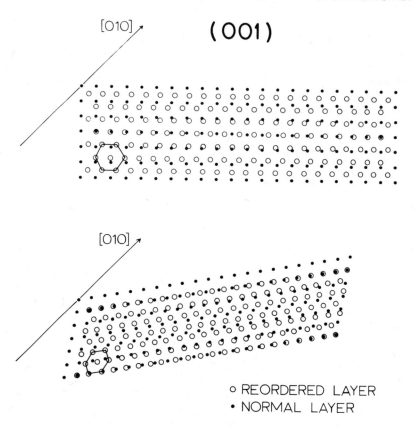

Fig. 6. Schematic representation of a normal (001) surface (solid circles) with a superimposed hexagonal layer (open circles) with a 5% contraction in the Au interatomic spacing. The upper view is along the [001] direction, and the lower is along the [$\bar{1}$01] direction.

Table I and compared to the model predictions. The difference between the measured yield of the reordered surface, 3.2, and that predicted by the model, 3.4, probably results from some small fraction of the reordered surface reverting to the normal (001) atomic arrangement as a result of small amounts of impurities picked up in the initial transfer to the PICS apparatus. Moreover, the placement of (5 x 20) domains in two orientations, needed to produce the four fold symmetry exhibited in Fig. 1b, will reduce the amount of surface arranged in the reordered structure. These measurements rather clearly rule out any possibility that the restructured surface region could be composed of more than one extra monolayer of the type shown in Fig. 6. They do not rule out the possibility that the structure might be somewhat less than a full monolayer. If future measurements

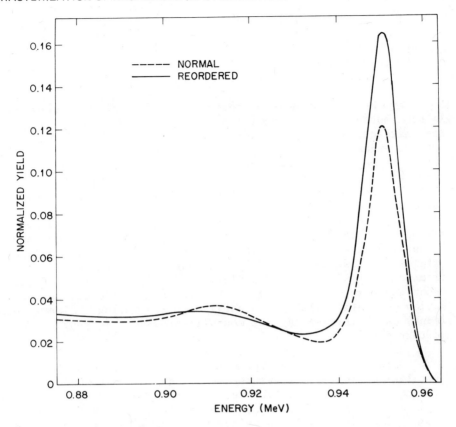

Fig. 7. Comparison of the Rutherford scattering spectra for 1.0 MeV ^{4}He ions incident parallel to the [$\bar{1}$01] direction of (001) Au single crystals with normal (dashed curve) and reordered (solid curve) surfaces.

with cleaner surfaces and greater precision should show less than a full monolayer, the nature of the structure would be very interesting to explore. However, at the present time it is to be concluded that the measurements are in excellent agreement with a reordered surface composed of one monolayer of hexagonal symmetry.

DISCUSSION

These measurements clearly illustrate the applicability of the PICS technique for surface investigations, and the potential for extension to other surface problems is obvious. This application also emphasizes that complementary techniques, such as LEED and AES, and accurate model calculations are necessary in order to extract credible results. This particular investigation was only possible because the Au surface is especially resistent to contaminants. Extension to other systems will require even better ultra high vacuums and *in situ* cleaning and characterizations.

For the present case of reordered surface characterization there is some advantage to using low energy (typically 1 KeV) ion scattering instead of the higher energies used here. The lower energies provide enhanced sensitivity to the top monolayer and to adsorbed impurity atoms.[34] On the other hand, the additional information provided by the higher energy measurements may prove invaluable in this and other investigations. In the event of more than one reordered layer the additional depth sensitivity is needed. It is also clear from close scrutiny of the measurements on the normal and reordered surfaces in Fig. 7 that the sub-surface yield oscillations are slightly altered by the presence of the reordered surface layer. The origins of these sub-surface yield peaks can be easily identified from computer simulation calculations[28] and application of the PICS technique using these peaks as sensors can provide sub-surface information not obtainable by low energy ion scattering. Clearly experiments in the two energy regimes will provide different but complementary information.

A final comment concerning the importance of computer simulation calculations in investigations such as these seems appropriate. As with most spectroscopic techniques in physics, our final information depends on comparing our measurements with model calculations. The measurements are sensitive to every perturbation of the single crystal lattice from fine details like thermal vibrations, mosaic structure and non-statistical equilibrium effects in channeling, to gross lattice disorder. Consequently, even if the assumed model is correct, the calculations must reflect this detail before one can extract quantitative information. These features are much more easily handled in computer simulation calculations than any other way.

REFERENCES

1. A. M. Mattera, R. M. Goodman, and G. A. Somorjai, Surf. Sci. 7, 26 (1967); G. A. Somorjai, Surf. Sci. 8, 98 (1967); H. B. Lyon, A. M. Mattera, and G. A. Somorjai, in Fundamentals of Gas-Surface Interactions, edited by H. Saltsburg, J. N. Smith and M. Rogers (Academic, New York, 1967), p. 102.

2. D. G. Fedak and N. A. Gjostein, Phys. Rev. Lett. 16, 171 (1966); Acta Met. 15, 827 (1967); Surf. Sci. 8, 77 (1967).

3. P. W. Palmberg and T. N. Rhodin, Phys. Rev. 161, 586 (1967); J. Chem. Phys. 49, 134 (1968).

4. P. N. J. Dennis and P. J. Dobson, Surf. Sci. 33, 187 (1972).

5. G. E. Rhead, J. Phys. F (Metal Phys.) 3, L53 (1973).

6. D. M. Zehner, T. S. Noggle, and L. H. Jenkins, Surf. Sci. 41, 601 (1974).

7. F. Grønlund and P. E. Højlund Nielson, J. Appl. Phys. 43, 3919 (1972).

8. P. W. Palmberg and T. N. Rhodin, J. Appl. Phys. 39, 2425 (1968).

9. D. S. Gemmell, Rev. Mod. Phys. 46, 129 (1974).

10. Atomic Collisions in Solids IV, edited by S. Andersen, K. Björkqvist, B. Domeij, and N. G. E. Johansson, Gordon and Breach Science Publishers (1972).

11. Proceedings of the International Conference on Ion Beam Surface Layer Analysis, Thin Solid Films 19, 1-420 (1973).

12. Applications of Ion Beams to Metals, edited by S. T. Picraux, E. P. EerNisse, and F. L. Vook, Plenum Press (1974).

13. Atomic Collisions in Solids, Vols. I and II, edited by S. Datz, B. R. Appleton, and C. D. Moak, Plenum Press (1974).

14. E. Bøgh, Proceedings of the International Conference on Solid State Physics Research with Accelerators, edited by A. N. Goland, USAEC BNC 50083 (C-52), September 25-28 (1967), p. 76.

15. B. R. Appleton and L. C. Feldman, Atomic Collision Phenomena in Solids, edited by D. W. Palmer, M. W. Thompson, and P. D. Townsend, North-Holland Publishing Co. (1970), p. 417.

16. J. H. Barrett, Phys. Rev. 3B, 1527 (1971).

17. W. K. Chu, E. Lugujjo, J. W. Mayer, and T. W. Sigmon, Thin Films 19, 329 (1973).

18. W. F. Van der Weg, W. H. Kool, H. E. Roosendaal, and F. W. Saris, Rad. Eff. 17, 245 (1973).

19. G. Della Mea, A. V. Drigo, S. Lo Russo, P. Mazzoldi, and S. Yamaguchi, Appl. Phys. Letters 26, 147 (1975).

20. E. Bøgh, Channeling, edited by D. V. Morgan, John Wiley and Sons Publishers, (1973), p. 435.

21. B. R. Appleton, T. S. Noggle, J. W. Miller, O. E. Schow, III, D. M. Zehner, L. H. Jenkins, and J. H. Barrett, Proceedings of the Third International Conference on Applications of Small Accelerators, edited by J. L. Duggan and I. L. Morgan, ERDA Conf. 74-1040-Pk (1974), p. 86.

22. D. M. Zehner, B. R. Appleton, T. S. Noggle, J. W. Miller, L. H. Jenkins, and O. E. Schow, III, J. Vac. Sci. Technol. 12, No. 1, 454 (1975).

23. S. Datz, C. D. Moak, T. S. Noggle, B. R. Appleton, and H. O. Lutz, Phys. Rev. 179, 315 (1969); T. S. Noggle, Nucl. Inst. Methods 102, 539 (1972).

24. L. H. Jenkins and M. F. Chung, Surf. Sci. 24, 125 (1971).

25. P. W. Palmberg and T. N. Rhodin, Phys. Rev. 161, 586 (1967); J. Chem. Phys. 49, 134 (1968).

26. J. P. Biberian and G. E. Rhead, J. Phys. F (Metal Phys.) 3, 675 (1973).

27. M. Kostelitz, J. L. Domange, and J. Oudar, Surf. Sci. 34, 431 (1973).

28. J. H. Barrett, Phys. Rev. Letters 31, 1542 (1973).

29. G. Moliére, Z. Naturf. 2a, 133 (1947).

30. T. S. Noggle and J. H. Barrett, phys. stat. sol. 36, 761 (1969); see particularly p. 772.

31. J. F. Ziegler and B. L. Crowder, IBM Research Report, No. RC-3551, September 23 (1971).

32. This has been defined on p. 1541 of Ref. 16. What was referred to there as effective number of surface layers is perhaps better named the effective number of surface atoms per row, as is being done here, in order to avoid confusion with the usage of monolayer in discussing surfaces.

33. J. F. Ziegler and W. K. Chu, Atomic Data and Nuclear Data Tables 13, 463 (1974).

34. E. Taglauer and W. Heiland, Surf. Sci. 47, 234 (1975).

DISCUSSION

Q: (R.J. MacDonald) Your results indicate almost exact agreement between surface area and geometrical area for these thin films. Would you care to comment?

A: (B.R. Appleton) The Au films grown by John Noggle have been well characterized by transmission electron microscopy and X-ray scattering. In particular his measurement of the mosaic structure are performed in air with considerable goniometric manipulation and so the 6'-9' FWHM mosaic structure I quoted would include this. The TEM measurements show no twinning and low dislocation density. These characterizations have been published.

PROTON CHANNELING APPLIED TO THE STUDY OF

THERMAL DISORDER IN AgBr

      M. Roulet, C. Jaccard, H. Huber
      Institut de physique de l'Université

      CH - 2000 Neuchâtel, Switzerland

ABSTRACT

The dechanneling rate of 100 keV protons along the <100> axis in AgBr single crystals is measured between 90 and 595 K. Its anomalous increase with temperature above 400 K is explained by thermally produced silver interstitials reaching a concentration of 0.06±0.02 at 695 K, with an activation energy of (0.65±0.10) eV.

1. INTRODUCTION

Channeling can be applied with success to silver bromide for the following reasons: the high atomic number gives a large scattering cross section, the structure is simple, good quality monocrystals are available and the structural disorder produces significant effects at high temperature. Moreover the detailed nature of this disorder is not known definitively, although extensive studies have been done since 1936. There is only a partial agreement between different methods of investigation of high temperature anomalies, such as electrical conductivity, specific heat, and thermal expansion. This suggests a new approach by use of the channeling effect.

In this paper we present first experimental data, which are related to the concentration of interstitial silver atoms; the results are then discussed and compared with those obtained by other methods.

2. EXPERIMENTAL RESULTS

Single AgBr crystals are cut with a face perpendicular to the <100> axis and cleaned by etching in a KCN solution. They are mounted on a goniometer, the temperature of which can be varied between 90 and 700 K, and bombarded with a beam of 100 keV protons, with a divergence smaller than 0.05°, a diameter of 0.3 mm and a current of 0.1-10 nA. Backscattered protons are detected at 150° with a cooled silicon surface barrier detector connected to a multichannel analyser. The energy resolution amounts to 4-6 keV FWHM, corresponding to a depth of 120-180 Å for AgBr. The total dose is measured with a fine mesh nickel grid intercepting the beam, and a second grid emits secondary electrons which neutralize the proton charge on the insulating sample. A dose of 10^{-2} protons/channel obtained by shifting the sample with respect to the beam prevents any significant damage.

The critical angle for axial channeling decreases from $(2.5\pm0.1)°$ at 90 K to $(2.0\pm0.1)°$ at RT. The minimum yield amounts to $(20.0\pm0.5).10^{-2}$ at RT but it shows a pronounced dependence of depth x and temperature T according to the law

$$1 - \chi(x,T) = [1 - \chi(0,T)] \; \exp[-<W(T)>_e x/d]$$

represented in Fig. 1 (d is the separation between the layers and $<W(T)>_e$ is the experimental dechanneling probability per atomic layer).

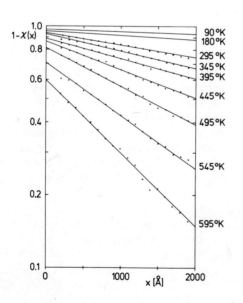

Fig. 1. Channeled normalized component versus depth between 90 and 595 K for 100 keV protons along the <100> direction in AgBr.

A thin silver film of $(140\pm20)\text{Å}$ evaporated on the surface increases the surface minimum yield from 3 to 32.10^{-2} at 90 K and from 6.5 to 42.10^{-2} at 295 K, whereas the dechanneling rate is increased from 0.2 to $0.7.10^{-4}\text{Å}^{-1}$ and from 1.0 to $2.0.10^{-4}\text{Å}^{-1}$, respectively.

3. DISCUSSION

At low temperature, the critical angle agrees closely with the value calculated by using Barrett's equation [1] but at high temperature, the agreement is not so good (the minimum yield is too large). The surface minimum yield is about two to three times larger than the values predicted by Barrett's semi-empirical formula [1], indicating possibly a slight surface disorder.

With the surface covered with a silver film of known thickness, the dechanneling cross section of silver atoms placed in it can be evaluated from the increase of the minimum yield, and it amounts to $(5.6\pm1.0).10^{-18}$ cm^2 at 90 K and $(6.9\pm1.3).10^{-18}$ cm^2 at 295 K. On the other hand the cross section can be calculated according to Lindhard [2] by using reduced energy and angle variables [3]. In this case we obtain the slightly smaller values $(3.8\pm0.4).10^{-18}$ cm^2 at 90 K and $(5.3\pm0.5).10^{-18}$ cm^2 at 295 K.

For silver atoms in the channels inside the crystal, the dechanneling cross section has to be corrected because of the inelastic energy losses of the temperature dependence of the critical angle and especially because the proton trajectories in the channels make with its axis an angle which is not zero but is distributed between zero and the critical angle. The last effect can be calculated by using Lindhard's formulas [2] and gives an increase by a factor of 6.

The experimental dechanneling probability per atomic layer $<W(T)>_e$, represented in Fig. 2, agrees at low temperature with the results of Kumakhov's statistical theory [4] of the vibrating crystal. Above 400 K, the anomalous increase can be attributed to the interstitial silver atoms required to explain anomalies of the specific heat or of the electrical conductivity. The difference between the curves a and b, together with the corrected cross section of the silver atoms, gives the concentration of the interstitials as a function of the temperature, which is reported in Fig. 3. The relevant parameters are given in Table I. The activation energy is (0.65 ± 0.10) eV and the concentration extrapolated to 695 K amounts to 0.06 ± 0.02 (melting point : 697 K).

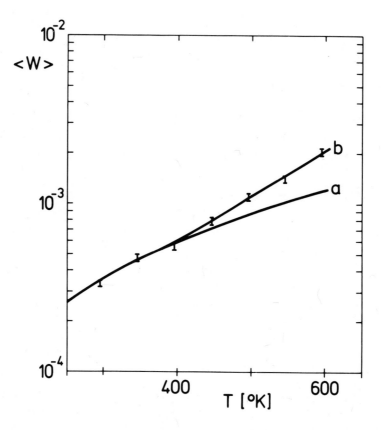

Fig. 2. Dechanneling probability per atomic layer. a) theoretical curve for a perfect vibrating crystal; b) experimental curve.

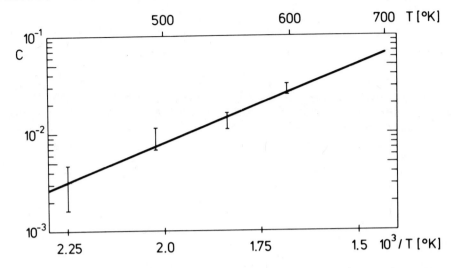

Fig. 3. Molar concentration of silver interstitials in AgBr.

Table I. Critical angle, dechanneling cross section, dechanneling probability and concentration of silver interstitials in AgBr.

T(K)	$\psi_c \times 10^2$	$\bar{\sigma}_d (10^{-17} cm^2)$	$<W>_e - <W>_t$	$c(T) \times 10^2$
445	3.15	4.4	0.82×10^{-4}	0.31±0.15
495	3.06	4.6	2.4	0.87±0.20
545	2.98	4.8	3.9	1.35±0.25
595	2.90	5.0	8.6	2.85±0.35

The measurements of the specific heat anomaly [5,6] yield a greater activation energy of 1.3 to 1.4 eV, as well as the analysis of the electrical conductivity [7,8,9,10,11]. The most recent value given by Aboagye and Friauf [11] is 1.13 eV. However, thermal expansion experiments by Lawn [12] indicate value of 0.59 eV. These discrepancies are probably due to the fact that the crude interstitial model is not correct, but has to be refined (e.g. with Hove's dumbbell model [13]) and that the different investigated properties are not affected in the same way by the details of the mechanism.

4. CONCLUSION

Silver bromide is adequate for channeling experiments, its surface can be prepared chemically so that the processes occurring in the bulk are not disturbed. Above 400 K, the parameters of Rutherford backscattering are significantly affected by the presence of silver interstitials in the lattice. A quantitative analysis gives their concentration as a function of temperature, and the corresponding activation energy agrees with the one obtained from thermal expansion, but is smaller by a factor of two with respect to the values obtained from electrical conductivity or from specific heat.

ACKNOWLEDGEMENT

We are indebted to Dr. P. Junod (Ciba-Geigy Laboratories) for supplying us with monocrystals and for valuable discussions, and to the Swiss National Foundation for the support of this work.

REFERENCES

1. J.H. Barrett, Phys. Rev. B, $\underline{3}$, 1527, 1971.
2. J. Lindhard, V. Nielsen and M. Scharff, Mat.-Fys. Medd. Dan. Vid. Selsk. $\underline{38}$, No 10, 1968.
3. L. Meyer, Phys. stat. sol. (b) $\underline{44}$, 253, 1971.
4. M.A. Kumakhov, Phys. Lett. $\underline{33A}$ (2), 133, 1970.
5. R.W. Christy and A.W. Lawson, J. Chem. Phys. $\underline{19}$, 517, 1951.
6. W. Jost and P. Kubachewski, Z.physik.Chemie, N.F.,$\underline{60}$, 69, 1968.
7. E. Koch and C. Wagner, Z.f.physik. Chemie, $\underline{B38}$, 295, 1938.
8. J. Teltow, Ann.Phys., Lpz., $\underline{5}$, 63, 1949.
9. M.D. Weber and R.J. Friauf, J.Phys.Chem. Solids, $\underline{30}$, 407, 1969.
10. P. Müller, Phys. stat. sol. $\underline{12}$, 775, 1965; $\underline{21}$, 693, 1967.
11. J.K. Aboagye and R.J. Friauf, Phys. Rev. B, $\underline{11}$, 1654, 1975.
12. B.R. Lawn, Acta Cryst. $\underline{16}$, 1163, 1963.

DISCUSSION

Q: (E. Rimini) The higher value of defects found by channeling measurements can be due to flux-peaking effect on the interstitials. Have you account for this effect ?

A: (C. Jaccard) If flux-peaking is taken into account, the effective dechanneling cross-section of the interstitials is increased, provided that they lie near the channel axis. However the concentration of the beam in the middle of the channel implies also that the angular distribution of the trajectories is peaked along the axis. This modifies the correction (sec. 3) by decreasing the effective cross-section. Therefore the two effects compensate each other.

NEON-ION IMPLANTATION DAMAGE GETTERING OF HEAVY METAL IMPURITIES IN SILICON

Shanghai Institute of Metallurgy

Chinese Academy of Sciences

Shanghai Institute of Nuclear Research

ABSTRACT

Rutherford backscattering, neutron activation analysis, secondary ion mass spectrometry and junction leakage measurement etc. were used to make a comprehensive study on the neon-ion implantation damage gettering effect of heavy metal impurities in silicon wafers. The energy of neon ions used was 100 keV. The dependence of gettering effect on target temperature and relative amount of residual damage were studied. The results of analysis indicate that heavy metal impurities in the damaged layer increase significantly after $900°C$ heat treatment, and that the junction leakage of diodes improves appreciably after damage gettering.

(Á) INTRODUCTION

In the last few years, the application of ion implantation to the fabrication of semiconductor devices has grown rapidly. Many new techniques and technologies emerged. Among them, ion implantation damage gettering of heavy metal impurities in silicon has received attention. Buck et al. measured the residual damage after annealing, the amounts and the depth distributions of Cu, Ni, Au and Fe in damaged layers by backscattering of He^+ ions. Hsieh et al. applied this new technique to the fabrication of silicon photodiode array camera targets. The damage gettering is applicable to other semiconductor device technologies where low minority-carrier life-time and soft junction can be problems. To make a more comprehensive understanding of the efficiency of damage gettering, we measured the damage induced by implantation and residual damage after annealing by Rutherford backscattering, the amount of heavy metal impurities in the damaged layer by neutron activation analysis and secondary ion mass spectroscopy (SIMS). In addition, we measured the quality of diodes fabricated on epitaxial Si wafers before and after damage gettering. The thickness of silicon wafers was chosen to be 200 μm to make provision that this technique might have led to practical applications to device fabrication processes. All implants were carried out in a direction titled 7° to the <100> axis.

(B) EXPERIMENTAL RESULTS

The experimental results of backscattering, diode damage-gettering and neutron activation analysis are given below:

(1) Backscattering measurements

The energetic ions used in these experiments were provided by a 180 keV ion implantation system designed and built in our country through self-reliance for both ion implantation and channeling effect studies. Backscattering experiments were carried out in a scattering chamber equipped with a two-axis goniometer. 160 keV proton beams were used in channeling effect measurements of the total amount and the depth distribution of the lattice disorder produced in Ne ion implantation. The energy spectra were measured with

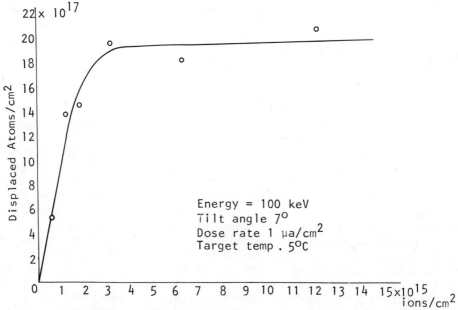

Fig. 1. The dose dependence of the lattice disorder for 100 keV Ne implantations.

a silicon surface barrier detector. The energy resolution of the system was about 7 keV(FWHM).

In calculating the depth distribution of lattice disorder, we made the assumption that the dechanneling probability increases linearly with depth. The stopping power of protons in silicon measured by Foster et al. was employed. Following results were obtained.
(a) The initial damage induced by 100 keV Ne ions implanted at a dose rate of 1 μA/cm^2 and the residual damage upon 900°C annealing for 60 min. were measured. As no measures were taken to cool down the target, the target temperature rose from 20°C to about 65°C during the implantation. The reproducibility of the experimental results are not satisfactory. Situation improved when measures were taken to keep the target at a fixed temperature.
(b) The temperature of the target samples was kept at 50°C during implantation. The dose dependence of the lattice disorder was measured (see Fig. 1). The critical dose to produce a completely disordered region occurred somewhere between 2×10^{15} ions/cm^2 and 3×10^{15} ions/cm^2. At a dose of 3×10^{15} ions/cm^2, the amorphous layer started to grow toward the surface. In addition, we measured the amounts of residual damage upon 900°C annealing for 60 min. They were 50 % and 24-34 % corresponding to doses of 1.2×10^{16} ions/cm^2 and 6×10^{15} ions/cm^2, respectively. Fig. 2 gives the backscattering spectra for a Si crystal implanted with 1.2×10^{16} Ne ions/cm^2 before and after annealing. Depth distributions of the initial and residual amounts of lattice disorder are given in Fig. 3, which show that the amorphous layer extends from 1000 Å to 3500 Å below the surface. The peak of residual damage occurred at a depth of about 1400 Å. For implantation doses less than 3×10^{15} ions/cm^2 only a small amount of residual damage was left behind after annealing. The backscattering spectra for a Si crystal implanted with 1.8×10^{15} ions/cm^2 before and after annealing are given in Fig. 4.

(2) Analysis of impurities in damage-gettered layers P-type, 8 - 10 Ω-cm resistivity, dislocation-free, <100> oriented silicon wafer of 40 mm diameter was lapped and polished on both sides to a final thickness of 230 μm. Half of the surface was masked by an Al foil when the Si wafer was implanted with 2×10^{16} 100 keV Ne ions/cm^2. The ion current was 1 μa/cm^2. An amorphous layer with a thickness of about 4000 Å was formed below the surface. A Cu or Au film of 32 mm diameter and about 1000 Å thick was evaporated on the back surface. Upon 900°C, 30 min annealing, the back surface which was rich in Cu or Au was removed by polishing. For Au contaminated sample another experiment for 300 min annealing was performed. The samples were then irradiated under a neutron flux of 8×10^{13} n/sec cm^2 for 10 hours, and etched in a compound solution of hydrofluoric acid and nitric acid. The solutions employed for each removal were retained and the ^{64}Cu and ^{198}Au activities in them counted. The

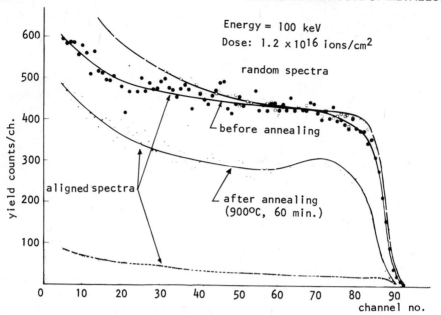

Fig. 2. Normalized energy spectra of backscattered 165 keV protons from a Si crystal.

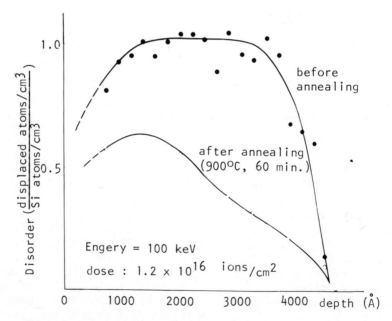

Fig. 3. Depth distribution of lattice disorder.

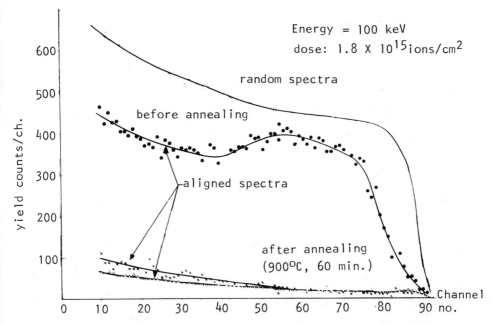

Fig. 4. Normalized energy spectra of backscattered 165 keV protons from a Si crystal.

amounts of Cu and Au in damage-gettered and undamaged regions were compared. For Cu, they were 2×10^{15} atoms/cm^2 and 8×10^{12} atoms/cm^2 respectively. The corresponding quantities for Au in the sample annealed for 30 min were 2.1×10^{13} and 1.5×10^{12} atoms/cm^2 and those in the sample annealed for 300 min were 1.6×10^{14} atoms/cm^2 and 6.7×10^{12} atoms/cm^2 respectively. Another compound solution with the following composition HF : HNO$_3$: HAC : H$_3$PO$_4$ = 5 : 100 : 60 : 30 was used as etching agent for further activation analysis. The wafer was sectioned by a chemical etching procedure and the activities in each section were counted. The measured depth distributions of Cu and Au in damaged layers are given in the following table.

In addition the depth distributions of impurities in damaged layers were measured through SIMS. The results are given in Fig. 5. Most of the impurity atoms were gettered within a thin layer of 1600 Å thick at the surface. The Cu concentration at the surface

Impurity	Annealing Conditions	Average Thickness of Each Layer	Amount of Impurity in Each Layer (atoms/cm^2)							
			1	2	3	4	5	6	7
Cu	900°C, 30 min.	1400 Å	5.0×10^{15}	1.0×10^{14}	9.4×10^{12}	2.9×10^{12}	$<10^{12}$	$<10^{12}$	$<10^{12}$
Au	900°C, 300 min.	1200 Å	4.4×10^{13}	7.2×10^{13}	6.7×10^{13}	1.3×10^{13}	1.3×10^{12}	1.4×10^{11}	$<10^{10}$

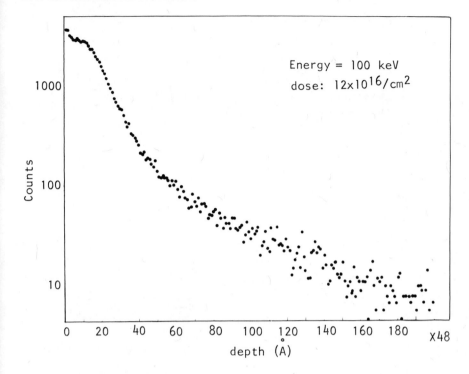

Fig. 5. Depth distribution of Cu impurity in damage - gettered layer through SIMS

was 1.5 orders of magnitude higher than that at a depth of 1600 Å and two orders of magnitude higher than that at a depth of 4000 Å. The foregoing results indicate that: (a) The total amount of copper (gold) gettered in the damage region is more than two orders of magnitude (one order of magnitude) larger than that gettered in the undamaged region. The efficiency of damage-gettering of heavy metal impurities like Cu or Au is significant. (b) The gettering efficiency of Cu is almost two orders of magnitude higher than that of Au. (c) The gettering efficiency of metals which are gettered slowly (like Au) is roughly proportional to the length of annealing time.

(3) Implantation damage-gettering of diode samples
(a) Gettering efficiency of diode samples contaminated with copper. Diodes with good qualities were fabricated with conventional technique on 200 μm thick epitaxial silicon wafers polished on both surfaces. These diodes were intentionally contaminated with copper through heat treatment in copper nitrate solution carried by oxygen

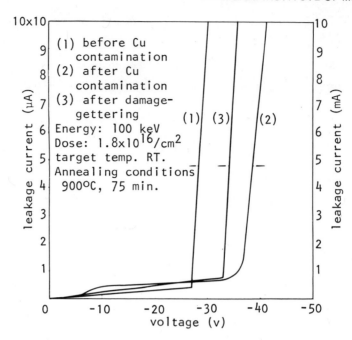

Fig. 6. Typical reverse characteristics of Cu contaminated diode samples.

at 1100°C for 30 min and then cooled down at a rate of 250°C/hr. After that the back surfaces were implanted with 1.8×10^{16} 100 keV neon ions/cm^2 at a rate of 1 μa/cm^2, followed by 900°C annealing in N_2 ambient for 75 min. The reverse characteristics of the diodes before and after damage gettering were measured. The results are shown in Fig. 6. The copper concentrations of a few samples were analyzed through SIMS before and after contamination and also after gettering (see Fig. 7).
(b) Damage-gettering of diodes with excessive reverse leakage current. Diodes were fabricated with conventional technique on 200 μm thick epitaxial Si wafers polished on both surfaces. Those diodes with excessive reverse leakage current were chosen and their back surfaces were implanted with 1.8×10^{16} 100 keV neon ions/cm^2 at a rate of 1 μa/cm^2 followed by 900°C annealing in N_2 ambient for 60 min. Their reverse characteristics were measured. The results are shown in Fig. 8.
(c) Summary.
(a) We have studied the lattice disorder of Si induced by 100 keV neon ions. At a dose rate of 1 μa/cm^2, the critical dose to pro-

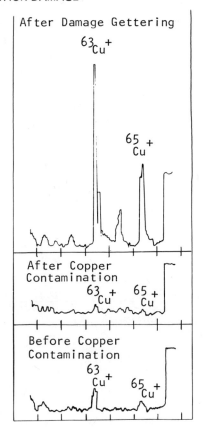

Fig. 7. Cu concentrations on the back surfaces of diode samples before and after Copper contamination and after damage - gettering through SIMS.

duce a completely disordered region occurs somewhere between 2×10^{15} ions/cm^2 and 3×10^{15} ions/cm^2. The amounts of residual lattice disorder induced by doses of 1.2×10^{16} ions/cm^2 and 6×10^{15} ions/cm^2, and followed by 900°C, 60 min. annealing are 50 % and 24-34 %, respectively. At a dose of 1.2×10^{16} ions/cm^2, the amorphous layer extends to a depth of 3500 Å from the surface. The peak of residual damage occurs at a depth of 1400 Å.
(b) For samples implanted with 100 keV neon ions to a dose of 2×10^{16} /cm^2 and at a dose rate of 1 μa/cm^2, the depth distribution of Cu impurity in the damaged layer measured by neutron activation analysis indicates that Cu impurity is gettered in a thin layer of 1400 Å thick at the surface. The results obtained through

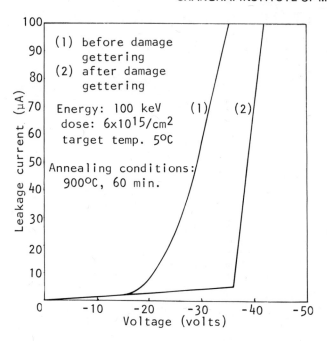

Fig. 8. Typical Reverse Characteristics of diodes after damage gettering.

SIMS indicate that Cu impurity is gettered at a depth of about 1600 Å from the surface.

(c) The results of damage-gettering experiments on diode samples contaminated with copper indicate that the reverse leakage current drops down by two orders of magnitude after annealing. For those diodes which were soft breakdown before annealing, the reverse leakage current drops down to less than 10 μa after gettering. The great leader of Chinese people Chairman Mao teaches us: "We must learn to look at problems allsidely, seeing the reverse as well as the obverse side of things. In given conditions, a bad thing can lead to good results and a good thing to bad results." Radiation damage induced by ion implantation was regarded as a troublesome phenomenon in semiconductor field. But it has been proved that this phenomenon is helpful in device improvements if it is used properly. Not only it can be employed to getter heavy metal impurities in semiconductor devices, but also it might have other practical applications to device industry.

Workers, cadres and research members worked in close coordination in performing the experiments. This work was carried out in socialist cooperation with other institutes. Neutron activation analysis and SIMS analysis were performed at Institute of Atomic

Energy, Chinese Academy of Sciences and Shanghai Bureau of Standards, respectively.

References

Buck, T.M., Poate, J.M., Pickar, K.A., and Hsieh, C.M. (1973) Surface Science 35, 362.

Foster, C., et al. (1972) Radiation Effect 16, 139.

Hsieh, C.M., Matthews, J.R., Seidel, H.D., Pickar, K.A., and Drum, C.M. (1973) Applied Physics Letters 22, 238.

Related Techniques

A NEW SCANNING ION MICROSCOPE

FOR SURFACE AND IN-DEPTH ANALYSIS

K. Wittmaack

Gesellschaft für Strahlen- und Umweltforschung mbH
Physikalisch-Technische Abteilung
D-8042 Neuherberg, Germany

ABSTRACT

The characteristics of a new UHV scanning ion microscope with quadrupole mass filter are described. The instrument provides ion images with a lateral resolution of about lo µm. Moreover stereoscopic imaging is achieved. High sensitivity surface and in-depth analysis can be carried out in the raster scanning mode. Edge effects are eliminated by means of an electronic aperture. Analysis of insulating specimens is accomplished by simultaneous bombardment with positive primary ions and low energy electrons. The geometry of the system allows combination with other analytical techniques.

I. INTRODUCTION

The techniques of ion beam surface layer analysis have been improved considerably during the last few years /1/. These methods include secondary ion mass spectrometry (SIMS)/2,3/, Auger electron spectroscopy /4/, low and high energy backscattering /5,6/, ion induced X-rays /7/ and photons and nuclear reactions /8/. Comparison of the techniques indicates /9/ that, although unique in certain applications, all of them suffer from limitations as to their general applicability in surface layer analysis. Usually the methods provide complementary information. To achieve meaningful results in all fields of application it is necessary, therefore, to combine at least two of the techniques. Equipment providing in situ analysis with different tools seems to be most promising in this respect.

Considering the various methods separately one can state that none of them has been optimized ultimately. In each case improvements with respect to sensitivity and depth resolution are possible. For some methods the present state of data interpretation is far from being satisfactory.

Another aspect of surface layer analysis is lateral resolution. Routine microanalysis has been accomplished previously only with SIMS /2,3,1o-13/. Standard SIMS instrumentation /1o-13/, however, comprises highly specialized machines the geometry of which does not allow in situ combination with other analytical tools. In this paper we present a new UHV scanning ion microprobe mass spectrometer which will allow simultaneous investigations by means of SIMS and other analysis techniques.

2. EXPERIMENTAL EQUIPMENT

The experimental arrangement is an improved version of a recently described UHV differential in-depth analyzer (DIDA) /14/. The main part of the set-up comprising target chamber, mass spectrometer, einzel lens and scanning equipment is shown in Fig. 1. The primary ion beam (1-15 kV, usually Ar^+) is supplied by a triode ion gun. A pressure-step between gun vessel and target chamber provides differential pumping. Prior to focusing the beam diameter is defined by suitable apertures placed on either side of the scanning plates. The einzel lens is designed such that focusing is achieved for midpotentials close to the acceleration potential.

The quadrupole mass filter used for secondary ion analysis is preceded by an energy filter /14,15/ which additionally serves as an accel-decel system to optimize the secondary ion intensity /14/. The target can be bombarded simultaneously with primary ions and low energy electrons (E ≤ 5oo eV). This procedure is used to compensate the positive charge in case that insulating specimens are analysed. Since the optimum secondary ion extraction (acceleration) voltage amounts to only 1oo V the deflecting action of the ion extraction field on the electron trajectory is small. It can be compensated easily by means of deflection plates placed at the exit of the electron gun. Moreover the deflector allows the electron gun to be positioned such that the target cannot "see" the filament. Contamination of the specimen by material evaporated from the heated filament can thus be avoided.

As can be seen from Fig. 1 the separation between target holder, energy filter and einzel lens is relatively large so that there is ample space to introduce additional equipment such as the electron gun. The large distance between einzel lens and target allows the focused beam to be raster scanned across a desired area of more than 1 mm^2. An unavoidable drawback of such a geometry is that spot dia-

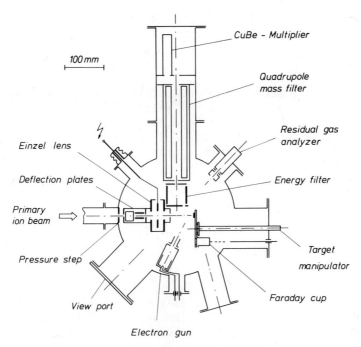

Fig. 1. Spectrometer section of the DIDA scanning ion microscope.

meters of the order of 1 µm can only be achieved with very small entrance apertures of the einzel lens (< 0.1 mm). This, however, reduces the beam current drastically. As a reasonable compromise we used an aperture with a diameter of 0.5 mm. At 10 keV Ar^+ currents of more than 1 µA could be fed through this aperture after optimizing the ion source operation for this purpose.

3. SYSTEM CHARACTERIZATION

3.1 Profiles of focused beams

Usually the minimum spot size of a focused beam is expected to be determined by beam transport characteristics and lens aberrations /16/. At high beam currents I and at a drift length L large compared to the minimum radius r_m, space charge expansion of the ion beam must be considered /17/. For a 10 keV Ar^+ beam of 0.5 µA the relative expansion due to this effect amounts to $r/r_m = 2$ at a reduced drift length $L/r_m = 10^3$ /17/. Assuming $r_m = 10$ µm this means that a drift length of only 10 mm will cause the beam to expand by a factor of 2. Since the distance between exit electrode of the einzel lens and target is about 6o mm in set-up of Fig. 1 pronounced space charge effects had to be expected. Therefore a detailed study of the spot diameter as a function of beam current and ion source parameters was carried out.

The investigations were accomplished using wire gauzes as profile monitors. Spot sizes of the order of the wire diameter or the wire separation can be determined easily from the amount of current transmitted through or collected by the mesh. Quantitative analysis was performed by assuming a Gaussian current density distribution in the spot, i.e.

$$j = (I/\pi r_s^2) \exp(-r^2/r_s^2). \qquad (1)$$

Eq. (1) provides an adequate definition of the spot radius r_s because a beam with uniform current density $I/\pi r_s^2$ and radius r_s contains the same total current I as the above Gaussian distribution.

An example for the variation of the current collected upon sweeping the beam across wires of the gauze is shown in Fig. 2. In that case beam size and mesh orientation were such that the current to neighbouring wires can be neglected. The profiles indicate a spot size much smaller than the wire separation. Moreover details of the current density distribution in the spot can be deduced. For routine measurements of the spot diameter, however, profile tracing is too much time consuming. E.g. when studying spot characteristics as a function of the ion source parameters it is sufficient to determine the minimum current I_{min} and the maximum current I_{max}. I_{min} is detected when feeding the beam through an opening in the mesh whereas I_{max} is collected when the beam hits the crossing point of two wires. Integrating Eq. (1) for the respective geometry the spot radius r_s can be deduced from either I_{min} or I_{max}. It was found that the spot radii thus determined are in reasonable agreement with one another. This is demonstrated in Fig. 3 where results are presented for optimum ion source operation. The consistency of the results indicates that this somewhat rigorous but rapid method of beam profile control is very adequate for such purposes.

As can be seen from Fig. 3 the spot size increases nearly linearly with beam current. Following the above estimate this effect must be attributed to space charge expansion in high density beams. A quantitative comparison with theoretical values /17/ is not possible because the convergence angle of the beam leaving the einzel lens is unknown. We use the results of Fig. 3 to conclude that with the present geometry spot diamenters of about lo µm and scanning ion images with good lateral resolution can be expected only at beam currents of the order of lo nA. Nevertheless the large beam current mode of operation will be very useful for high sensitivity in-depth analysis. It allows the sputtered area to be as large as

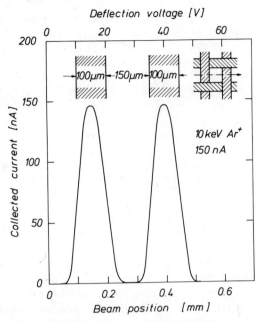

Fig. 2. Current collected on a steel gauze as a function of the beam position. Wire position and dimension are indicated as well as the trace of the beam on the mesh.

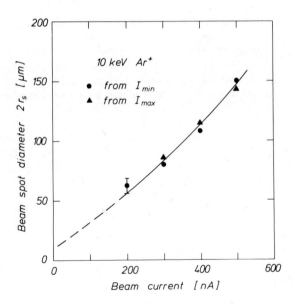

Fig. 3. Minimum diameter of the focused beam as a function of the beam current.

1 mm^2. The number of secondary ions emitted per depth interval is thus up to three orders of magnitude larger than in conventional ion microprobe mass analyzers /1o-13/. This is achieved while maintaining a reasonable sputtering rate $\dot{x}$. For a mean current density of 1 μA/mm^2, a sputtering yield $S = 1$, and a target density $n = 6 \times 10^{22}$cm^{-3} one finds $\dot{x} = 1$ Å/s.

3.2 Raster scanning ion imaging

Ion images were achieved in the manner already applied in magnetic type mass spectrometers /11-13/: The focused primary ion beam is raster scanned over the surface of the specimen. The quadrupole mass filter is tuned to the desired mass number and the appropriate secondary ions are detected by an electron multiplier. The amplified output signal of the detector is used to modulate the intensity (z axis) of an oscilloscope the x,y deflection of which is synchronized to the primary ion raster. The oscilloscope intensity distribution thus reflects the lateral surface distribution of the element of interest. By continous bombardment successive images at desired depth intervals can be obtained which provide threedimensional characterization of the sample.

The steel gauze used for spot size measurements (Sect. 3.1) was taken as a specimen. Examples of ion images obtained from the gauze are presented in Fig. 4. The mesh structure is shown schematically in Fig. 4a. Fig. 4b provides a characterization of the lateral resolution and the depth of focus. The image clearly exhibits the braiding pattern of the gauze. The stereoscopic imaging capabilities demonstrated in Fig. 4b are similar to those of a scanning electron microscope, except for the reduced lateral resolution which can be estimated to amount to about 1o μm. This is supported by further test runs on a copper mesh with a wire diameter of 3o μm /18/.

The image in Fig. 4b was obtained at a low beam current of 1o nA. As expected from the results of Sect. 3.1 the lateral resolution decreases rapidly with increasing beam current. This is demonstrated in Figs. 4c and 4d. The loss in resolution is in good agreement with the spot size measurements.

Note that for all the examples presented the area imaged amounted to 1 x 1 mm^2. As can be seen from the images there is no loss in resolution across this area. Further measurements indicated that this is true even for areas as large as 1.5x1.5 mm^2 (which corresponds to the upper limit of sweep capability in the present set-up at a primary ion energy of 1o keV).

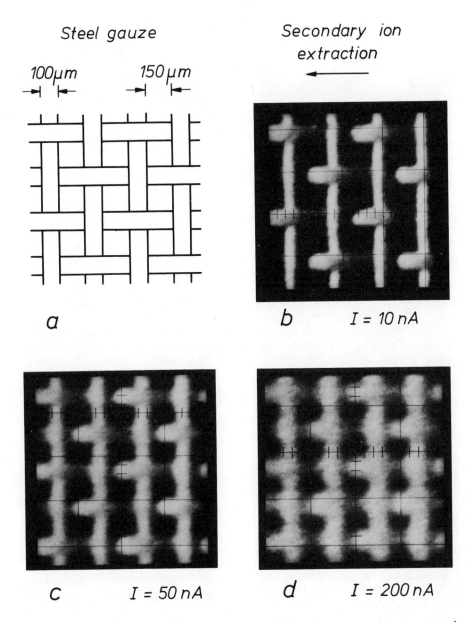

Fig. 4. (a) Braiding pattern of the steel gauze used as a specimen. (b) - (d) Scanning ion microprobe images obtained from the gauze. Primary ions: 10 keV Ar^+. Secondary ions: $^{56}Fe^+$. Parameter is the primary beam current.

3.3 In-depth analysis

Depth profiling with ion microprobes operated in the raster scanning mode offers the advantage to get rid of implications faced in analysis with large area beams. In the latter case edge effects can only be avoided with specially prepared specimen structures /19,20/. In the raster scanning mode contributions from the rim of the area under investigation can be eliminated by use of an electronic aperture: The scaler which counts the mass-analyzed secondary ions is gated electronically by signals derived from the scan voltage power supply so that only ions from a pre-selected central part of the sputtered area are counted.

In this way multilayer structures have been investigated with high depth resolution. In particular the instrument was used to study interface composition in epitaxial films prepared by chemical vapour deposition on insulating substrates /21/. The analysis could only be carried out with the help of the charge compensating electron gun described in sect. 2. The method was found to work well even with raster scanned primary ion beams.

3.4 Combination with other analysis techniques

Although the microprobe mass analyzer described was designed not specifically for a combination with other surface analysis techniques it is obvious from Fig. 1 and the performance characteristics of the mass spectrometer that other instruments can be added easily. Optimum geometry for combined use of two techniques can be achieved by slight changes in the design of the target chamber. In many cases this will allow not only in situ but also simultaneous investigation by different techniques.

4. CONCLUSIONS

The new scanning ion microscope described has been shown to be a very useful analytical tool. With respect to versatility it is superior to other secondary ion mass spectrometers. The most important properties of the instrument are the capability to provide (i) stereoscopic imaging with reasonable lateral resolution and (ii) in-depth profiling with high sensitivity and high sputtering rate. Conducting as well as insulating specimens can be analyzed under UHV conditions ($<10^{-8}$ Torr). Improvements with respect to lateral resolution can be obtained by reducing relevant dimensions such as the diameter of the einzel lens entrance aperture.

REFERENCES

/ 1/ J.W. Mayer and J.F. Ziegler, Ed., Proc. Int. Conf. Ion Beam Surface Layer Analysis, publ. in Thin Solid Films 19 (1973).
/ 2/ H.W. Werner, Vacuum 24, 493 (1974).
/ 3/ H. Liebl, J. Vac. Sci. Technol. 12, 385 (1975).
/ 4/ J.M. Morabito, Thin Solid Films 19, 21 (1973).
/ 5/ R.E. Honig and W.L. Harrington, Thin Solid Films 19,43 (1973).
/ 6/ W.K. Chu, J.W. Mayer, M.-A. Nicolet, T.M. Buck, G. Amsel and F. Eisen, Thin Solid Films 17, 1 (1973).
/ 7/ J.A. Cairns, A.D. Marwick and I.V. Mitchell, Thin Solid Films 19, 91 (1973).
/ 8/ G. Amsel, T.P. Nadai, E. d'Artemare, D.Davied, E. Girard and J. Moulin, Nucl. Instr. Meth. 92, 481 (1971).
/ 9/ J.W. Mayer and A. Turos, Thin Solid Films 19,1 (1973).
/1o/ A.J. Socha, Surface Sci. 25, 147 (1971).
/11/ C.A. Evans, Jr., Anal. Chem. 44, 67A (1972).
/12/ H. Liebl, Anal. Chem. 46, 22A (1974).
/13/ B.F. Phillips, J. Vac. Sci. Technol. 11, 1o93 (1974).
/14/ K. Wittmaack, J. Maul and F. Schulz, Proc. Sixth Int. Conf. Electron and Ion Beam Science and Technology, R. Bakish, Ed. (The Electrochemical Soc., Princeton 1974) p. 164.
/15/ K. Wittmaack, J. Maul and F. Schulz, Int. J. Mass Spectrom. Ion Phys. 11, 23 (1973).
/16/ R.L. Seliger and W.P. Fleming, J. Appl. Phys. 45, 1416 (1974).
/17/ R.G. Wilson and G.R. Brewer, Ion Beams (J. Wiley & Sons, New York 1973).
/18/ K. Wittmaack, submitted to Rev. Sci. Instrum.
/19/ F. Schulz, K. Wittmaack and J. Maul, Rad. Effects 18,211(1973).
/2o/ K. Wittmaack, F. Schulz and B. Hietel, Proc. Fourth Int. Conf. Ion Implantation in Semiconductors, S. Namba, Ed. (Plenum Press, New York 1975) p. 193.
/21/ Ch. Kühl, M. Druminski and K. Wittmaack, Proc. Third Int. Conf. on Thin Films, to be published.

SPUTTERING OF THIN FILMS IN AN ION MICROPROBE

W.O.Hofer and H.Liebl

Max-Planck-Institut für Plasmaphysik, 8046 Garching

bei München, EURATOM Association, Germany

ABSTRACT

The method of measuring sputtering yields by the SIMS technique and its main complications are discussed. One of them, cone formation, can often be avoided by a proper choice of the substrate, while the other, secondary ion yield enhancement, must be dealt with by tracing ions characteristic of the transition at the interface. The method has been applied to an investigation of the dependence of the titanium sputtering yield on oxygen coverage.

INTRODUCTION

A relatively simple method of measuring sputtering yields of metals by secondary ion mass spectroscopy (SIMS) has been described by Benninghoven /1/. It is essentially based on the determination of the time it takes to sputter through a thin film of known thickness which has been deposited on some substrate. From the current density and the time till break-through, it is straightforward to calculate the number of ejected target atoms per incident ion. This method is especially advantageous when the sputtering yield is close to or less than unity, since then with weight loss techniques the retention of the primary ions in the target has to be known accurately /2/. Another attractive feature of this method is its applicability to a wide variety of targets.

EXPERIMENTAL

The experiments described in the following were performed with an ion microprobe described previously /3/. The primary beam is raster scanned over a defined target area and, by virtue of an electronic aperture /4/, only ions from the center part of the erosion crater are counted. Ions emitted from the rim of the bombarded area would smear out the depth profile and thus deteriorate the timing precision in sputtering yield measurements. The size of the rastered area (85 x 85 μm^2) was determined in a scanning electron microscope. This measurement constitutes the largest source of experimental uncertainty.

The vacuum deposition of the thin films was performed under UHV conditions at deposition rates greater than 10 Å/s. An oscillating quartz crystal was used for the thickness determination.

RESULTS AND DISCUSSION

Figure 1 shows the time variation of the secondary ion signals for a titanium-vanadium interface. From these symmetrical

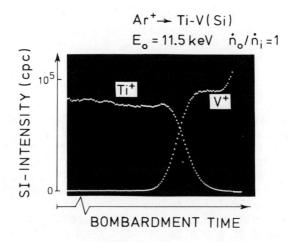

Figure 1. Secondary ion intensities near the interface for a 1000 Å Ti-layer deposited on V. The Ti$^+$ signal drops from 90 % to 10 % within 13 % of the total time.

profiles an unambiguous determination of the break-through time can be made, giving a sputtering yield for Ti with 11.5 keV Ar^+ ions of S = 1.42 + 0.05 atoms per ion for a clean surface. This ideal situation is, however, the exception rather than the rule, the main obstacles being cone formation and secondary ion yield enhancement at the film-substrate interface. We have observed cone formation on a great variety of multilayer targets and an example is presented in Fig.2

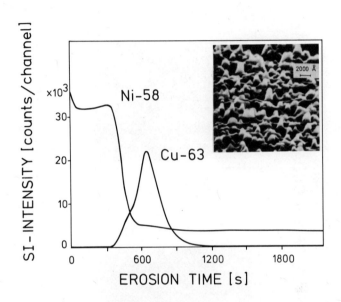

Figure 2. Secondary ion intensities for a 1000 Å Ni film deposited on a 1000 Å Cu layer (substrate Si). Strong cone formation is observed as shown in the scanning electron micrograph.

While this effect can be avoided in depth concentration profiling by using reactive primary ions /5/, one does not have this option when the sputtering yield for a given projectile-target combination is to be determined. Thus, the only alternative left is to try to find a substrate whose sputtering yield is similar to that of the film, and whose atomic mass is not too different.

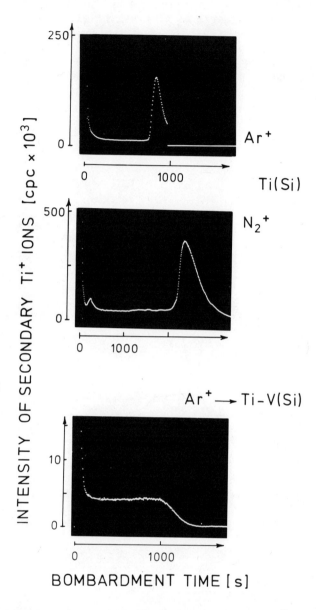

Figure 3. Intensity of secondary Ti^+ ions for a 1000 Å Ti film deposited on a Si-substrate, upper two figures directly, lower figure with 400 Å vanadium buffer layer in between. Despite the strong enhancement with N_2^+ bombardment the enhancement at the interface is still predominant.

The other obstacle - the ion yield enhancement - is demonstrated in Fig.3. The decrease of the secondary ion signal is overcompensated here by the enhancement, rendering it difficult to specify the mean break-through time. It was found possible, however, to pin down the point of break-through by tracing additional secondary ions characteristic of the sample, either molecular ions or impurity ions from the interface.

We applied this method in investigations of the oxygen influence on the sputtering yield of Ti (Fig.4), the main reason being to devise an alternative approach to the problem of oxide sputtering. Compared with bombarding bulk oxide samples we see one major advantage: the collision cascade in the target is that of the monoelemental base metal and all that is changed by surface oxidation is the surface binding energy.

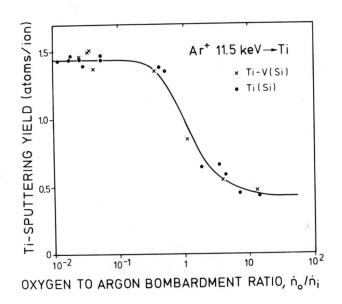

Figure 4. Ti sputtering yield vs. oxygen-to-argon flux ratio.

With increasing oxygen-to-ion flux ratio the Ti sputtering yield drops to about one-third of the original value. One has to keep in mind, however, that assuming TiO_2 is being formed, for each Ti atom ejected, two oxygen atoms are also removed, thus bringing up the total sputtering yield to that of pure Ti again. From the surface binding energies one would expect a sputtering yield lower by $S(TiO_2)/S(Ti) = E_b(Ti)/E_b(TiO_2) = 0.7$. The somewhat higher yield found experimentally is probably due to the surface oxide decreasing the fraction of channeled projectiles and thus increasing the energy deposited in the region near the surface.

The decrease in differential sputtering yield occurs around $\dot{n}_0 \approx \dot{n}_i$, which is at much lower exposures than expected from static oxygen exposure experiments /6/. This might have favourable technological consequences in view of the first wall problem of a fusion reactor /7/.

ACKNOWLEDGEMENTS

We gratefully acknowledge experimental help from G.Staudenmaier in the first phase of the experiments, the preparation of the targets by H.Kukral and the technical assistance of A.Schlamp.

REFERENCES

/1/ A.Benninghoven, Z.Angew. Physik, 27 (1969) 51

/2/ A.J.Summers, N.J.Freeman and N.R.Daly, J.Appl.Phys. 42 (1971) 4774

/3/ H.Liebl, Int.J.Mass Spectr.Ion Phys. 6 (1971) 401

/4/ W.O.Hofer, H.Liebl, G.Roos and G.Staudenmaier, Int. J.Mass Spectr. Ion Phys., in press

/5/ W.O.Hofer and H.Liebl, Applied Physics, in press

/6/ A.Müller and A.Benninghoven, Surf.Sci. 41 (1974) 493

/7/ J.Kistemaker in Course on the Stationary and Quasi-Stationary Toroidal Reactors p 281, Erice-Trapani 1972, EUR 4999e.

Xe^+ ION BEAM INDUCED SECONDARY ION (Si^+) YIELD FROM Si-METAL INTERFACES

T.NARUSAWA, T.SATAKE and S.KOMIYA

ULVAC Corporation, Chigasaki, Kanagawa, JAPAN

A.SHIMIZU, M.IWAMI and A.HIRAKI

Department of Electrical Engineering, Osaka University, Suita, Osaka, JAPAN

In connection with the phenomenon of Si low temperature migration in Si-Metal(M) systems, the present authors have measured Xe^+ ion($\sim$1.5keV) beam induced secondary Si^+ ion yield from Si-M(especially Si-Au) interfaces by secondary ion mass spectroscopy(SIMS) technique. The yield is about ten times higher than that from the ordinary Si crystal surface. The origin of this high Si^+ ion yield is proposed to be due to the Si metallic phase existing in the interface region. The metallic phase of Si is assigned by Auger electron spectroscopy, soft X-ray spectroscopy and nuclear magnetic resonance techniques.

1. INTRODUCTION

In recent years, Si low temperature migration in Si-Metal(M) systems has been studied in many laboratories using He^+ ion backscattering technique[1], Auger electron spectroscopy[2,3] and other methods to indicate that Si-M interfaces control the phenomena. For example, Hiraki et al.[4] of Osaka University and Tu et al.[5] of IBM independently have proposed that at Si-M interfaces the chemical state of Si is metallic rather than ordinary covalent state to form Si-M metallic bonding and since this Si-M bonding is far weaker than covalent Si-Si bonds, the ejection of Si atoms from the interface takes places at low temperature to induce the above mentioned phenomenon of low temperature Si migration. The foundation of the proposal by Hiraki et al. is, as explained briefly later, based upon studies of valence electronic states of the metallic Si in Si-Au systems[6], which is one of interesting Si-M systems ever studied, by several techniques such as Auger Electron Spectroscopy(AES)[2], Soft X-ray Spectroscopy(SXS)[7] and Nuclear Magnetic Resonance(NMR)[8].

Recently, present authors have measured Xe^+ ion($\sim$1.5keV) beam induced secondary Si^+ ion yield from the metallic Si by Secondary Ion Mass Spectroscopy(SIMS) technique and found that the yield was about ten times higher than that from the ordinary Si crystal to suggest that the SIMS technique may also be used to check the above proposal, since there are several reports[9,10] to show that secondary ion yield depends rather sensitively on the chemical states. The main purpose of the present paper is to report the result of the SIMS measurement of secondary Si^+ ion yield from Si-Au interface.

2. METALLIC STATE of Si EXISTING at Si-M INTERFACES

From thermodynamical point of view[11], any interface between two materials tends to become a diffuse interface with a certain thickness ℓ and a composition gradient in order to minimize interfacial energy and the phase or electronic state of diffuse interface are different from the equilibrium phase indicated by the phase diagram. In Si-Au phase diagram, for instance, there is no equilibrium solid compound or silicide phase, but the diffuse interface region, if formed, as shown in Fig.1, must be some kind of Si-Au solid compound or metastable(or non-equilibrium) silicide. Therefore, the electronic or chemical nature of Si in this diffuse interface must be different from that of pure Si.

To investigate the presence of diffuse interface in Si-Au system, Hiraki et al. employed AES technique. In this study, onto a clean surface of a Si substrate a Au film was deposited at temperature lower than 50°C in high vacuum($\sim 10^{-7}$ Torr) and then Si(LVV) spectra, where V represents a valence energy state, were

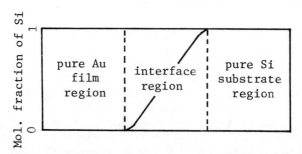

Fig.1 A schematic diagram of Si(substrate)-Au(thin film) diffuse interface region.

taken from both the substrate(pure Si) and the interface. As is
shown schematically in Fig.2, the Si spectrum of the interface region had two peaks at 90 and 95eV and clearly differed from that of
pure Si which had a single peak at 92eV to indicate that, as was
expected, the electronic state of Si in the interface was quite
different from that of pure Si and therefore the Si-Au interface
was diffuse instead of sharp.

With regard to the double-peaked Si spectrum, it was proposed
from the AES study[12] of "vapor-quenched" thin films of metastable
Si-noble metal alloys(Si-Ag, Si-Au and Si-Cu) that the spectrum
could be assigned to correspond to metallic Si from following
reason. Si, as well as Ge, become metallic in liquid state.
In analogy with the quenching of metallic Ge[13], Si liquid containing a certain amount of noble metal(Ag, Au and Cu) was rapidly quenched to room temperature by the vapor quenching method onto
non-reactive metal plates inside the AES spectrometer specimen-
chamber, to freeze as metallic Si. The role of adding the noble

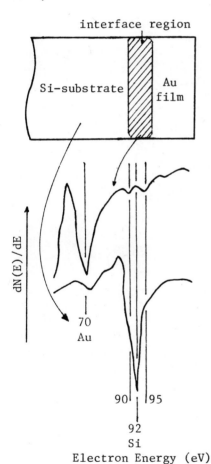

Fig.2

Si(LVV) Auger spectra from substrate Si and Si-Au interface. Notice that the former has a single peak(92eV) and the latter has a double-peak (90 and 95eV).

metal was thought to be necessary to stabilize the metallic Si by preventing the formation of covalently bonded Si which is semiconductor rather than metal. As seen in Fig.3, exactly the same double-peaked Si(LVV) spectra were observed for all the vapor-quenched Si-noble metal films with metal concentrations of more than 70at.%. But with smaller metal concentration the Si(LVV) spectra were the same as that of pure Si with a single peak at 92eV to support the above assignment of the double-peaked spectrum to Si metallic state because it is readily understandable that metallic Si could not be stabilized in these metal-depleted specimens.

Further and more direct support to the assignment was obtained from both SXS[7] and NMR[8] studies of metal-rich alloy specimens.

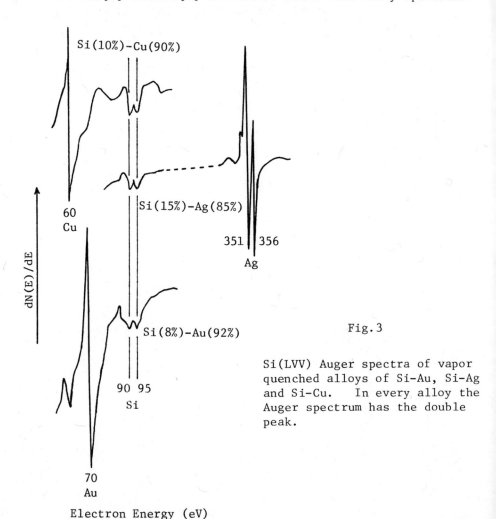

Fig.3

Si(LVV) Auger spectra of vapor quenched alloys of Si-Au, Si-Ag and Si-Cu. In every alloy the Auger spectrum has the double peak.

Xe+ ION BEAM INDUCED SECONDARY ION (Si+) YIELD

Si K_β-emission spectra were measured both from Si-Cu specimens whose AES spectra of course had the double-peaked structure as seen in Fig.3 and from pure Si specimens for purpose of comparison. As a consequence of the selection rule, the present K_β-emission gives information only about Si 3p band or, more accurately speaking, local density of states of 3p electrons at the Si sites. As shown in Fig.4 both K_β-spectrum and 3p-band of Si in metal rich Si-Cu alloy are quite different from those of pure Si: namely, from the density of states at the Fermi-energy(E_F) it is clear that Si is metallic in the alloy, whereas in the pure specimen very little, if any, electrons are present at E_F corresponding to semiconductor state of Si.

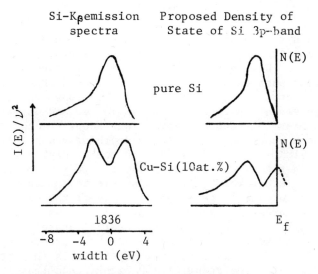

Fig.4 Si K_β-emission(soft X-ray) from pure Si and Si(10%)-Cu(90%) alloy. The former has a single peak and the latter has a double peak.

By NMR technique it is possible to identify metallic Si by observing a spin lattice relaxation time(T_1) of ^{29}Si nuclear spins (I=1/2) at low temperature. In metals, especially whose conduction electrons at Fermi surface have s-like character(or wave function), the relaxation time(T_1) is mainly determined by the Fermi-type contact interaction and T_1 at temperature T is expressed by the following equation:

$$(T_1 T)^{-1} = (64\pi^3 k/9\hbar)(\gamma_e \gamma_N \hbar^2)^2 [\rho_s(0, E_F)]^2$$

where $\rho_s(0,E_F)$ is the energy distribution of the density of s electrons at the ^{29}Si nucleus(r=0) having energy of E_F, k is the

Boltzman constant and γ_e and γ_N are gyromagnetic ratios of electron and nucleus, respectively. From the equation it is understood that in metals, since $P_s(0,E_F)$ is almost independent of T, T_1T=constant and T_1 must be far shorter than in semiconductor state whose $P_s(0,E_F)$ is almost zero at low temperature. In fact, the measured T_1 in metal-rich Si-Cu alloys obeyed the relation of T_1T=constant and T_1 was of the order of 10sec at T=1.4K which is more than 2 orders of magnitude shorter than the reported T_1[14] measured in the degenerate Si semiconductor containing 10^{19} P-donor-impurities per cm^3.

3. Xe^+ ION BEAM INDUCED SECONDARY ION(Si^+) YIELD from METALLIC Si

It is rather popular that SIMS(Secondary Ion Mass Spectroscopy) is used for the analysis of surface layers, thin films and interfaces. In this method the sample is bombarded with a beam of primary ions such as Ar^+, Xe^+ and O_2^+. As a result of the bombardment, neutrals, positive and negative ions are sputtered away. These secondary ions are mass selected and detected in a mass spectrometer to produce informations about the composition of the sample as a function of sputtered depth or depth informations. Although the SIMS is recognized as a method of high detectivity, to obtain a quantitative information by this technique is sometimes difficult. In other words, it is possible to get quantitative informations about the target specimen only if the secondary ion yield from the target is known. However, the secondary ion yield is often strongly dependent upon the chemical state(or bonding nature) of the constituents of the target to introduce difficulties. One of the reported examples[9] on this effect is that the Al^+ ion yield in SiC is about a factor of 100 higher than that from pure Al due to the difference of bonding nature of Al in both matrices and consequently the effect of this kind is called as "matrix effect". This matrix effect is ascribed[10] mainly to an increase in the degree of ionization, though the detailed mechanism of the effect is not yet known.

Then it is an interesting question to see the difference in secondary Si^+ ion yields, if any, from pure Si and from metallic Si, since it is clear from the preceding chapter that the chemical (or bonding) nature of Si is different from each other.

The home made SIMS-AES apparatus[15] by ULVAC Research Laboratory was used for the present study. By this apparatus performing simultaneous SIMS and AES analyses together with measurements of sputtering rates of the specimens was possible. As primary ions Xe^+ beam of $\sim$1.5keV were used. The specimen chamber of the spectrometer was at first evacuated down to 10^{-9} Torr and then Xe gas of 10^{-5} Torr was introduced in order to avoid the oxygen ions(O_2^+) to mix with Xe^+ ions. Because O_2^+ ions, if exist in the bombarding beam, are known to modify greatly the secondary ion yields[16].

The metal rich alloys of Si-Au and Si-Cu and pure Si specimens were investigated. Identifications of pure(or semiconductor) Si and metallic Si were done through the Si(LVV) AES spectra in the same manner as was described already. As was expected, the secondary Si^+ ion yield from metallic Si differed from that from semiconductor(or pure) Si and about a factor of 10 higher yield was observed in metallic Si. Although the mechanism of this enhanced Si^+ ion yield from metallic state of Si is not known, this fact suggests that SIMS can possibly be used for the study of Si-M interfaces where Si metallic state is expected to exist.

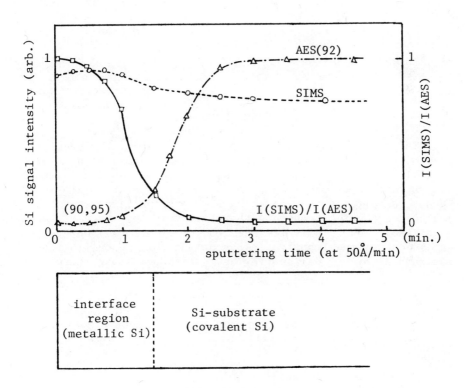

Fig.5 Si^+ yield[I(SIMS)] and Si(LVV) intensity[I(AES)] from Si(substrate)-Au(thin film) system(only interface and substrate regions). The ratio of I(SIMS)/I(AES) is also shown. Notice that in the interface region Si is metallic which can be identified by the double peaked AES[Si(90 & 95)] spectra.

4. ENHANCED SECONDARY Si^+ ION YIELD from Si-Au INTERFACE

A Si wafer of (111) orientation, (10mm wide)×(15mm long) and 0.1mm thick, was mounted on the sample holder inside the spectrometer and cleaned by Xe^+ ion bombardment until no trace of SiO_2 on the wafer surface could be detected by AES. Then pure Au was deposited to thickness of approximately 200Å in ultrahigh vacuum[17] and was then exposed at ~25°C for ~10min in Xe gas at 10^{-5} Torr, as was done in the case of SIMS measurement of metallic Si, to obtain depth profiles of the sample[Au(film)-Si(substrate)] from both SIMS and AES analysis. In Fig.5, are shown the intensities (in arbitrary units) of both SIMS and AES of Si only near the interface region(interface and substrate) together with their ratios [denoted by I(SIMS)/I(AES)] as a function of sputtering time(min); the Xe^+ ion beam(~1.5keV and current density of 2.5μA per cm^2) used could sputter off 50Å Cu per min. For the identification of Si in the interface and substrate, the already mentioned AES spectra from Si(LVV) transition[Si(90 & 95eV) and Si(92eV)] were used.

It is clear from Fig.5 that Si^+ ion yield from the interface is more than a factor of 10 higher and this is consistent with the already proposed picture that at Si-Au interface the chemical state of Si is metallic.

5. CONCLUDING REMARK

The present authors have employed several techniques for the possible identification of metallic Si which is displayed in Table.1. They believe that these techniques can provide more understandings of phenomenon of Si low temperature migration, because despite of many works on Si-M systems[1] done mostly by He^+ ion backscattering technique there are very few study of the interface itself, especially its chemical(or electronic) nature which may be one of the important clues of the phenomenon[4,5] and can not be done by He^+ backscattering. Although among these tech-

Table 1. Techniques used for the identification of metallic Si.

°AES[Si(LVV)]
SXS[Si-K_β emission]
NMR[T_1 of ^{29}Si-nuclei]
°SIMS[secondary Si^+ yield]

° : good for interface study

niques, SIMS is thought only to give informations about the composition of the target specimen, as does He$^+$ ion backscattering, through mass analysis, the authors would like to emphasize its high capability of detecting the chemical(or bonding) state of the constituent of the target by the measurement of secondary ion yields. Then for the direct study of interface itself, including its chemical state, SIMS as well as AES(or combination of both) can be one of the prospective tools: SXS and NMR become powerless when the interface thickness is thin as that in the case of Au deposition onto Si substrate which is less than 100Å[17].

The present authors would like to express their thanks to Dr.T.HAYASHI(ULVAC Corporation) and Professor K.KAWABE(Osaka University) for their warm and continuous encouragement toward the present study.

REFERENCES

[1] J.A.Borders and S.T.Picraux, Proc. IEEE, Vol.62(1974)1224.
[2] G.Y.Robinson, Appl. Phys. Letters, Vol.25(1974)158.
[3] T.Narusawa, S.Komiya and A.Hiraki, Appl. Phys. Letters, Vol.20 (1972) 272.
[4] A.Hiraki, A.Shimizu and M.Iwami, Extended Abstract of 147th Meeting of Electrochemical Society, Toronto, 1975, p.265.
[5] K.N.Tu, ibid, p.267.
[6] A.Hiraki, E.Lugujjo and J.W.Mayer, J. appl. Phys., Vol.43 (1972) 3643.
[7] K.Tanaka, M.Matsumoto, S.Maruno and A.Hiraki, Appl. Phys. Letters, Vol.27 (1975).
[8] A.Hiraki, K.Shuto, A.Shimizu, M.Iwami, M.Matsumura, T.Kohara and K.Asayama, J. Phys. Soc. Japan (to be published).
[9] H.W.Werner and A.M.DeGrefte, Surface Science, Vol.35 (1975) 458.
[10] H.W.Werner, Develop. Appl. Spectroscopy, Vol.7A, Eds. E.L. Grove and A.J.Perkins (Plenum, New York, 1969).
[11] J.W.Cahn and J.E.Hilliard, J. Chem. Phys., Vol.28 (1958) 258.
[12] A.Hiraki, A.Shimizu, M.Iwami, T.Narusawa and S.Komiya, Appl. Phys. Letters, Vol.26 (1975) 57.
[13] B.S.Stritzker and H.Wuhl, Proc. 12th Conf. Low Temp. Phys., Kyoto, 1971, p.339.
[14] W.Sasaki, S.Ikehata and S.Kobayashi, J. Phys. Soc. Japan, Vol.36 (1974) 1377.
[15] S.Komiya and T.Narusawa, J. Vac. Sci. Technol., Vol.11 (1974) 312.
[16] S.Komiya, T.Narusawa and T.Satake, ibid, Vol.12 (1975) 361.
[17] T.Narusawa, S.Komiya and A.Hiraki, Appl. Phys. Letters, Vol.23, (1973) 389.

SURFACE ANALYSIS OF ION BOMBARDED METAL FOILS BY XPS

E. Henrich, H.J. Schmidt

Institut für Heisse Chemie

75 Karlsruhe, Postfach 3640

ABSTRACT

Polycrystalline gold and nickel foils were sputtered with 45 or 60 keV Cs^+ or I^+ ions. The surface concentration of the implanted ions at saturation was analysed by X-ray photoelectronspectroscopy. Information concerning the concentration profiles near the surface can be detected by taking the spectra at different photoelectron take-off angles. The determination of absolute concentrations is still somewhat limited by the present state of the art of quantitative X-ray photoelectronspectroscopy. However, the change of relative concentrations seems especially suited to obtain the influence of various irradiation parameters like total dose, dose rate, temperature, angle of beam incidence etc. The absolute sputtering ratios at saturation which can be determined under certain conditions from the absolute surface concentrations, show deviations of ±20 % in the same and ±50 % in different runs, indicating the possible influence of some additional irradiation parameters. Because of elemental and chemical sensitivity, XPS seems to be especially useful for the investigation of irradiated multicomponent targets.

The surface properties of metals can be changed systematically by bombarding the surface with different ions of variable energy (1-3). The maximum implant concentration which can be obtained, is related in some way to the sputtering ratio. The aim of the following experiments was to see if such a correlation between the sputtering ratio and the surface concentration, analysed by X-ray photoelectronspectroscopy (XPS), can be found.

I. SATURATION CONDITIONS DURING SPUTTERING

The sputtering ratio S is the mean number of atoms removed by an impinging ion. Therefore, a solid surface is eroded by ion bombardement if the sputtering ratio is greater than one. When the diffusion of the stopped ions out of the surface or into the unirradiated part of the target can be neglected, a saturation is reached after some time, where the number of bombarding ions in the target is at maximum and remains constant.

During such steady state conditions, each bombarding ion removes one previously implanted ion plus S-1 target atoms, preferably from the immediate surface region. The surface concentration of the projectiles should be S-1 (4,5). Without diffusion in the damaged region during and after bombardment, the concentration profiles at saturation are relatively flat near the surface, since only a negligible amount of ions is stopped directly behind the surface (Fig. 1).

Therefore, the surface region can be treated as a nearly homogeneous binary alloy, whose composition may be analysed by XPS without disturbance from an inhomogeneous interior of the sample.

Diffusion and especially radiation enhanced diffusion within the damaged region is able to distort the concentration profiles. Accumulation or reduction of the implant near the surface may cause steeper concentration profiles in this region. The information concerning the concentration profiles in such inhomogeneous samples can be extracted by taking the X-ray photoelectron spectra at different photoelectron take-off angles.

Without going into detail some necessary aspects of quantitative X-ray photoelectronspectroscopy of homogeneous and inhomogeneous samples are discussed in the following section.

II. X-RAY PHOTOELECTRONSPECTROSCOPY (6)

Principles

Soft monochromatic X-rays are used to eject core electrons of atoms. Especially suitable are the K_α-X-rays of Al or Mg with an energy of 1487 or 1254 eV and a halfwidth of about 1 eV. The kinetic energy of the photoelectrons corresponds to the X-ray energy minus the electron binding energy. It is a characteristic value for each element, thus providing the basis for elemental analysis. Changes in the chemical surroundings of an element (e.g. oxidation state) may cause a chemical shift up to several eV. Therefore, some chemical information can be obtained, too.

An electron spectrometer is used to analyse the energy of the photoelectrons. The electrons escaping from the sample without energy loss can be seen as definite peaks whereas the electrons which have suffered energy loss are forming the background of the spectrum.

Information Depth

Because of these energy loss events, the intensity contribution to the photoelectron peaks diminishes exponentially with increasing distance from the sample surface. Therefore, XPS is a surface sensitive method with an information depth of only some 10 Å. Fig. 2 shows the mean free escape depth λ for electrons with different kinetic energies (7). Between 100 - 1500 eV λ shows an approximate $\lambda \sim \sqrt{E}$ dependence. Different compounds may have different escape depths (8). The mean free path of the kilovolt X-rays is several orders of magnitude larger. Thus the X-rays are essentially unattenuated over the escape depth, which simplifies intensity considerations.

Intensity (9,10)

The intensity or peak area J_i of a photoelectronpeak i in the spectrum of a homogeneous sample with a clean surface is given by: $J_i \sim \varepsilon_i \cdot \lambda_i \cdot c_i \cdot S_i$;

ε_i photoionisation cross section of the X-rays for electron i
λ_i mean free path of the photoelectron i
c_i concentration of element producing electron i
S_i spectrometer constant for kinetic energy of electron i

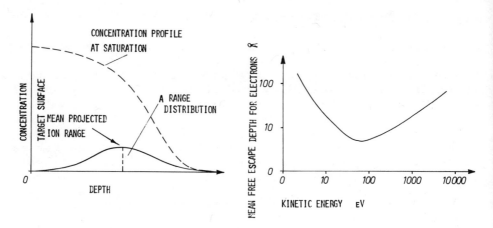

Fig. 1 (left): Concentration profile of implant at saturation
Fig. 2 (right): Mean free electron escape depth

Surface roughness factors and other small corrections are neglected. In an inhomogeneous sample λ_i and c_i change within the information depth. The usual carbon contamination layer on the samples reduces the intensity by additional energy loss processes.

The relative intensity of two different elements in a homogeneous binary sample with a clean surface can be expressed by: $J_A/J_S = \varepsilon_A \lambda_A c_A S_A / (\varepsilon_S \lambda_S c_S S_S)$.

For a stoichiometric compound AS the elemental concentrations are the same and the relative intensities are: $J_A/J_S = \varepsilon_A \lambda_A S_A / (\varepsilon_S \lambda_S S_S)$.

If element S is taken as the standard element the ratio J_A/J_S is called the relative sensitivity. Because the spectrometer constants are still included the relative sensitivities are spectrometer dependent and experimental sensitivity determinations are necessary.

But even if elemental sensitivities are determined experimentally some possible error sources remain. These result from surface reactivity and contamination and from additional chemical effects connected with satellite peaks.

Satellite peaks occur if there is more than one final
state, resulting from the removal of a core electron.
The intensity of the main peak is transferred partly
to the satellite. Such phenomena caused by multiplett
splitting in paramagnetic atoms, electron shake-up and
shake-off processes and configuration interaction are
greatly affected by the chemical environment. Therefore,
the transfer of relative sensitivities from one system
to another may be limited especially for solids, where
the satellite structure is often obscured by the background.

Inhomogeneous Samples (11)

Sample inhomogeneities including contamination and surface reaction layers can be detected experimentally by
changing the information depth with the photoelectron
take-off angle (Fig. 3). Since varying the surface to
bulk contribution changes the relative peak intensities,
the spectra of inhomogeneous samples change with the
photoelectron take-off angle. The complicating effects
due to diffraction and reflection of electrons and X-
rays in emission or incidence at very low angles to the
sample surface can be neglected experimentally for take-
off angles between $20 - 70°$.

III. EXPERIMENTAL

The targets were commercial polycristalline 4x12 mm gold
or nickel foils. They were sputtered with 45 or 60 keV
J^+ or Cs^+ ions in an oil free vacuum of 10^{-7} Torr. A one
msec triangular beam sweep in x- and y-direction was
applied to get an uniform irradiation of the target area.
During sweeping, the beam density was a few $\mu A/cm^2$. The
original beam current density changed with the beam
shape from one run to another and was about one order
of magnitude higher. Within experimental error the saturation doses were found to be about 10^{17} ions/cm^2.

After irradiation the samples were transported under
pure argon to the sample chamber of an AEI-ES-200B photoelectron spectrometer equipped with a Mg-Kα X-ray gun.
At a total pressure of 10^{-9} Torr successive scans of the
most intense photoelectron lines of the target material,
the ion, carbon and oxygen were made; peak integration
was done graphically.

For inhomogeneity control the spectra were taken at
different photoelectron take-off angles. The intensity
of the carbon and oxygen lines reflect the extent of

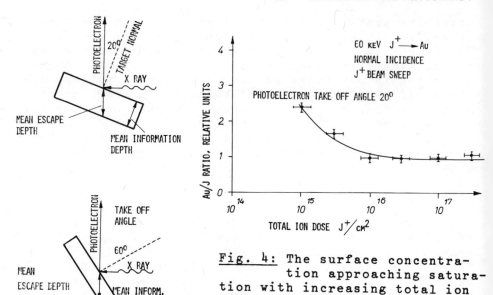

Fig. 3: Change of the information depth with the photoelectron take-off angle

Fig. 4: The surface concentration approaching saturation with increasing total ion dose

contamination and surface oxidation. Some samples were rejected where contamination or oxidation (Ni) was excessive.

IV. RESULTS AND DISCUSSION

The experimental target/ion ratio approaches a constant saturation value with increasing ion dose; the 60 keV $J^+ \rightarrow$ Au system is shown in fig. 4. For total ion doses exceeding about $5 \cdot 10^{16}$ ions/cm^2 saturation conditions are obtained within experimental error. For all the following experiments the total ion dose was $>10^{17}$ ions/cm^2.

Fig. 5 shows the relative Au/J ratio at different angles of iodine ion incidence on the target. The curve represents the well known influence of the beam incidence angle on the sputtering ratio.

The effect of lowering the XPS information depth by changing the photoelectron take-off angle can be seen

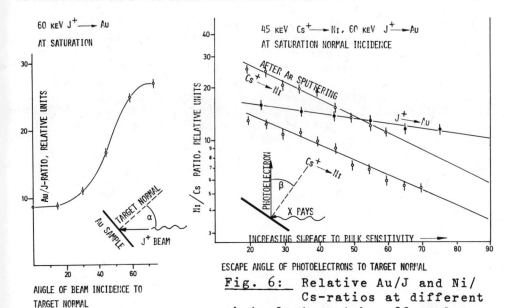

Fig. 5: Relative Au/J-ratio at different angles of beam incidence

Fig. 6: Relative Au/J and Ni/Cs-ratios at different photoelectron take-off angles

in fig. 6. The relative Au/J ratio is only slightly affected by changing the surface to bulk sensitivity, indicating a relatively flat increase of the iodine concentration towards the surface. According to fig. 1, this will be expected for systems, where the mean projected ion range is of the same order of magnitude as the XPS information depth.

The Cs-concentration in the 45 keV $Cs^+ \rightarrow$ Ni system increases considerably towards the target surface. The steeper concentration profile is caused at least partly by the lower mean projected Cs^+ ion range (about 30 Å), which is less than half the 60 keV J^+ range in gold.

High total doses or dose rates may cause additional effects by radiation enhanced diffusion (12) which may result in the movement of the implants to or away from the surface, thus distorting the original concentration profiles.

Removing several monolayers of the Ni foil by "etching" with 400 eV Ar^+ ions, shows the expected lower Cs-concentration in the interior of the sample.

The immediate surface concentration of the implant can be obtained by graphical extrapolation. Within experimental error a linear extrapolation seems to be consistent with our data, if the logarithmic target/ion ratio is plotted versus the photoelectron take-off angle. This may be different in other ion-target combinations or for different irradiation conditions, which may influence the concentration profiles. For an irradiation run, the extrapolated target/ion surface ratio agreed within ±20%, but differed up to ±50 % for different runs, which is about 5 - 10 times the experimental XPS accuracy. This may be connected with differences in the target foils or with difficulties reproducing the primary beam shape in different runs.

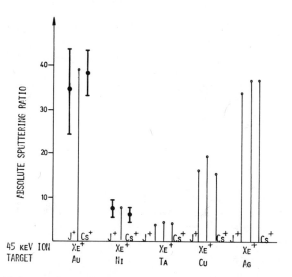

Fig. 7: Sputtering ratios at saturation and normal incidence for 45 keV J^+, Xe^+ and Cs^+ ions in different targets [13] (● this work)

With the help of the elemental photoelectron sensitivities which were determined experimentally the absolute sputtering ratios S at normal incidence and saturation can be calculated from the extrapolated immediate surface concentration S-1. These values are shown in fig. 7 and can be used only with all the restrictions concerning quantitative photoelectron spectroscopy mentioned before.

According to O. ALMEN and G. BRUCE (13) the sputtering ratios show maxima around the noble gases and are similar for neighbouring elements. This is shown for Ta, Au and Ag-targets irradiated with 45 keV J^+, Xe^+ and Cs^+-ions. The same trends should be true also for Au or Ni targets, when only the 45 keV Xe^+ sputtering ratios are known experimentally. Our sputtering ratios for the 45 keV Cs^+ and I^+ ions are the mean values of several runs and confirm the expected trends.

CONCLUSIONS

These preliminary results and some additional studies with other target ion combinations support the following conclusions:

1. For target ion combinations, where normal diffusion can be neglected and the mean projected ion range is larger than the information depth, a determination of the absolute sputtering ratio S at saturation can be made if the surface concentration S-1 is determined by XPS. Contrary to other methods, it is neither necessary to know the weight loss of the target nor the exact total ion dose. The obtainable accuracy depends largely on the present state of the art of quantitative XPS, which may become better in the future.

2. Relative changes of the sputtering ratio at saturation caused by a variation of the irradiation parameters like total ion dose, dose rate, energy, angle of beam incidence or target temperature can be obtained easily and with sufficient accuracy even without knowing the elemental sensitivities.

3. Inhomogeneities near the sample surface caused by contamination, surface reactions, diffusion or implantation profiles can be observed by changing the photoelectron take-off angle.

4. Because of its elemental and chemical sensitivity, XPS may become especially useful for the observation of multicomponent targets and the chemical aspects of irradiation.

ACKNOWLEDGEMENTS

We thank Prof.Dr.F. Baumgärtner for his support and encouragement, Prof.Dr.G.K. Wolf and his group for the irradiations, and Miss A. Wolff for the XPS-measurements.

REFERENCES

1) R. BEHRISCH
   Ergebnisse exact. Naturwiss. 35, 295 ff.

2) G. CARTER, J.S. COLLIGON
   in "Ion Bombardement of Solids", Chap. 7

3) G. WILSON, G.R. BREWER
   "Ion Beams With Applications to Ion Implantation"
   Wiley & Sons, 1973

4) H.J. SMITH
   Radiation effects 18 (1973) 55,65

5) G. CARTER, J.S. COLLIGON, J.H. LECK
   Proc.Phys.Soc. 79 (1962) 299

6) K. SIEGBAHN
   "Electron Spectroscopy for Chemical Analysis", Uppsala University, Sweden 1967

7) G. ERTL, J. KÜPPERS
   "Low Energy Electrons and Surface Chemistry", Weinheim 1974

8) T.A. CARLSON, G.E. McGUIRE
   J. Electronspectrosc. 1 (1972-73) 161

9) W.J. CARTER
   ORNL-TM 4669

10) W.J. CARTER, G.K. SCHWEITZER
    "Experimental evaluation of a simple model for quantitative Analysis in X-ray photoelectron spectroscopy",
    Int.Conf.on Electron Spectroscopy, Namur, Belgium 1974

11) C.S. FADLEY, R.J. BAIRD, W. SIEKHAUS, T. NOVAKOV, S.A.L. BERGSTRÖM
    J. Electronspectrosc. 4 (1974) 93

12) R.S. NELSON
    in S.T. PICRAUX, E.P. EerNISSE, F.L. VOOK
    "Applications of Ion Beams to Metals", N.Y. 1974, 221

13) O. ALMEN, G. BRUCE
    Nucl.Instr.Meth. 11 (1961) 257 and 279

CHEMICAL REACTION ENHANCEMENT AND DAMAGE RATE OF SURFACE LAYER BOMBARDED WITH INERT ION BEAMS

T. Tsurushima and H. Tanoue

Electrotechnical Laboratory

5-4-1 Mukodai, Tanashi, Tokyo 188, Japan

ABSTRACT

Experimental investigations on chemical dissolution enhancement of single crystal garnets bombarded with inert ions have been carried out. Relations between the damage rate and the dissolution enhancement are discussed. A typical dissolution rate corresponding to the damage rate of the order of 10 eV/Å^3 is approximately 10 Å/sec in phosphoric acid at room temperature, which shows that the dissolution rate of the ion bombarded surface layer becomes $10^2 \sim 10^3$ times larger than that of unbombarded crystals. The dissolution enhancement, which is closely related to the depth distribution of the damage rate, vanishes at the particular depth where the damage rate falls to $0.5 \sim 0.005$ eV/Å^3, depending on the ion mass and chemical reaction temperature. The effect of the damage recovery during ion bombardment and chemical etching on the dissolution enhancement is negligible, at least at room temperature.

INTRODUCTION

An energetic ion coming into crystalline material makes many violent collisions with host atoms, displacing them from their lattice sites, and produces a disordered region around the path along which it slows down. At high doses, the disordered regions can overlap each other to form a highly damaged layer at the surface. In many cases, the damaged surface layer thus produced can be dissolved selectively in a weak solvent which hardly dissolves the undamaged substances [1,2]. This type of selective dissolution enhancement is realized simply by bombarding the substances with inert gas ions typically with doses of the order of 10^{15} ions/cm^2.

This paper presents some experimental data on such an ion-bombardment-enhanced-selective-dissolution of garnets and discusses relations between the dissolution enhancement and the damage rate of the surface layer.

EXPERIMENTAL PROCEDURE

Selective ion bombardments were carried out using masks made from Mo-sheets with simple test patterns photoetched in advance. Ion beam currents were set between 0.5 and $2\mu A/cm^2$, which correspond to $3 \times 10^{12} \sim 1.2 \times 10^{13}$ ions/$cm^2 \cdot$sec in dose rates. After the bombardment, each sample was given a series of etchings in phosphoric acid (H_3PO_4) of specified temperature. Measurement of the chemically etched step depth was made precisely by coating the sample surface with vacuum-deposited aluminum (approximately 500 Å in thickness), and using a multiple beam interferometer which had an accuracy of about ±25 Å ($\sim \lambda/200$). Prior to this series of experiments, etchings of unbombarded garnets in H_3PO_4 were done, and the dissolution rate was found to be less than 100 Å/day at room temperature and of the order of 0.1 Å/sec at 80 °C.

RESULTS AND DISCUSSION

Typical experimental results for the amount of dissolution (expressed as the etch depth) at two different temperatures for

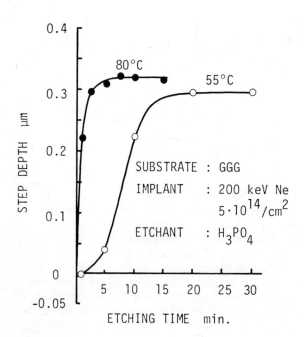

Fig. 1 Dissolution of ion-bombarded GGG in H_3PO_4.

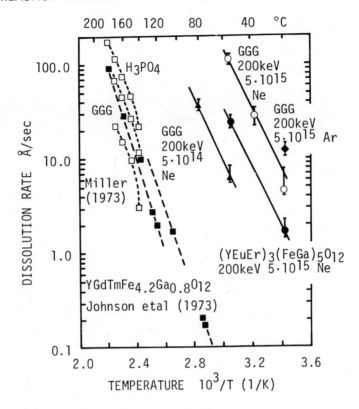

Fig. 2 Dissolution rate vs. reciprocal temperature.

$Gd_3Ga_5O_{12}$ (GGG) bombarded with 5×10^{14} Ne^+/cm^2 at 200 keV are shown in Fig. 1. Important features of these results are: (i) The dissolution rate reaches its maximum at a depth between 0.1 and 0.2 μm, which corresponds to the peak of the damage rate, and the amount of dissolution saturates almost completely within a certain period which depends on the temperature of chemical reaction. (ii) The final amount of dissolution also depends on the reaction temperature. This will be discussed again later.

In Fig. 2 the peak dissolution rates estimated from Fig. 1 are plotted as functions of the reciprocal temperature. It also includes the results of the same kind of experiments for GGG bombarded with 5×10^{15} Ne^+/cm^2 at 200 keV, 5×10^{15} Ar^+/cm^2 at 200 keV, and for a magnetic garnet $(YEuEr)_3(FeGa)_5O_{12}$ bombarded with 5×10^{15} Ne^+/cm^2 at 200 keV, together with those for unbombarded magnetic and non-magnetic garnets [2,3]. It must be noted that the dissolution rate of ion bombarded surfaces of garnets

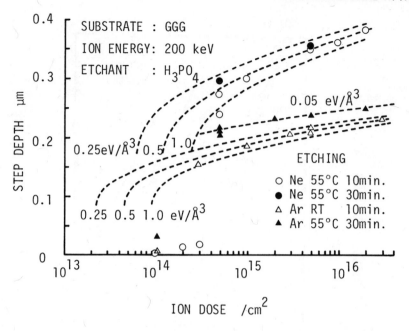

Fig. 3 Amount of dissolution vs. ion dose.

becomes $10^2 \sim 10^3$ times as large as that of unbombarded garnets at room temperature and at elevated temperatures up to at least 80 °C.

Fig. 3 shows some experimental results of the final amount of dissolution, or the final etch depth, vs. ion dose. The depth increases by 10 ∼ 20 % with the increase of one order of magnitude of the ion dose, if the dose exceeds an effective critical value to realize this kind of dissolution enhancement. The critical doses are approximately 10^{14} and $3 \times 10^{14}/cm^2$ in the cases of 200 keV Ne and Ar bombardments, respectively.

It is interesting to compare the obtained experimental data with the theoretical damage rates [4]. Several damage rate curves for GGG bombarded with inert ions have been calculated as shown in Fig. 4, in which the damage rate is expressed as the energy deposition rate per incident ion.

In the bombardment of 5×10^{15} Ne$^+$/cm^2 at 200 keV, the peak value of the deposited energy density is about 20 eV/Å^3, and the dissolution rates corresponding to this deposited energy density at room temperature and at 55 °C are approximately 10 Å/sec and 100 Å/sec, respectively. Using the results shown in Fig. 4, contours of the equi-density of the deposited energy have been

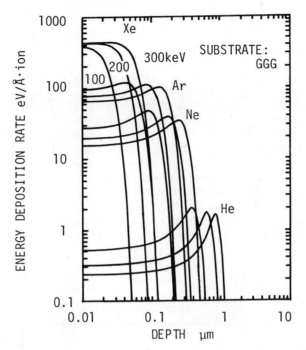

Fig. 4 Energy deposition rate in GGG for a variety of incident ions.

obtained theoretically as functions of the ion dose. Those are shown in Fig. 3 as several dashed lines. The experimental data for the final etch depth (total amount of dissolution) of a common etching condition are mostly on a common curve, which suggests that the damage rate of the particular depth at which the dissolution enhancement vanishes does not depend on the ion dose, while slightly depends on the etching conditions. In other words, the dissolution is self-limiting, and the final etch depth is determined simply by the deposited energy density which makes the surface layer soluble under the existing etching conditions. As just suggested in Fig. 1 (Ne on GGG) and Fig. 3 (Ar on GGG), the reaction temperature coefficient of the final etch depth is not more than 10 Å/°C.

Dependences of the amount of dissolution on ion masses are shown in Fig. 5. Apparently, light ions provide larger etch depths than heavy ions. However, the critical dose of the dissolution enhancement for light ions becomes considerably larger than for heavy ions. If we consider the cases of 5×10^{14} ions/cm^2 at 200 keV bombardments and 55 °C H_3PO_4 etchings, ions heavier than Ne can make the surface of crystalline garnet soluble into the etchant. The critical dose of O^+ or N^+ for the same etching condition looks slightly larger than 5×10^{14}/cm^2 but smaller than 2×10^{15}/cm^2, and those of protons and α-particles, larger than 2×10^{16}/cm^2.

In Fig. 5, theoretically obtained depths corresponding to three specified energy deposition rates: 10, 1 and 0.1 eV/Å·ion (these result in the deposited energy density: 0.5, 0.05 and 0.005 eV/Å^3, respectively, if the bombardments are 5×10^{14} ions/cm^2 at 200 keV) are given as dashed curves. Experimental data show that, in the present etching conditions, the dissolution enhancement vanishes at the depth where the deposited energy density is approximately equal to 5 meV/Å^3 for heavy ions, and 0.5 eV/Å^3 for medium mass ions. The annealing effect on the dissolution enhancement has been examined as shown in Fig. 6. Remarkable influences are observed after 600 °C or higher temperature annealings, if the surface layer is heavily damaged. In the case of light damage,

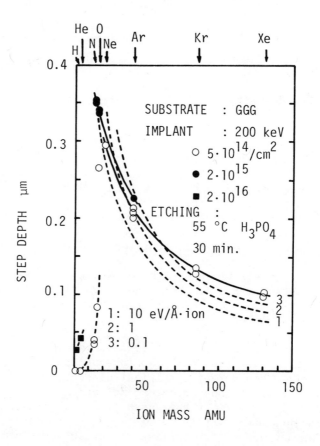

Fig. 5 Amount of dissolution vs. ion mass.

however, annealing at a lower temperature, e.g. 200 °C, begins to effect the amount of dissolution in the same etching condition.

In order to make clear the effect of defect recovery at room temperature on the dissolution enhancement phenomena, we attempted

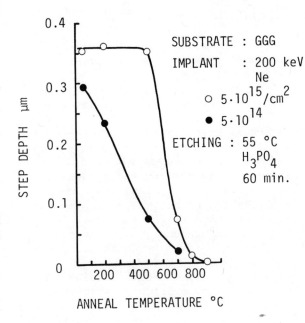

Fig. 6 Effect of annealing on the amount of dissolution.

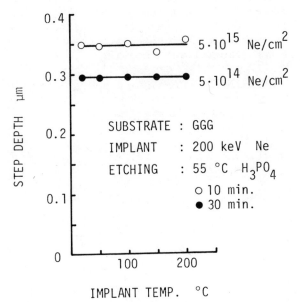

Fig. 7 Effect of temperature during bombardment on the amount of dissolution.

an etching a long while (more than 10^3 hours in which the samples had been kept at room temperature) after ion bombardment at room temperature and at several elevated temeratures up to 200 °C. Results are given in Fig. 7, which shows that the influence of the defect recovery to the dissolution enhancement is negligible, at least at room temperature.

CONCLUSIONS

Experimental investigations were carried out to characterize the chemical dissolution enhancement effect in single crystal garnets bombarded with energetic inert ions.

The dissolution rate of the ion bombarded surface layer in phosphoric acid becomes $10^2 \sim 10^3$ times larger than that of unbombarded crystals. The dissolution is self-limiting, and the final etch depth (the amount of dissolution) is determined mainly by the ion bombardment conditions. A typical room temperature dissolution rate corresponding to the damage rate (deposited energy density) of the order of 10 eV/Å^3 is about 10 Å/sec. The final depth corresponds to the position where the energy density deposited by the incident ions is approximately equal to 5 meV/Å^3 for heavy ions, and 0.5 eV/Å^3 for light ions. The temperature coefficient of the final depth is estimated to be less than 10 Å/°C. The depth increases by 10 $\sim$ 20 % with the increase of one order of magnitude of the ion dose. The ion dose must exceed an effective critical value to realize the dissolution enhancement. The critical doses are about 10^{14} and 3×10^{14}/cm^2 in the cases of 200 keV Ne and Ar bombardments, respectively. Light ions generally provide larger etch depths than heavy ions, if the ion dose exceeds its critical value. The influence of the defect recovery at room temperature to the dissolution enhancement is negligible. Remarkable influences are observed, however, after 600 °C or higher temperature annealings.

REFERENCES

[1] J. F. Gibbons, E. O. Hechtl, and T. Tsurushima, Appl. Phys. Lett. 15 (4) 117 (1969)

[2] W. A. Johnson, J. C. North, and R. Wolfe, J. Appl. Phys. 44 (10) 4753 (1973)

[3] D. C. Miller, J. Electrochem. Soc. 120 (12) 1771 (1973)

[4] T. Tsurushima, and H. Tanoue, J. Phys. Soc. Japan 31 (6) 1695 (1971)

Ion Induced X-Ray Spectroscopy

PROGRESS IN THE DESCRIPTION OF ION INDUCED X-RAY PRODUCTION;

THEORY AND IMPLICATION FOR ANALYSIS

Finn Folkmann

GSI, Darmstadt, Germany

The production of characteristic x-rays induced by light ions is well understood. X-rays generated by heavy ions are more complex and exhibit several interesting features. Background radiation arises from bremsstrahlung of secondary electrons and directly of the incident ions. Compton scattering of nuclear γ-rays and special heavy ion background reactions are mentioned. Sensitivities in trace element analysis are limited to 10^{-6}-10^{-7} in concentration, 10^{-9}-10^{-16}g in absolute amounts and 10-1 μm in lateral resolution. The secondary electron bremsstrahlung may be reduced using very thin samples and its angular distribution makes backward detection angles favourable.

1. INTRODUCTION

Measurement of characteristic x-rays, emitted during bombardment of samples with energetic ions, has for the last few years been applied to trace element analysis of small samples[1]. This kind of work has preferentially been made with solid state x-ray detectors for rapid multielement analysis.

When atoms are bombarded with ions there is a high probability that inner shell electrons are ejected, and by the subsequent filling of a vacancy by an outer electron the energy gained by the transition may be radiated as a characteristic x-ray. Its energy depend quadratically on the Z value of the emitting atom ($E_{K_\alpha}=Z^2 \cdot 0.010$ keV) and the energy of the x-ray thus tells which atom was originally hit. The ion energies considered range from 0.5 to 10 MeV/nucleon and most applied work has been made with 1-4 MeV protons. The x-ray detectors are usually Si(Li), Ge(Li), or intrinsic Ge detectors working between 1 and 100 keV. This means that we can observe x-rays from elements with Z≥10 and the detector resolution allows us to separate these elements clearly helped by the with Z increasing distance between the x-ray lines of adjacent elements. The higher resolving power with increasing Z using these energy dispersive x-ray spectrometers contrasts the situation for Rutherford backscattering (and by the way also for Bragg crystal spectrometers, which separate the x-ray lines better at low energies).

The spectrum from a semiconductor detector, as shown in fig. 1, grows very rapidly and gives information on the elements present in the sample. The experimental setup for such measurements is very simple with the detector most often perpendicular to the ion beam and is discussed in detail in refs. 2 and 1. Fig. 1 shows that characteristic x-ray peaks of high intensity are produced from a broad range of elements, and the excitation mechanisms and the peaks are discussed in chapter 2. The perspective of ion-induced x-radiation is that the background is low compared with other modes of x-ray excitation, but from fig. 1 it is seen that it is not negligible. The background encountered is mentioned in chapter 3 and the consequences it has for the analytical sensitivity outlined in chapter 4. The last two chapters analyse two possible ways in which the dominant bremsstrahlung background may be reduced. The reduction in very thin targets is estimated and the importance of the angular distribution pointed out.

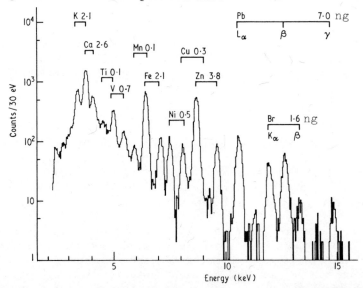

Fig.1 X-ray spectrum from routine analysis of air-born dust collected on a 30 µg/cm² thick polystyrene foil and bombarded with 3 MeV protons. Characteristic x-rays are seen and the amount present of different elements is given in 10^{-9} g.

2. CHARACTERISTIC X-RAYS

The chance that characteristic x-rays are emitted from an element which we want to study is very high due to the fact that we are dealing with atomic processes. The cross section is often of the order of kb and is for incident protons and light ions determined by the Coulomb interaction between the projectile and each of the atomic electrons to be ejected. In the binary encounter approximation (BEA) the cross section for this process can be easily calculated classically and is also valid quantum mechanically. The BEA calculation of Garcia et al.[3] of the cross section for producing vacancies in the K-shell of various elements Z is shown in fig. 2 as the solid curve, which is seen to be in agreement with the experimental points from $15<Z<70$ and protons of different energies. When the projectile of charge Z_p and mass $A_p M$ has the energy E_p the BEA result for the K vacancy production of fig.2 and ref. 3 obeys the scaling relation

$$\sigma_K(E_p) = Z_p^2 \cdot \frac{1}{U_K^2} \cdot f(\frac{m_e \cdot E_p}{A_p M \cdot U_K}) \qquad (1)$$

It depends on the projectile parameters and Z (through the binding energy $U_K = U_K(Z)$ of the ejected electrons) via a function f of one single parameter $\eta = m_e \cdot E_p/(A_p M \cdot U_K)$. Here m_e is the electron mass, M the proton mass and $\sqrt{\eta}$ then the velocity of the projectiles divided by the mean velocity of the bound electrons. In eq. (1) K can for other electron shells be substituted by L (or M) with the same function f as an average over the shell, except for a multiplication up to the total number of electrons in the shell of

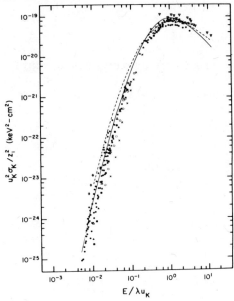

Fig.2 K vacancy production for proton bombardment of a variety of elements with different energies. When plotted as function of $E/\lambda U_K$, where λ is the proton-electron mass ratio, E the proton energy and U_K the K binding energy of the atom, the data follow a universal curve. The full drawn curve is calculated in the binary encounter approximation (BEA) according to ref.3, from which the figure is taken, and obeys the scaling of eq.(1). The dashed curve is the similar plane wave Born approximation result (PWBA).

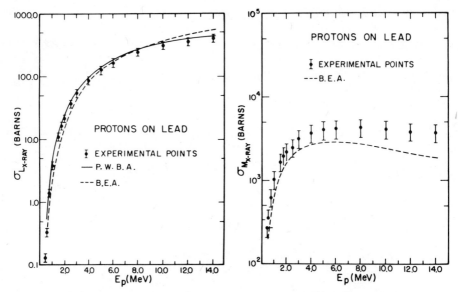

Fig.3 L and M x-ray emission cross sections for protons of energies 0.5-14 MeV on Pb (from ref. 8). Theoretical results are given for BEA (dashed curve) and for the L cross sections also for PWBA (full curve).

interest (e.g. with a factor 4 = 8/2 for a full L shell). X-ray emission cross sections are obtained from vacancy production values as in eq.(1) simply by multiplication with a fluorescene yield[3], depending on Z and accounting for the x-rays seen for one original vacancy.

BEA is only one method of calculation and other approximations can be used. Plane wave Born approximation (PWBA) results[4,5] are shown as the dashed curve in fig. 2. Various refinements can be added to this approach, as in the paper of Basbas, Brandt and Laubert[6], which treats the perturbations for light incident ions due to Coulomb deflection, increased binding energy and polarization. In the semiclassical approximation (SCA) is emphasized the impact parameter dependence[7] and the importance of the Coulomb deflection of the projectile, which is of interest for low velocities ($E_p/A_p \lesssim 0.5$ MeV). These theories give a satisfactory explanation of x-ray production cross sections for impact of light ions. This is well known for K x-rays as illustrated by fig. 2 and is also reflected by measurements for higher shells. Fig. 3 shows experimental and theoretical results[8] for L and M x-rays from Pb as function of the proton energy. The characteristic x-rays are divided in several groups, e.g. for the L x-rays in L_α, L_β and L_γ corresponding to different transitions and in fig. 4 is displayed both the total cross section for Au L x-rays by proton impact and the intensity ratios L_α/L_β and L_α/L_γ.

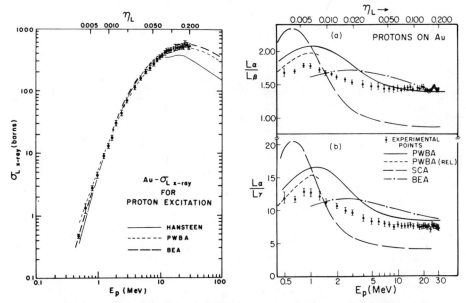

Fig.4 Total L x-ray emission cross sections, L_α/L_β and L_α/L_γ ratios for protons of energies 0.5-30 MeV on Au (from ref.9). Theoretical values are given for BEA, PWBA and the semi-classical approximation (SCA, Hansteen[7]).

Most x-ray work has been made with protons, but also many experimental cross sections and comparisons with theory exist for heavier incident ions. From the BEA scaling in eq.(1) we get the cross section for a fixed target element (Z, U_K) for ions (Z_p, A_p) of energy E_p related to that of equal velocity protons (of energy E_p/A_p) by

$$\sigma_x(\text{ion}, E_p) = Z_p^2 \cdot \sigma_x(\text{proton}, \frac{E_p}{A_p}) \qquad (2)$$

This relation applies both to the x-ray emission and to the vacancy production cross section and is also expected from PWBA and SCA calculations. In fig. 5 is shown target x-ray cross sections from Li ions incident on various targets directly compared with the BEA values of eq.(1) and fig. 2. It is seen that the experimental results are typically within 20% of the theory and that deviations mainly occur (down to a reduction of 50%) for low incident velocities. These conclusions hold for most light ions.

For heavy ions, however, the simple description above has to be modified as several other mechanisms participate in the production process and affect the appearance of the observed characteristic x-rays. The charge state of the incident ion is a parameter of importance for the detailed magnitude of the x-ray emission cross

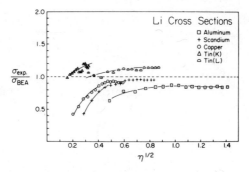

Fig.5 Cross section for production of characteristic x-rays for ^{6}Li-ion bombardment of Al, Sc, Cu and Sn relative to BEA values (from ref. 10). The cross sections are shown as function of $\eta^{1/2}$, the ratio of the projectile velocity to target orbital electron velocity ($\eta = m_e E_p / A_p MU_K$ was used as parameter in fig.2 and in eq.(1)). For the data E_p = 4-34 MeV.

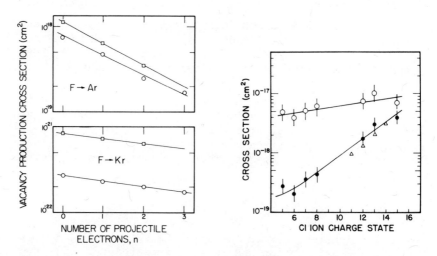

Fig.6 Ar and Kr K shell vacancy production cross sections for F-ion bombardment of gas targets, shown as function of the number of electrons of the incident F ion (from ref. 11). The charge state dependent cross sections are given for 36 MeV (circles) and 48 MeV (squares) ^{19}F ions.

Fig.7 Ne x-ray (filled circles) and Auger-electron (open circles) cross sections for 50 MeV Cl bombardment of a Ne gas as function of the Cl charge state (from ref.12). The figure shows that the ratio between the two ways of filling a vacancy depends strongly on the charge state of the ion, i.e. that the fluorescence yield is not constant.

section. In fig. 6 is shown the target K vacancy production cross section for Ar and Kr gas targets with incident F ions of different charge states, and it is seen to be a function of the number of electrons carried by the F ion (from 0 to 3) as well as of the F energy. This dependence is a consequence of interactions between the two heavy collision partners. Effective in sudden collisions is the capture of target electrons into empty orbitals of the projectile, but the smooth variation of the cross section with the charge state calls also for explanations in terms of rearrangement of the atomic shells during the collision. The x-ray emission cross sections are also sensitive to changes in fluorescene yields, observed in heavy ion collisions. For photon or proton excited inner shell vacancies the filling under emission of x-rays is always in a constant ratio to the filling by ejection of Auger electrons. But for heavy incident ions this is not always the case as seen in fig.7, where the x-ray emission cross sections from Ne are shown for 50 MeV Cl bombardment together with the higher cross section for Auger electron emission. The probability of filling a vacancy by x-ray emission (i.e. the fluorescene yield) gets higher with higher charge state of the projectile, and thereby with higher probability of ejecting atomic electrons, cf. fig. 6. With fewer electrons in outer shells it is understandable that the

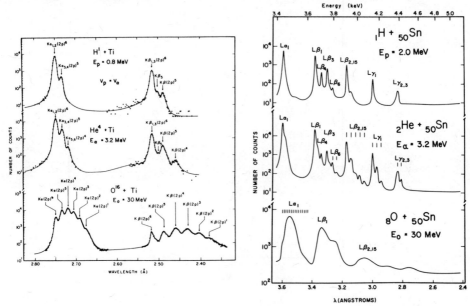

Fig.8   Fine structure of characteristic x-ray spectra for p, α and ^{16}O bombardment of the same target. The spectra are measured with a Bragg crystal spectrometer. Left is shown the result for the K_α and K_β groups from Ti with indication of the no of electrons in the L shell (from ref.3) and right for the L_α, L_β and L_γ groups of Sn (from ref.13).

decay channel via emission of such electrons as Auger electrons will decrease in importance. For applications with solid target, however, charge state effects are not likely to be seen as an average charge state of the projectile will be established in the front part of the target.

The strong ionization of heavy ions, expressed by the Z_p^2 term from the Coulomb interaction in eq.(2), has also a significant effect on the characteristic x-ray lines themselves. Additional vacancies in outer shells will shift the x-ray energies as clearly observed with a crystal spectrometer in fig. 8, where x-ray spectra from Ti and Sn are shown excited by incident protons, alphas and oxygen ions. With heavier ions in fig. 8 the satellite lines for a partially filled L shell prevail the fine structure of the K_α and K_β lines of Ti. For the L lines of Sn the individual transition lines are not resolved, but a clear increase of the width of the main components is seen due to the higher degree of ionization with heavier ions. By observation of characteristic x-rays with a solid state detector of worse resolution in the same way an increasing width and a shift to higher energies is seen for

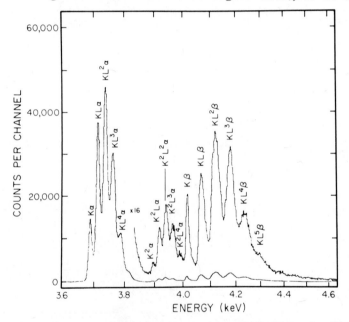

Fig.9 Fine structure of the K x-rays of Ca from bombardment with 48 MeV ^{16}O ions, observed with a crystal spectrometer (from ref.14). Each line is labeled by the electron vacancies in the initial state. Hyper-satellite lines (with two K vacancies) lie between the two groups of normal satellite lines (one K vacancy and various L vacancies) built upon the K_α and K_β lines of Ca.

excitation with heavy ions, complicating the elemental analysis a little. A good separation of the individual satellite lines is seen in the crystal spectrometer spectrum of fig. 9 for the K x-rays of Ca with incident 48 MeV ^{16}O ions[14]. Fig. 9 also shows hyper satellite lines from two initial K vacancies and it can be mentioned that Wölfli et al.[15] have recently observed x-rays from filling simultaneously two K-vacancies with two electrons. These vacancies were produced in nearly symmetric heavy ion collisions (Ni,Fe)→(Ni,Fe) and the energy of the x-rays was approximately the double of the normal one electron transition energies, as seen in fig. 9.

The effects mentioned here may change the estimate of eq.(2) for the x-ray emission cross section of heavy ions by factors of 10, but an even higher increase of the cross section may be expected from electron promotion via molecular orbitals (MO) transiently formed during the collision[3]. In low energy collisions between heavy ions MO excitation is responsible for increase in the x-ray cross sections many orders of magnitude[3] and also in the energy regime around 0.5 MeV/amu and higher, vacancies may be brought into the collision by the projectile and via coupling between MO's transferred to the target atom. For near-symmetric heavy ion collisions K vacancy sharing as described by Meyerhof[16] explains a strong increase in the observed cross sections for target atoms close in Z to the projectile. In fig. 10 is for 47 MeV I ions shown the vacancy production yield, and the target excitation (circles) is enhanced a few orders of magnitude around $Z_2 \sim 53$ above the exponentially decreasing cross sections on both sides, determined by the Coulomb excitation, eq. (2). X-ray cross sections for

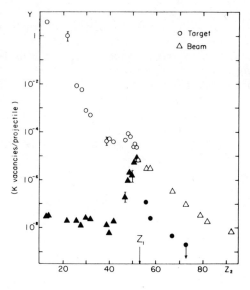

Fig.10 K yields from thick targets bombarded with 47 MeV I ions versus target atomic number (from ref.16). An enhanced yield of target x-rays (circles) is apparent for target atoms close to the projectile due to molecular orbital (MO) excitation. This effect is even more striking if yields from the higher Z collision partner (closed symbols) is considered.

symmetric systems as high as Pb on Pb have later been analysed[17] and in ref. 18 is compiled absolute values of cross sections for Cl, K, Ni, Br, Kr, I, Xe and Pb ions incident on a range of atoms (summed for target and projectile).

The selective MO excitation by heavy ions of K vacancies in atoms close in Z to the projectile is one perspective of using heavy ions to create x-rays instead of protons or light ions. As seen from fig. 10, however, also the projectiles are highly excited and emit x-rays, so the importance for applied work is presently not obvious and must be clarified by a more careful study of the background. Another perspective is the general high value of the cross sections, eq. (2), but as shown in the next two chapters this fact has only very meager consequences.

3. CONTINUOUS X-RADIATION

To observe x-rays from trace elements in a sample it is essential to know the background over which the characteristic peaks are to be seen. The continuous background radiation normally arises from the interaction of the ion beam with the most abundant elements of the matrix (the bulk material of the sample) and can be assigned to the following processes:

    SEB   Bremsstrahlung of secondary electrons
    PB   Projectile bremsstrahlung
    CS   Compton scattering of γ-rays
    REC   Radiative electron capture
    MO   Quasi-molecular radiation

To these radiation continua originating from the sample (and for CS produced in the detector) come background contributions from characteristic lines of elements in the sample and of the projectile or from tails of these lines, created in the detector. For light ions the dominant background contribution at low radiation energies comes from SEB. With 2 MeV proton bombardment of C and Al matrices the radiation has been measured[19] and shown in fig. 11 to be due to SEB and PB. As proton induced x-rays prevail in analytical work these two continua always present in that type of measurements will be commented before turning to CS and to the two special heavy ion continua REC and MO.

Secondary electrons are ejected when the incident ions hit and ionize the atoms of the matrix. They are produced with high probability with kinetic energies up to

$$T_m = \frac{4 \cdot m_e}{M} \cdot \frac{E_p}{A_p} \qquad (3)$$

which is the maximum energy transfered to a free electron in a collision with an ion (Z_p, A_p) of energy E_p, and with a strongly decreasing likelyness with higher kinetic energies (which can be obtained from the fraction of hardly bound atomic electrons). Then these electrons later undergo collisions in the matrix and are slowed down over short distances they have a certain chance of emitting bremsstrahlung quanta being accelerated in the nuclear electric field of the matrix atoms. The resulting SEB show the same main spectral shape as the original secondary electron energy spectrum decreasing rapidly in intensity above the photon energy T_m, eq. (3). This fall over many orders of magnitude can be seen from the spectra of fig. 11, where T_m = 4 keV. It must be emphasized that SEB is closely related to the production of characteristic x-rays, as the first part of the process, the ejection of secondary electrons, is essentially the vacancy production (in the matrix atoms) and can be calculated by PWBA and BEA[4,3,19]. It implies that SEB has the same scaling with projectile parameters as was given in eq. (2) and that the spectral shape remains the same, according to simple scaling, as long as the ions have the same velocity. The scaling of eq.(2) has for the SEB background to be taken for the spectrum from the matrix atoms and for the characteristic x-rays for different trace elements of interest. Although these latter cross sections grow high for heavy ions, exactly the same applies

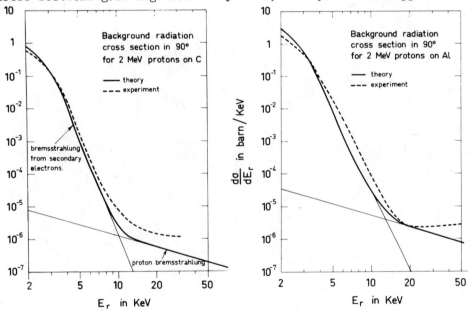

Fig.11 Background cross sections for 2 MeV proton bombardment of a carbon and an aluminum matrix (from ref.19). The experimental background (dashed line) in a Si(Li) detector is dominated by the secondary electron bremsstrahlung and the proton bremsstrahlung shown by full-drawn theoretical curves.

to the background, and for the same velocity of incident ions (same E_p/A_p) the ratio is the same. From the nature of this dominant background (and of the x-ray production) it is a good starting point to compare analytical situations for ion beams of equal velocity.

In the high energy region of the spectra in fig. 11, where the SEB has gone to small values the background observed can be explained by the bremsstrahlung directly of the protons in the field of the matrix nuclei. This dipole radiation is proportional to $(Z_p/A_p-Z/A)^2$ where (Z,A) denotes the matrix and will thus be very small when the ions have the same charge to mass ratio as the matrix (e.g. $1/2$)[19]. The high energy spectral region is, however, for ions of $Z_p/A_p \simeq 1/2$ filled by CS due to huge excitation cross sections for nuclear γ-rays, which are also important for protons of higher energies [20,1]. The hope of using the PB cancellation may be justified in special situations, but generally a CS background is present, at least as intense as the proton bremsstrahlung and will represent the background at the highest observed x-ray energies.

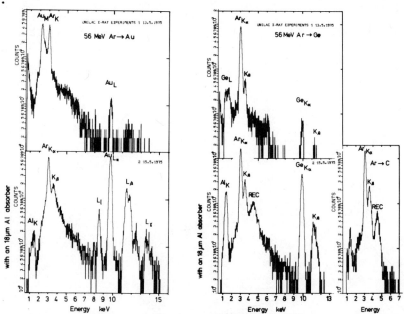

Fig.12 X-ray spectra from 1.4 MeV/amu Ar-ion bombardment of Au, Ge and C measured at GSI, Darmstadt. Characteristic M and L x-rays from Au, L and K x-rays from Ge and in all cases the K x-rays from Ar are seen. The latter are accompanied by a radiative electron capture (REC) continuum, which is clearly seen as a peak for the C and Ge spectra. For the lower spectra an Al absorber between the target and the Si(Li) detector suppresses the intense low energy radiation.

For heavy ion bombardment several other continua occur additionally, and the most important contributions to the background will be mentioned. In fig. 12 is shown spectra from 56 MeV Ar bombardment of Au, Ge and C targets. Besides the target x-rays characteristic x-rays from the projectile are seen and a little higher the REC continuum, which for Ge and C has a peak-like appearance. This radiation is emitted when electrons of the target are captured into vacancies (here K) of the projectile. It has a maximum at an energy which is higher than the binding energy of the excited projectile by $T_m/4$ and a width which is proportional to the projectile velocity and the mean velocity of the outer electrons of the target. For the C target the REC peak is thus seen to be narrower than for Ge and in the case of Au it is not recognizable as a peak due to contributions from many shells of Au, some of which are strongly bound. REC continua are also seen in fig. 13 for S bombardment of C and Al higher than the S K x-rays, but in these cases another continuum prevails at even higher photon energies. It is a band of quasi-molecular radiation (MO) emitted during the collision by electrons filling vacancies in transiently formed molecular orbitals. As function of the internuclear distance the tightest bound K orbital, to which the transitions go, approaches the binding energy of the united atom E_u at the shortest distances, but the MO radiation seen in fig. 13 extends to even higher energies due to dynamical broadening caused by the finite interaction time. In fig.14 is shown a number of spectra from Br ions incident on Ni,

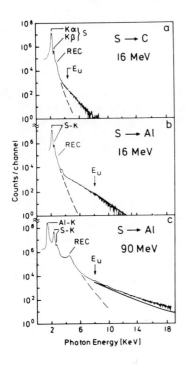

Fig.13 Background continua from sulfur ion bombardment of C and Al (from ref.21). The radiative electron capture (REC) lies higher than the characteristic K x-rays of S and extends a tail (dashed continuation) to higher energies. The continuum at high energies is a quasimolecular radiation band (MO) decreasing exponentially with photon energy also above the binding energy E_u of the united atom.

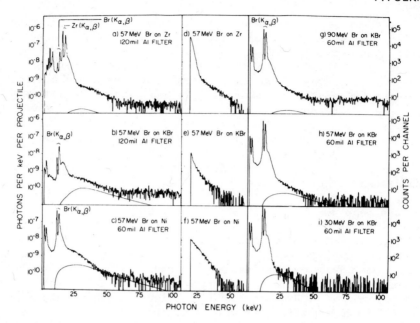

Fig.14 Background continua from Br ion bombardment of Ni, KBr and Zr at high radiation energies (from ref.22). The full-drawn curve is the calculated projectile bremsstrahlung yield which accounts for the high energy radiation of c), h) and i). Compton scattering of γ-rays produces in the Ge detector the flat continua of a), g) and b) and MO contributions might be seen in the spectra d), e) and f) where the first two contributions are subtracted.

KBr and Zr targets taken with a Ge detector with special interest to the radiation at the highest photon energies 25-100 keV. From the spectra the radiation above the characteristic x-rays and REC may be interpreted as PB, CS and MO, but the situation is not very clear-cut. A quantative analysis of the background as is possible for the simple proton induced continua SEB and PB is presently still difficult to make in detail for the heavy ion induced background continua.

4. SENSITIVITY IN X-RAY ANALYSIS

The analytical use of ion-induced x-rays has mainly been based on proton beams or use of He ions, as they are cheapest and also seem to offer the best sensitivities. We will therefore here primarily for proton induced x-rays mention the limitations which the strongest involved physical processes impose on the obtainable sensitivity. We will base our treatment on calculations of the x-ray yields from trace elements and of the bremsstrahlung background as described in ref. 1, using many procedures from ref. 19. In fig.

Fig.15 Calculation of x-ray yields in a detector from 2 MeV proton bombardment of a C matrix. Characteristic total K and L x-ray intensities are shown for the K_α or L_β photon energy for Z of the target indicated along the curve (for 1 ng/cm² of the element). Bremsstrahlung background yields per 200 eV are given from 1 mg/cm² C. A total dose of 100 µC protons is assumed and a solid angle of the detector of 0.038 sr.

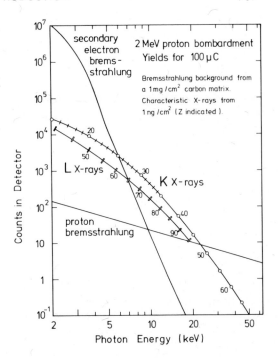

15 we have displayed the yield we will expect in a detector from these calculations, when we set the detector solid angle to $\Delta\Omega$ = 0.038 sr and bombard a 1 mg/cm² C matrix with 100 µC protons of energy 2 MeV (e.g. a 150 nA beam in 11 min). The bremsstrahlung background can directly be compared with the experimental results in fig. 11. Fig. 15 also shows the number of characteristic x-rays one would observe from 1ng/cm² of various trace elements with Z indicated along the curves (15<Z<69 for K x-rays and 40<Z<95 for L x-rays), i.e. for a concentration of 1 ppm relative to the matrix. The similarity of the K and L x-ray curves is due to the fact that the ionization probability mainly depends on the binding energy of the electrons and that the closely related characteristic x-ray energy is used as abscissa (at the position where it has to be seen over the background).

As criterion for observing an x-ray peak we employ

$$N_T \geq 3\sqrt{N_B} \qquad (4)$$

where N_T is the characteristic x-ray counts from the trace element and N_B the number of background counts within the FWHM of the peak as seen with a Si(Li) detector (150 eV at 5.9 keV) according to eq. (1) of ref. 1. With various proton energies but otherwise the same assumptions as for fig. 15 we have used eq. (4) to find the minimum concentration of different trace elements which can be de-

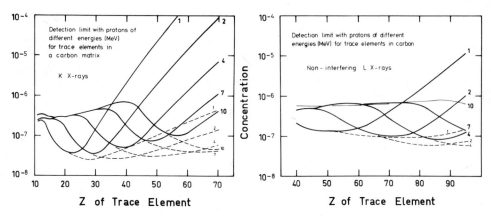

Fig.16 Detection limits calculated according to eq.(4) for protons of different energies (stated in MeV for the curves). Concentrations of trace elements (by weight) which can be seen with a Si(Li) detector of solid angle 0.038 sr are given for bombardment with 100 μC protons of a 1 mg/cm² C matrix. Left results for K x-rays and right for L x-rayx from the trace element. For scaling of the detection limits to other run-parameters see eq.(5) and text.

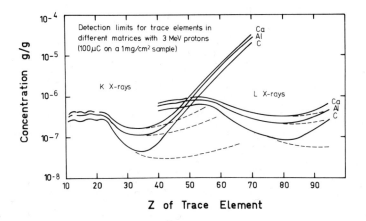

Fig.17 Detection limits for 3 MeV protons incident on 1 mg/cm² of C, Al and Ca matrices. Results are shown for both K and L x-rays covering trace elements from Ne to Am. Caption else as for fig.16. The dashed continuation of the curves give the concentration which is observed if only the SEB background is present.

tected in a 1 mg/cm^2 carbon matrix. In fig. 16 the resulting detection limits are shown as function of Z of the trace element. K x-ray results are given to the left and L x-ray values to the right. Calculations are depicted for 1, 2, 4, 7 and 10 MeV protons so the region of optimal sensitivity is visualized as function of this external variable. The sensitivity at low Z following one curve (fixed proton energy) is determined by SEB, and if only this background had been present, the dashed contribution at higher Z would had applied. The full curve at high Z reflects the importance of the proton bremsstrahlung background. In fig. 17 the same representation of the concentration sensitivity as in fig. 16 is shown for one proton energy (3 MeV, often used in analytical work)but for three different matrices, C, Al and Ca, each of thickness 1 mg/cm^2. The results of figs. 16 and 17 can be scaled to other experimental parameters using the dependence of the minimum concentration of a certain element Z_T

$$\text{Conc}(Z_T, \text{param}) = C(Z_T) \cdot \frac{\sqrt{\Delta E}}{\sqrt{\Delta \Omega \cdot t_M \cdot N_p \cdot Z_p^2 \cdot C_{abs} \cdot \varepsilon_d}} \quad (5)$$

where the constant $C(Z_T)$ can be read from the figures according to the detector resolution ΔE (eq. (1) in ref. 1), detector solid angle $\Delta \Omega (=0.038$ sr$)$, matrix thickness $t_M (=1$ mg/cm$^2)$, no of projectiles $N_p (=6.24 \cdot 10^{14})$ of charge $Z_p (=1)$, internal detection efficiency $\varepsilon_d (=1)$ and absorption factor $C_{abs} (=1)$. Eq.(5) can be used for one fixed proton energy, but applies also to the same energy per nucleon for other incident ions with proper insertion of their nuclear charge Z_p due to the scaling relation, eq.(2) for the x-ray production and SEB.

Concentration		$10^{-6} - 10^{-7}$		
Absolute amounts	for matrix	conc.	beam	area
10^{-9} g	1 mg/cm^2 C	10^{-6}	defocused	1 cm^2
10^{-12} g	100 µg/cm^2 C	10^{-6}	focused	1 mm^2
10^{-16} g	100 µg/cm^2 C	10^{-6}	micro	(10µm)^2
Range (effective)	for matrix	with protons	of full range	
2 mg/cm^2	carbon	2 MeV	8 mg/cm^2	
4 mg/cm^2	carbon	3 MeV	16 mg/cm^2	
Lateral resolution	for beam	in the range		
10 mm	defocused	20 - 5 mm		
0.5 mm	focused	2 - 0.1 mm		
4 µm	micro	10 - 1 µm		

Table 1. Practical sensitivities in ion-induced x-ray analysis.

The conclusion from figs. 16 and 17 is that there is a strong limitation in the concentration which can be measured with ion-induced x-rays bounding it to values down to the $10^{-6}-10^{-7}$ level (allowing experimental conditions a little worse than in the calculations). However, sensitivities can also be expressed in other ways. E.g. the absolute amount observed is equal to $\text{Conc}(Z_T,\text{param}) \cdot t_M \cdot \text{Area}$, where Area is the beam area hitting the sample, and is stated in table 1 for typical analytical situations to lie between 10^{-9}g and 10^{-16}g. The depth probed in the sample depend for a thick target on the range of the projectiles and is typically a quarter of this distance, i.e. a few mg/cm^2. The lateral resolution of the beam varies normally between 20 and 0.1 mm (defocused - focused), but can with a strongly focused microbeam get down to a working range of 10^{-4} µm keeping a high intensity[1]. With careful construction of carbon slits collimation down below 1 µm is achievable but with an extreme low beam intensity[23]. Although we have considered low Z matrices one can also easily detect trace elements in a higher Z matrix with good sensitivity[24].

For heavy ions the concentration sensitivity can be related to that for protons of the same velocity (energy per nucleon) due to the scaling of eqs. (5) and (2) as long as only the SEB background is important. In table 2 are results shown with different criteria for comparison. For the energy loss is assumed a charge squared dependence like in eq. (2) $dE/dt(\text{ion}) = Z_p^2 \cdot dE/dt(\text{proton})$ and thereby for a constant fractional energy loss $(\Delta E/E)$ $t_M(\text{ion}) = t_M(\text{proton}) \cdot A_p/Z_p^2$. The equilibrium charge Z_p in these formulae is for light ions equal to the nuclear charge Z_p and for heavy ions somewhat lower, depending on the energy (eq. $\sim$20 for 1.4 MeV/amu Xe).

Criterion (restriction)	constraint for eq.(5)	variable	Concentration limit for equal velocity ions
Equal counting rate	$N_p Z_p^2$	N_p	$\text{Conc}(\text{ion}) = \text{Conc}(\text{proton})$
Equal energy loss	$N_p Z_p^2$	N_p	$\text{Conc}(\text{ion}) = \frac{Z_p}{Z_p} \text{Conc}(\text{proton})$
Equal number of projectiles	N_p	-	$\text{Conc}(\text{ion}) = \frac{1}{Z_p} \text{Conc}(\text{proton})$
Equal fractional energy loss	$t_M \cdot Z_p^2/A_p$	t_M	$\text{Conc}(\text{ion}) = \frac{Z_p}{Z_p} \frac{1}{\sqrt{A_p}} \text{Conc}(\text{pr.})$

Table 2. Detection limits of ions compared with that of equal velocity protons for different criteria. Z_p^2 scaling of the cross sections, eq.(2), is assumed.

DESCRIPTION OF ION INDUCED X-RAY PRODUCTION

The analysis in this sections was based on theoretical calculations which seem to be in agreement with the observed background but careful measurements with more detailed comparisons are needed. One can either from the practical point of view measure the background from respresentative matrices under analytical conditions or try to follow the physics to a level of higher precision. For the study of the SEB background we[25] are trying this last approach at GSI in Darmstadt measuring directly the electron spectra with special interest to the differences, which occur when heavy ions are used as projectiles instead of light ions as protons.

5. ESCAPE OF ELECTRONS FROM A VERY THIN TARGET

The most important contribution to the background is the bremsstrahlung of secondary electrons and it is worth while to estimate in which cases one could reduce this radiation. One obvious possibility is to make the target so thin that we can cut the two step process into pieces, i.e. by letting the electrons escape from the target without making much bremsstrahlung. To estimate when this happens the essential quantity is the stopping power of the electrons, which by Bethe's formula is

$$\frac{dE_e}{dt}(E_e) = -P_B \cdot \frac{Z}{A \cdot E_e} \cdot \ln\frac{1.166 \cdot E_e}{J} \tag{6}$$

where $P_B = 7.85 \cdot 10^4$ keV$^2 \cdot$cm^2/g and an approximation for the mean ionization potential is $J = Z \cdot 11.5$ eV, Z and A being the charge and mass numbers of the matrix atoms with which the electrons of energy E_e are interacting.

From experimental results[26] and theoretical calculations, e.g. in the BEA[27], it is known that the ejected secondary electrons of the highest energies are strongly forward peaked, as is illustrated in fig. 18. To determine the critical thickness t_c for which important escape of secondary electrons occur, we will then simply assume that the electrons go straight forward. We will now study the steep part of the radiation spectra (see figs. 11 and 15) where the intensity of the secondary electrons[19] goes with $\sim E_e^{-10}$. If the electrons now leave the target after having been slowed down to an energy where the intensity was originally double so high we will get half the amount of radiation, i.e. for $\Delta E_e \sim -0.07 \cdot E_e$. Setting $dE_e/dt = -0.07 \cdot E_e/\Delta t$ in eq.(6), $\Delta t = 0.5 \cdot t_c$ and using $Z/A \sim 1/2$ we get with transformation to the radiation energy $E_r \sim 0.97 \cdot E_e$

$$t_c = \left(\frac{E_r}{1\text{keV}}\right)^2 \cdot \frac{1}{\ln\frac{1.2 \cdot E_r}{Z \cdot 0.0115\text{keV}}} \cdot 3.8 \text{ µg/cm}^2 \tag{7}$$

Matrix \ E_r	1keV	2keV	5keV	10keV	20keV	50keV	100keV
C	1.3	4.3	21	74	260	1400	5090
Al	1.8	5.5	26	87	300	1580	5680
Fe	2.7	7.3	32	103	347	1790	6340

Table 3. Critical escape thickness in µg/cm², according to eq.(7) for various radiation energies E_r. Secondary electrons will escape and radiate only half the bremsstrahlung they produce in a thick target in a matrix of the given thickness.

Target thicknesses lower than this critical value t_c will then exhibit an important reduction of the background radiation of energy E_r due to escape of secondary electrons. In table 3 is t_c given according to eq.(7) for radiation energies from 1 to 100 keV in matrices of C, Al and Fe.

For energies $E_r < T_m$ the estimate, eq. (7) is too low because of the less rapid variation with E_e of the electron spectra which however here are emitted in a wider forward cone. On the other hand eq. (7) is an overestimate for low E_r, as electrons of energies in the 1 keV region and lower will undergo strong angular scattering during their slowing down, i.e. they will curl up within small distances of the target and there emit all the bremsstrahlung (isotropically).

6. ANGULAR DISTRIBUTION OF THE RADIATION

Characteristic x-rays from trace elements in the target are normally isotropic. For L and M x-rays this is not obvious, but it has in several cases been established to a few per cent by measurements. However, for characteristic x-rays of heavy ion projectiles themselves large anisotropies have been observed in some cases, but this has only little importance for applied work. Most interest is then to be paid to the angular distribution of the background radiation, as was initiated in ref. 19. The angular distribution of the proton bremsstrahlung was in ref. 19 found to have a minimum in 90° but deviated not much from isotropy. The Compton scattering of γ-rays has not been separately investigated, but the angular distribution follows that of the primary nuclear γ-rays, for which often anisotropic emission patterns are expected and known from nuclear physics studies. Of interest for heavy-ion induced x-ray analysis is that REC has very close to a $\sin^2\theta$ dependence and that several of the MO continua also have a $\sin^2\theta$ term, being peaked around 90°.

For the most dominant background contribution, the bremsstrahlung of secondary electrons (SEB), the angular distribution can be estimated from the degree to which the electrons are forward peaked (see fig. 18), which is simply calculated with the BEA theory[27], combined with the angular distribution of the electron bremsstrahlung[28] and the angular spread of the electrons being slowed down before they radiate. For high electron energies ($E_e \gtrsim 10$ keV) the electrons are strongly peaked and also continue to proceed forward, being slowed down a little, so in this case we will expect something like the angular distribution of incident electrons shown in fig. 19. Fig. 19 gives the angular distribution of the bremsstrahlung from 5 and 50 keV electrons incident on Al calculated at radiation energies 4 and 40 keV by Tseng and Pratt[28]. The distributions are sharply peaked away from the beam axis and due to the retardation effect the peak of maximum intensity is moving to forward angles with increasing electron energy. Although this angular distribution is expected to be smeared somewhat out by the initial angular emission of secondary electron, it should clearly be visible in the high energy part of the spectra (in ref. 19 a mean increase of the yield in 90° over the isotropic radiation was assumed from this argument). However, for low radiation energies both the less peaked emission of secondary electrons and the later angular deflection of the electrons mentioned at the end of chapter 5 will strongly tend to wipe out the emission pattern, peaked at 90°, towards isotropy.

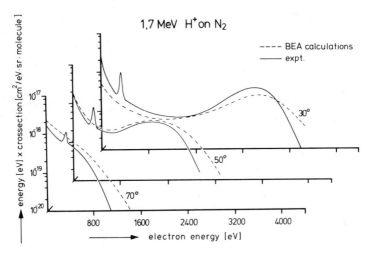

Fig. 18 Energy distribution of secondary electrons from 1.7 MeV protons on N_2 (from ref.25). The experimental cross sections are from ref.26 and the BEA calculations are made according to ref.27.

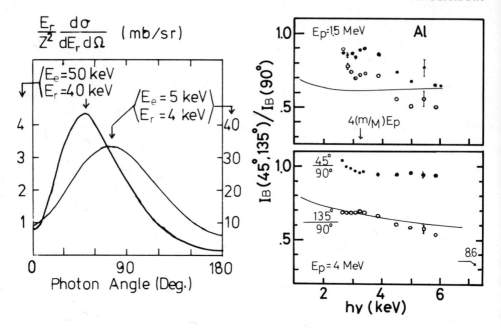

Fig.19 Angular distribution of electron bremsstrahlung (from ref.28). For electrons of 50 and 5 keV is shown the calculated angular distribution of the emitted photons with energy equal to 80% of the electron energy.

Fig.20 Experimental anisotropies of the background radiation from bombardment of Al with 1.5 and 4 MeV protons (from ref. 29). The observed intensities in forward ($45°$) and backward ($135°$) are divided by the $90°$ intensity. The radiation is not symmetric around $90°$ and has the lowest intensity at backward angles.

Angular distributions of SEB have been measured by Ishii et al[29] and their results shown in fig. 20 for the yield in 45º and 135º divided by the yield in 90º with 1.5 and 4 MeV protons incident on Al. As expected a clear forward peaking of the radiation is seen for the radiation for the highest energies, and at least for 1.5 MeV protons the results approach isotropy below ~ 3 keV. However, a similar trend should have been exhibited by the 4 MeV data, but at least the data show that SEB is really anisotropic and peaked at angles smaller than 90º relative to the beam direction.

From these studies of the SEB background it is evident that the background can be reduced significantly (more than a factor 2) moving the detector to backward angles, making this geometry more favourable for analytical work than the traditional set-up with the detector in 90º.

References

1. F. Folkmann, J. Phys. E: Sci. Instr. $\underline{8}$, 429 (1975)
2. T.A. Cahill, New Uses of Ion Accelerators, ed. J.F. Ziegler (Plenum Press, New York, 1975) p.1
3. J.D. Garcia, R.J. Fortner and T.M. Kavanagh, Rev. Mod. Phys. $\underline{45}$, 111 (1973)
4. E. Merzbacher and H.W. Lewis, Handbuch der Physik $\underline{34}$, ed. S. Flügge (Springer Verlag, Berlin, 1958) p. 166
5. B.H. Choi, E. Merzbacher and G.S. Khandelwal, Atomic Data $\underline{5}$, 291 (1973)
6. G. Basbas, W. Brandt and R. Laubert, Phys. Rev. $\underline{A7}$, 983 (1973)
7. J.M. Hansteen, O.M. Johnsen and L. Kocbach, Atomic Data and Nuclear Data Tables $\underline{15}$, 305 (1975)
8. C.E. Busch, A.B. Baskin, P.H. Nettles, S.M. Shafroth and A.W. Waltner, Phys. Rev. $\underline{A7}$, 1601 (1973)
9. S.M. Shafroth, G.A. Bissinger and A.W. Waltner, Phys. Rev. $\underline{A7}$, 566 (1973)
10. F. Hopkins, R. Brenn, A.R. Whittemore, J. Karp and S.K. Bhattacherjee, Phys. Rev. $\underline{A11}$, 916 (1975)
11. S.J. Czuchlewski, J.R. Macdonald and L.D. Ellsworth, Phys. Rev. $\underline{A11}$, 1108 (1975)
12. D. Burch, N. Stolterfoht, D. Schneider, H. Wieman and J.S. Risley, Phys. Rev. Lett. $\underline{32}$, 1151 (1974)
13. D.K. Olsen, C.F. Moore and P. Richard, Phys. Rev. $\underline{A7}$, 1244 (1973)
14. D.K. Olsen and C.F. Moore, Phys. Rev. Lett. $\underline{33}$, 194 (1974)
15. W. Wölfli, C. Stoller, G. Bonani, M. Suter and M. Stöckli, Phys. Rev. Lett. $\underline{35}$, 656 (1975)
16. W.E. Meyerhof, Phys. Rev. Lett. $\underline{31}$, 1341 (1973)
17. W.E. Meyerhof, R. Anholt, T.K. Saylor and P.D. Bond, Phys. Rev. $\underline{A11}$, 1083 (1975)
18. L. Brown, G.E. Assousa, N. Thonnard, H.A. Van Rinsvelt and C.K. Kumar, Phys. Rev. $\underline{A12}$, 425 (1975)
19. F. Folkmann, C. Gaarde, T. Huus and K. Kemp, Nucl. Instr. Meth. $\underline{116}$, 487 (1974)
20. F. Folkmann, J. Borggreen and A. Kjeldgaard, Nucl. Inst. Meth. $\underline{119}$, 117 (1974)
21. H.-D. Betz, F. Bell, H. Panke, W. Stehling, E. Spindler and M. Kleber, Phys. Rev. Lett. $\underline{34}$, 1256 (1975)
22. C.K. Davis and J.S. Greenberg, Phys. Rev. Lett. $\underline{32}$, 1215 (1974)
23. R. Nobiling, Y. Civelekoglu, B. Povh, D. Schwalm and K. Traxel, Nucl. Instr. Meth. to be published
24. M. Ahlberg, R. Akselsson, D. Brune and J. Lorenzen, Nucl. Instr. Meth. $\underline{123}$, 385 (1975)
25. F. Folkmann, K.O. Groeneveld, R. Mann, G. Nolte, S. Schumann and R. Spohr, Z. Phys. A to be published
26. L.H. Toburen, Phys. Rev. $\underline{A3}$, 216 (1971)
27. T.F.M. Bonsen and L. Vriens, Physica $\underline{47}$, 307 (1970)
28. H.K. Tseng and R.H. Pratt, Phys. Rev. $\underline{A3}$, 100 (1971)
29. K. Ishii, S. Morita and H. Tawara, Phys. Rev. $\underline{A}$ to be published

DISCUSSION

Q: (W. Reuter) You calculated the detection sensitivity for impurities in a thin carbon matrix. For the more typical case, i.e. a thick target of atomic number 30 one would expect a considerable reduction in the detection sensitivities. Have you made any calculations for this condition?

A: (F. Folkmann) Yes, and the bremsstrahlung background follows the trend with Z_M indicated in fig. 17 for the target. For a thick target the same trend applies and as shown in a later paper today (7.6) the high Z_T part will be somewhat higher. In total I will estimate that a thick high Z_M matrix gives an optimal sensitivity, which is one order of magnitude worse than for a thin $Z_M = 13$ matrix. To this comes the characteristic X-ray peaks from the matrix, which is a severe problem in a semiconductor detector. They can be reduced with a critical absorber (e.g. Cr for a Fe matrix, ref. 24) so one will still get reasonable sensitivities even below the major matrix component, but 1-2 orders of magnitude worse than the Al thin target values. Another way to overcome this last difficulty is to use a Bragg crystal spectrometer to observe the X-rays.

K-SHELL IONIZATION OF BORON INDUCED BY LIGHT IONS BOMBARDMENT

K.Kawatsura[*], K.Ozawa[*], F.Fujimoto[**], and M.Terasawa[***]

[*] Japan Atomic Energy Research Institute, Tokai, Japan
[**] College of General Education, University of Tokyo, Tokyo, Japan
[***] Toshiba R & D Center, Tokyo Shibaura Electric Co., Ltd., Kawasaki, Japan

ABSTRACT

Absolute cross sections have been measured for the single and double K-shell ionization of boron atom by hydrogen and helium ions bombardment over the energy range from 0.5 to 2.0 MeV. Characteristic x-ray spectra of boron and boron nitride produced by ion bombardment were obtained by using a flat crystal spectrometer. Compared to boron, the K and K^2 lines from nitride were at energies a few eV lower. The measured cross sections are compared with those calculated using the plane-wave Born approximation (PWBA) and the binary-encounter approximation (BEA). The result for the single ionization is in good agreement with that by BEA, while the result for the double ionization deviates from the calculated one.

INTRODUCTION

Recently, the inner shell ionization of atoms by ion bombardment has been studied with high resolution by many workers. It has been revealed that multiple inner shell ionization in ion-atom collision process induces many satellites and hypersatellites. In the case of light ion bombardment such as hydrogen and helium ions, the ionization process can be explained by direct Coulomb excitation. The cross section for single K-shell ionization can be predicted with some success by theoretical calculations using the plane-wave Born approximation (PWBA) or the binary-encounter approximation (BEA). For the double K-shell ionization process, Olsen and Moore

[1] studied oxygen bombardment on calcium at high energy. However, the resulting cross section was in poor agreement with theoretical calculations. Recently, Nagel et al. [2] obtained K^2 line energies for the elements Be through Ne by heavy ion bombardment, and compared successfully with theoretical calculations.

Here, we report the energy dependence of the single and double K-shell ionization processes using 0.5 - 2.0 MeV/amu hydrogen and 0.125 - 0.5 MeV/amu helium ion bombardment on a thick polycrystalline boron target. The measured cross sections were compared with the available theoretical calculations. Characteristic x-ray spectra from boron nitride were also measured and the difference of spectra between boron and boron nitride is explained by the effect due to the chemical bonding of nitrogen atom to boron.

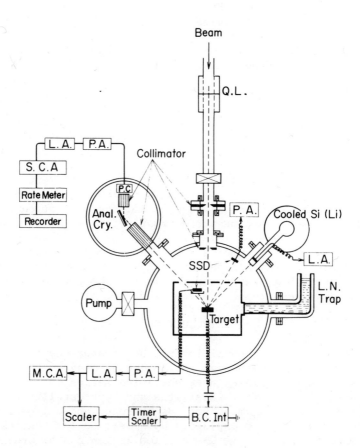

Fig. 1. Schematic diagram of the apparatus used for x-ray spectrum measurement.

EXPERIMENTAL

The experiment was performed with a 2-MV Van de Graaff accelerator of JAERI using protons and singly ionized helium ions of several energies between 0.5 and 2.0 MeV. Figure 1 shows a schematic diagram of the apparatus used for x-ray spectrum measurement. A thick target of boron or boron nitride was bombarded vertically with the ion beam. The vacuum at the target was about 5×10^{-7} torr. The resulting characteristic x-rays were measured at 45° take-off angle by a Bragg spectrometer with a flat crystal of the lead-stearate soap film consisting of 100 layers (2d spacing = 100 Å) and with Soller slits. The detector was a gas-flow proportional counter with a thin stretched polypropylene window. A gas mixture of 25 % argon and 75 % methane was used because this ratio showed better energy resolution for measuring low energy x-rays than any other mixture of these two gases [3].

RESULTS AND DISCUSSION

Figure 2 shows the spectra observed from thick boron and boron nitride targets bombarded by 1.5 MeV helium ion. The energy scale was calibrated by measuring separations of boron, carbon and nitrogen K x-ray lines with proton bombardment and using the x-ray energy data

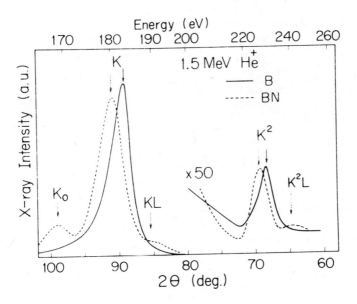

Fig. 2. Boron K x-ray spectra for 1.5 MeV helium ion bombardment on boron (solid curve) and boron nitride (dashed curve).

of Bearden [4]. The x-ray emission spectrum from boron target consists of two peaks. The intense peak at the energy of 183.0 eV is due to K x-rays of boron atoms. The other weak one at 230.8 eV can be identified as radiation from boron atoms with double K-shell vacancies and this energy is in good agreement with other measurements and theoretical calculations [2]. On the other hand, the x-ray emission spectrum from boron nitride is rather complicated. The energies of the two main peaks, K and K^2, are 180.7 and 228 eV, respectively and they are lower by a few eV in energies than those of boron as the effect of chemical bonding of nitrogen atom to boron. The main peaks have a satellite in the high energy side of each peak and these satellites were also observed in the case of heavy ions bombardment, where they were very strong. They are identified as KL and K^2L x-rays of boron [5]. Another faint peak at 169 eV is also observed. In this case, it can be presumed that the 2s electron of nitrogen fills up the single vacancy of 1s level of boron [6]. Table 1 summarizes the energy of the recognizable peaks in both spectra with the result of electron bombardment [7] for comparison. In the case of hydrogen ion bombardment, the peak due to the double K-shell ionization could not be observed above the background level.

In order to obtain absolute cross sections from observed spectra and a non-dispersive intensity, the following process is carried out:

Thick target yield Y(E) is obtained from the directly measured x-ray yield by correcting for incident flux, geometry factors, window absorption, reflection efficiency of analysing crystal and detector efficiency. The x-ray production cross section can be calculated from a thick target yield by the relation

$$\sigma_x(E) = \frac{1}{n}\left(\frac{dY}{dE} S(E) + \frac{\mu \cosec(\pi/4)}{\rho} Y(E)\right) \qquad (1)$$

Table 1

Energy values (eV) of K lines of boron

	BN			B		
	He	H	e & X	He	H	e & X
K_0^1	169	169	169.0			
K^0	180.7	181.7	180.1	183.0	183.4	182.5
KL	189		190.0			
K_2^2	228			230.8		
K^2L	240					

where S(E) is the stopping power for the incident ion in the target material, μ/ρ is the target absorption coefficient for the characteristic radiation and n is the density of target atoms. The stopping powers were interpolated from the data of Northcliffe and Schilling [8]. The absorption coefficients were taken to be 3350 and 57000 cm^2/g for the single and double K-shell ionization x-rays, respectively, from the data of Henke and Elgin [9]. The value of n was taken to be 5.57×10^{22} g^{-1}. The derivative of the yield function was obtained analytically from least-squares polynomial fits to the experimental yield curve. The ionization cross section σ_i is related to σ_x by $\sigma_i = (1/\omega) \sigma_x$, where ω is K-shell fluorescence yield whose value is 7.1×10^{-4} given by Dick and Lucas [10] and the same value is used for the double K-shell ionization though the conventional fluorescence yield value for the single ionization may not be applicable to the case of the double ionization. The uncertainty in the absolute ionization cross sections is within a factor of 2.

Figure 3 shows the experimental cross sections for the single and double K-shell ionization of boron atom by hydrogen and helium ion bombardment. For the single ionization, the cross section varies from 1.3×10^{-18} to 2.5×10^{-18} cm^2 in the present energy range and agrees very well with the theoretical calculations of direct Coulomb excitation for K-shell electrons predicted by PWBA and especially by BEA. On the other hand, the cross sections for the double K-shell ionization are one order of magnitude less than those for the single K-shell ionization and the energy dependence of the cross section is much stronger than those for single ionization. Then, the position of maximum cross section is located near the position of the single ionization case. The measured cross section is 4.8×10^{-20} cm^2 at 125 keV/amu and rapidly increases with increasing energy and reaches a maximum value, 2.6×10^{-19} cm^2 at about 375 keV/amu. For higher energy, it decreases sharply with incresing energy. The lower solid curve in Figure 3 shows the ionization cross sections with double K-shell vacancies calculated by BEA derived by Hansen [11]. He has developed a semiempirical calculation based on the BEA which can be useful in estimating the multiple ionization cross section. According to his treatment, the ratio of the gross ionization cross section σ_g to the calculated one σ_c is given by $\sigma_g/\sigma_c = 1 + (N-1) \bar{P}/2$ provided that the average probability of ionization per electron, $\bar{P}$, is much less than unity. In this case, $\sigma_g = \sigma_1 + 2\sigma_2$ and $\sigma_c = \sigma_1 + \sigma_2$, where σ_1 and σ_2 are the the cross section for the single and double K-shell ionization, respectively. Then, σ_2 can be calculated by $\sigma_2 = \bar{P}\sigma_1/(2-\bar{P})$. The probability $\bar{P}$ has been semiempirically derived as follows:

$$P \approx \frac{7 \times 10^{-5} z_1^2 z_2^2}{U^2 n^3 (\ell + \frac{1}{2})} \left(\frac{\bar{v}}{v_1}\right) \frac{\sigma_1(v_1^2/\bar{v}^2)}{\sigma_1(v_1^2/\bar{v}^2=1)} \qquad (2)$$

with Z_1 and Z_2 the charges of projectile and target atom, respectively, U is the binding energy in keV, n and ℓ are the principle and angular momentum quantum numbers, and $v/v_1 = (MU/mE)^{1/2}$. The energy dependence of the measured cross sections for the double K-shell ionization is similar to that of the calculated ones, but it is smaller by a factor 3 to 5. In respect with the ratio of σ_2 to σ_1, Olsen and Moore [1] have measured for calcium K x-rays by oxygen ion bombardment at very high energy, i.e., 24.0 - 48.0 MeV. This energy range corresponds to 2.3×10^{-1} to 4.4×10^{-1} of $E/\lambda U$. They did not give the absolute cross sections and the energy range examined was different from that of the present experiments, so we cannot compare directly both results. However, it seems reasonable that σ_2/σ_1 gives much the same value at about 4.0×10^{-1} of $E/\lambda U$ in both measurements. In either case, the measured cross sections for the double K-shell ionization are smaller than the calculated ones.

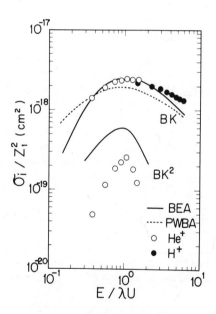

Fig. 3. Boron K-shell ionization cross sections in proton and helium ion bombardment : open circles, helium ion ; solid circles, proton ; dashed curve, PWBA ; solid curve, BEA.

In summary, the absolute cross sections of both single and double K-shell ionization have been measured for the first time. Moreover, it has been observed that the energy of maximum cross section for the double ionization is nearly equal to the energy of the single ionization case. However, the measured cross sections for the double K-shell ionization are less than that of the calculated ones by BEA. The deviations between the measured and calculated cross sections seem to arise from neglect of binding energy change and distortion effect during the collisions suggested by Olsen and Moore [1].

ACKNOWLEDGMENTS

The kind operation of Van de Graaff staff is gratefully acknowledged.

REFERENCES

[1] D.K.Olsen and C.F.Moore, Phys.Rev.Lett. 33, 194(1974).
[2] D.J.Nagel, A.R.Knudson and P.G.Burkhalter, Vacuum UV Radiation Physics (Pergamon Press, 1974).
[3] M.Terasawa, Thesis (University of Tokyo, 1974), unpublished.
[4] J.A.Bearden, Rev.Mod.Phys. 39, 78(1967).
[5] K.Ozawa, K.Kawatsura, F.Fujimoto and M.Terasawa, to be published
[6] D.W.Fischer, Advances in X-Ray Analysis Vol.13, p.159(Plenum Press, New York, 1970).
[7] R.C.Ehlert and R.A.Mattson, Advances in X-Ray Analysis Vol.9, p.456(Plenum Press, New York, 1966).
[8] L.C.Northcliffe and R.F.Schilling, Nucl.Data Tables A7, 233 (1970).
[9] B.L.Henke and R.L.Elgin, Advances in X-Ray Analysis Vol.13, p.639(Plenum Press, New York, 1970).
[10] C.E.Dick and A.C.Lucas, Phys.Rev. A2, 580(1970).
[11] J.S.Hansen, Phys.Rev. A8, 822(1973).

EFFECT OF CHANNELLING ON IMPURITY ANALYSIS BY CHARGED PARTICLE

INDUCED X-RAYS

P.B. Price, B.E. Cooke, G.T. Ewan and J.L. Whitton†

Department of Physics, Queen's University, Kingston, Ont.
† Chalk River Nuclear Laboratories, Atomic Energy of
   Canada Limited, Chalk River, Ont.

There are two important background effects limiting the sensitivity of impurity analysis in solids by charged particle X-rays. A general limitation is imposed by the continuous background produced by δ-ray bremstrahlung while a relatively high yield of X-rays from the host material can mask adjacent smaller signals. In this report we shall consider only the effect of the δ-ray bremstrahlung.

This can be drastically reduced when single crystals are used. The ordered lattice allows use of the channelling effect which steers the charged particles away from the nuclei and tightly bound electrons. This reduces the yield of knock-on electrons arising from close-encounter collisions.

The net result of the channelling effect is to reduce the background continuum and allow observation of impurity levels that are normally masked by the background. Two types of impurity distributions are influenced
1) those lying on the immediate surface of the crystal and
2) those distributed throughout the bulk.

Several authors[1,2,3,4] have shown how the channelling effect can be utilized in this way for elastic scattering analysis of impurities of mass heavier than the matrix. The advantage of the channelling effect applied to charged particle induced X-rays is that it allows analysis of impurities lighter than the matrix. The application of this technique has been illustrated by Chemin et al[5] for the location of P and S implants in Ge. It is, therefore, a complementary technique to the elastic scattering analysis.

The effects reported here were observed during a study of specimens of GaAs implanted with 40 keV ^{34}S ions along the most open (<110>) crystal direction. This direction was chosen to allow a high fraction of the incident ions to penetrate deeply in the crystals and to increase the chance of them occupying substitutional sites. The implantations were done at 150°C to encourage in-situ annealing of the ion bombardment induced damage[6].

For the analyses, the crystals were first aligned by standard backscattering techniques[7]. Typical χ_{min} values (ratio of aligned to random yields near the surface) were 0.06 with 2 MeV He ions. The emitted X-rays were detected with a Kevex X-ray detector[8], filtered with 150 μm of Be to screen out the Ga(L) X-rays, i.e. only the As(L) X-rays and the K X-rays from both elements are seen to arise from the matrix.

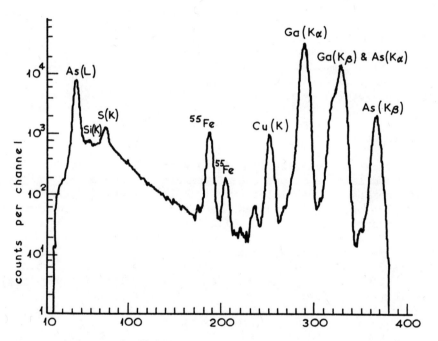

Fig. 1. X-ray spectrum from S implanted GaAs for a 2 MeV He$^+$ beam incident in a random direction.

Figure 1 shows the X-ray spectrum observed when the crystal was oriented in a random direction. The background observed in the 1-5 keV region is due to δ-ray bremstrahlung produced by

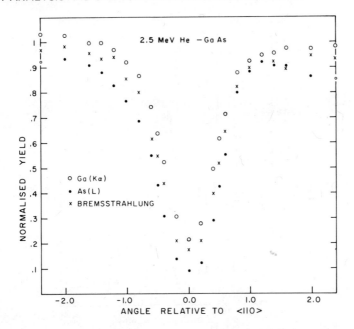

Fig. 2. Normalized yield curves for As(L), Ga(Kα) and bremsstrahlung for tilt scan across the <110> in GaAs with 2.5 MeV He$^+$ ions.

knock-on electrons from collisions of He ions with bound electrons. These are close-encounter processes and would be expected to show channelling effects similar to those for other close encounter processes such as characteristic X-ray production or Rutherford backscattering.

Figure 2 shows typical dip curves for the As(L) X-ray peak, the Ga(K) X-ray peak and the bremsstrahlung in the energy range 1.5 -4 keV. All show the well-known dips associated with the channelling process. This suggests that the bremsstrahlung production is indeed associated with a close-encounter collision and can be reduced by orienting the crystal so that the charged particles are channelled down a crystal axis. If surface impurities are being studied which will not be affected by this channelling the reduction in background increases the sensitivity.

This is illustrated in Figure 3 which shows the low energy portion of the X-ray spectrum in a random and aligned direction. The spectra were taken for the same total integrated current and can be directly compared.

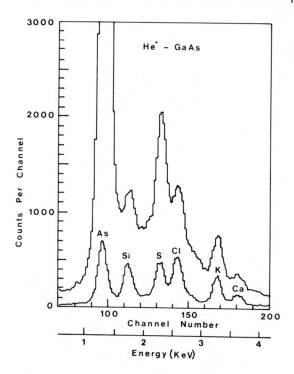

Fig. 3. X-ray spectra for 2.0 MeV He$^+$ ions incident in random and aligned on a <110> GaAs crystal implanted along the <110> direction with 40 keV ^{34}S ions.

Signals arise from the As of the GaAs matrix and from the implanted S; these can be considered as bulk distributions. Si (from pump oil) and the remaining signal peaks from Cl, K and Ca are purely surface distributions. The latter three are most likely deposits from the various laboratory reagents used in rinsing the crystal. Similar deposits have been noted previously[4] by elastic scattering analysis of silicon crystals.

On the upper, random spectrum, the high continuum background makes integration of the peak areas very difficult. Improvement is seen when the crystal is aligned (lower spectrum), in particular, the Si, Cl, K and Ca peaks are more pronounced although having the same integrated area (the slight increase in the Si yield is due to build-up of this contaminant from the pump oil during the analysis).

Integration of the areas under these surface contaminant peaks in the aligned spectrum shows the minimum detectable level of Cl, K and Ca impurities to be $\sim 1 \times 10^{13}$ cm^{-2}, i.e. $\sim 1/200$ of an atom layer.

The reduction in the intensity of the As(L) X-ray line in the aligned direction is the well-known channelling effect. If the implanted S atoms were all substitutional they should show the same reduction in intensity except for a small correction associated with the different mean depths from which different energy X-rays are observed. The observed reduction in intensity is less than expected and our atom location analysis[9] shows ~ 60% of the S atoms lying in substitutional (therefore shielded) positions so the yield from the S atoms is lower also.

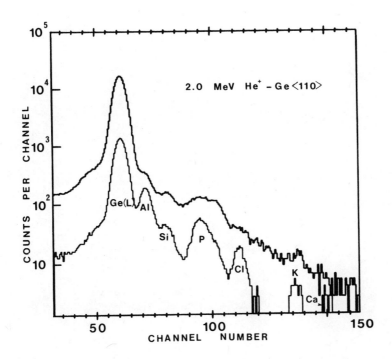

Fig. 4. X-ray spectra for 2.0 MeV He$^+$ ions incident in random and aligned directions on a <110> Ge crystal.

In Figure 4 we show similar results for a <110> Ge crystal. The δ-ray continuum is lowered due to the channelling effect and impurity peaks that are masked in the random spectrum show up very clearly in the aligned spectrum.

TABLE I

Minimum detectable concentrations for 2.0 MeV He$^+$ estimated from measured concentration /N_S where $N_S = N_P/3N_B^{1/2}$

ELEMENT	ORIENTATION	MINIMUM CONCENTRATION (10^{13} atoms cm^{-2})	
		GaAs (20µC)	Ge (4µC)
Aℓ	R		-
	A		500
Si	R	48	250
	A	24	100
S (implanted)	R	4	
	A	3	
P	R		100
	A		30
Cl	R	1.5	-
	A	.9	2
K	R	1	3
	A	.5	1
Ca	R	1	-
	A	.6	1

R: - Random Orientation
A: - Aligned <110>
-: - Not detected

Since the visibility of the signal is limited by statistical fluctuations in the background, a suggested requirement for the recognition of impurities is[10] that

$$N_P \geq 3N_B^{1/2}$$

where N_P is the number of counts in the peak and N_B is the background measured at F.W.H.M. Table I lists the minimum concentration of impurities estimated from the spectra in Figs. 3 and 4 according to the criterion $N_S = N_P/3N_B^{1/2} = 1$ (where N_B in this case is the full background). For Cl, P, K and Ca the minimum detectable sensitivity is estimated to be about 10^{13} atoms cm^{-2}. Since the

Ge crystal is unimplanted (i.e. less damaged), the increase in sensitivity on alignment is greater than for GaAs. The measure of the sensitivity is dependent on the integrated current and the result is seen in the comparatively larger sensitivity values obtained in all cases for Ge (4µC) as compared with GaAs (20µC).

In these experiments 150 µm of Be were used to attenuate low energy X-rays. This gives an attenuation of $\sim 10^{-2}$ at 1 keV. As a result the sensitivity to light elements, e.g. Aℓ and Si is relatively poor as indicated in the Table. In general, though backgrounds can be reduced by up to a factor of 10, the use of the criterion $N_B \geq 3N_B^{\frac{1}{2}}$ implies increases in sensitivity of only $\sqrt{10} \approx 3$ for the same integrated currents.

It should be noted that in making analyses for surface impurities, it is only when the analyzing beam is in a random direction to the crystal that the real concentration is measured relative to the bulk material. If channelling is used to reduce the background due to δ-ray bremstrahlung the results should be corrected for the channelling effect on the bulk material. The channelling effect causes the yield to be lowered when the impurities are in (shielded) substitutional positions or enhanced when the impurities are in (open) interstitial positions (the flux peaking effect[7]).

REFERENCES

(1) W.D. Mackintosh and J.A. Davies, Anal. Chem. 41, 26A (1969).

(2) W.D. Mackintosh, J. Rad. Anal. Chem. 16, 421 (1973).

(3) R.L. Meek, T.M. Buck and C.F. Gibbon, J. Electrochem. Soc. 120, 1241 (1973).

(4) R.L. Meek, J. Electrochem. Soc. 121, 172 (1974).

(5) J.F. Chemin, I.V. Mitchell and F.W. Saris, J. Appl. Phys., 45, 532, 537 (1974).

(6) J.L. Whitton and G.R. Bellavance, Rad. Effects 9, 127 (1971).

(7) D.S. Gemmell, Rev. Modern Phys. 46, 1 (1974).

(8) Kevex-ray spectrometer, Kevex Analytical Division, Burlingame, California 94010.

(9) B.E. Cooke, G.T. Ewan and Whitton, Rad. Effects (to be published).

(10) F. Folkmann, J. Sci. Instr. $\underline{8}$, 429 (1975).

DISCUSSION

Q: (W.L. Brown) Have you looked at the channelling angular dependence of different parts of the bremsstrahlung spectrum?

A: (P.B. Price) No. We have looked only at the whole (1.5 - 4 keV) spectrum.

DEPTH PROFILING WITH ION INDUCED X-RAYS

L. C. Feldman and P. J. Silverman

Bell Laboratories

Murray Hill, New Jersey 07974 U.S.A.

I. INTRODUCTION

The backscattering of MeV particle is a demonstrated and successful technique in the near surface analysis of solids. Similarly, light-ion induced X-rays have been shown to be an important technique for trace element analysis. The principal advantage of the X-ray technique is the elemental discrimination, while in the backscattering case it is depth resolution. Of course, the scattering technique provides mass discrimination, but since the energy signal contains both the mass and depth information, ambiguities may arise in a number of important cases.

As the energy dependence of the X-ray production cross section is known and the stopping power of He^+ ions is understood, one can, in principle, obtain an elemental depth profile by observing the change in X-ray yield while varying the He^+ trajectory. The change in trajectory can come about by varying either the beam energy or the angle between the beam and sample surface. In this paper we describe the use of the latter technique to extract elemental depth profiles.

In a previous report[1] we have described the calibration of this technique, determined the depth resolution and used the qualitative features to explore the composition of oxide layers on gallium arsenide. In this paper we shall describe in detail two more cases involving GaAs. This is a particularly significant substrate since it is an important semiconductor while its two constituent masses are too close to be easily resolved by backscattering. We shall show that the technique is rather direct in supplying qualitative information on the near surface profile; however, the experimental errors propagate in such a manner that accurate quantitative profiles are difficult to extract. In a number of recent papers Pabst has also considered the mathematical problems described here.[2]

II. FORMALISM

The total yield, Y_i, of characteristic X-rays from element i imbedded in a solid may be written as

$$Y_i = \int_0^{t_f} N_i(t) \, \sigma_i[E(t)] e^{-\mu_i t} \, dt \qquad (1)$$

where σ_i is the cross section for excitation of the X-ray by an ion with energy E, t_f is the range of the ion, μ_i is the appropriate absorption coefficient, and $N_i(t)$ is the profile of the element at depth t. As noted in Fig. 1, t is measured along the

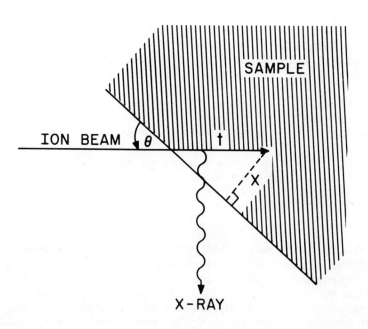

Fig. 1   Geometry for the depth profiling technique.

particle trajectory while usually one is interested in $N_i(x)$, the profile measured parallel to the surface normal. If we assume that there is no variation in N_i other than that in the x-direction then by a change of coordinates

$$x = t \sin \theta$$

Y_i may be written

$$\sin \theta \, Y_i(\theta) = \int N_i(x) \sigma_i(x/\sin \theta) e^{-u_i/x/\sin \theta} \, dx. \quad (2)$$

By measuring the yield at a variety of angles one can use Eq. (2) to extract the profile $N_i(x)$.

In the work described here the absorption factor is small and will be omitted for the remainder of the discussion. Obviously, Eq. (2) can be written as a sum

$$\sin \theta \, Y_i(\theta) = \Sigma \, N_i(x) \, \sigma_i(x/\sin \theta) \, \Delta x \quad (3)$$

which is the way it has been treated in this analysis. For different angles θ Eq. (3) forms a set of simultaneous linear equations in which the $Y_i(\theta)$ are the measured yields, the cross sections, σ_i, are the known coefficients and the $N_i(x)$ are to be extracted. Alternately, one can formulate the problem by fitting the data for a given form of $N_i(x)$. We note that in the trivial case of $N_i(x)$ = constant the yield is constant.

Implicit in this analysis has been the transformation from $\sigma_i(E)$ to $\sigma_i(t)$. For trace elemental analysis this transformation depends on an accurate knowledge of the cross section for X-ray production and the stopping power. For film analysis, however, the stopping power may not be known since the composition is not known. The stopping power may be estimated from the backscattering spectrum and the integral X-ray yield. For film analysis this is an intrinsic difficulty in the technique.

In order to give the reader a feel for $\sigma(t)$, Fig. 2 shows this function for excitation of Ge K X-rays by 1.0 MeV He incident on Ge. Since 90% of the X-rays are created in a thickness $t_{crit} \sim 8000$ Å, one can obtain a depth resolution of about 900 Å at the 10% accuracy level in GaAs by tilting the sample to angles of $\sim 7°$.

III. RESULTS

1. $Al_{.67}Ga_{.33}As$

As a first example we consider the concentration profile of Al doped GaAs. The sample was nominally doped to 2/3 Al to a depth of 1.7 μ. The questions to be answered concern the actual concentration of Al and the uniformity. In this particular case the As-L X-rays fall very close in energy to the Al-K and hence we used the Ga-K to profile the Ga concentration in the near surface region of the sample.

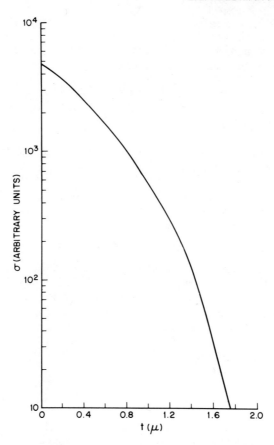

Fig. 2 The K X-ray production cross section versus thickness in germanium for 1.0 MeV incident He$^+$ ions.

The results of this profile are shown in Fig. 3. In this figure the vertical scale is established by comparison with a pure GaAs sample. The yield data show that the extracted concentration is .32 ± .01 over the depth sampled. The yield showed no significant variation over the angles sampled and one must conclude that the concentration is constant. These results are borne out by optical measurements[3] which assert that the Al concentration could not vary by more than 10% without producing a significant broadening of the emission line. This broadening is not observed. As an indication of the sensitivity of the techniques we show in Fig. 3 the expected yield data for a profile of Al as indicated in the insert. This type of profile would show up as a large variation in the extracted concentration but as a small variation in the angular dependence. From the shape of the $\sigma(t)$ function it is clear that this type of profile

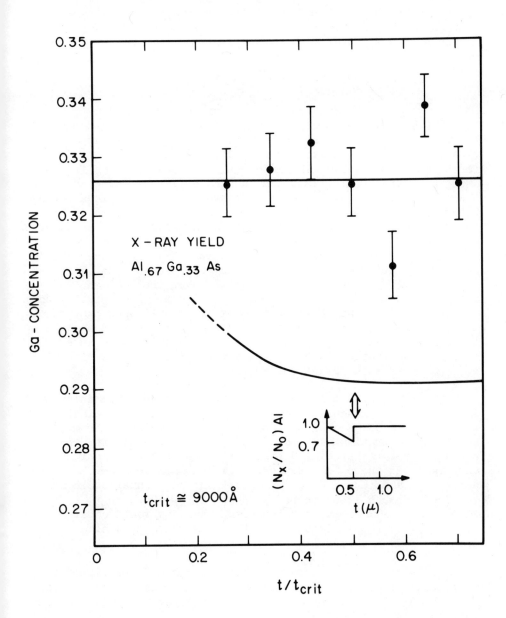

Fig. 3 The Ga concentration in Al doped GaAs as a function of sin θ or t/t_{crit}. The lower curve shows the expected yield for a hypothetical Al concentration shown in the insert.

variation is more difficult to differentiate from a uniform distribution, than a concentration profile of opposite slope. In this case we concluded that the Ga concentration is .32 ± 0.1 and the Al concentration varies less that ~20% over the explored depth.

2. Pt-GaAs Films

The motivation for this study is in understanding the metallization schemes of GaAs. Such metallization systems are important in choosing contacts for GaAs devices and the metal layers for Schottky Barriers. The interest in Pt GaAs derives from a study by Poate et al.[4] in which the properties of Au, W and Pt on GaAs were measured by Rutherford scattering. The backscattering spectrum of Pt-GaAs before and after annealing is shown in Fig. 4. In order to understand the solid state chemistry that has occurred one is desirous of understanding at least the elemental composition and the chemical composition near the surface. This information is not evident from the backscattering spectrum alone. Figure 5 shows a series of spectra from reacted (heat treated) and unreacted Pt-GaAs films at 10° and 45° incidence. The spectra show the Ga and As Kα and Kβ lines and the Pt-L X-ray spectrum. It is clear from the direct spectra that the films are rich in Ga at the surface. With appropriate stripping techniques we have extracted the intensity of the Ga and As lines for a series of spectra taken at a variety of angles.

The results of this analysis for the reacted film is shown in Fig. 6. The intensity, on the vertical scale, is relative to the intensity of Ga or As observed in a pure GaAs spectrum. Thus the Ga concentration, is ~.6 of the Ga concentration in GaAs, averaged over the probing depth, while the As concentration is very low. In these intensities, corrections have been made for absorption and fluorescence effects. The horizontal scale defines the depth probed in terms of sin θ or "t/t_{crit}" where t_{crit} ~ 5000 Å for Pt-Ga or Pt-As compounds. Thus one would conclude that integrated over the first 800 Å, the film contains a high concentration of Ga and very little As.

It is not difficult to anticipate one simple profile that is consistent with the integral data. A pure Pt-Ga layer of thickness ~1500 Å followed by a Pt-As layer of approximately the same thickness is qualitatively correct and suggested by the backscattering spectrum in Fig. 4. In this model there will be a larger Ga yield at grazing incidence while the ion-trajectory is entirely in the Pt-Ga layer; as the angle increases and the He trajectory penetrates to the Pt-As layer the As yield takes a dramatic increase and the Ga yield shows a step decrease. Of course, the

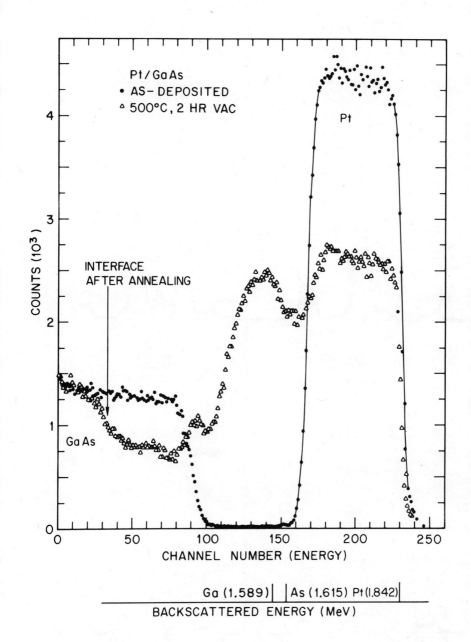

Fig. 4 Backscattering spectrum using 2.0 MeV He ions from the as-deposited (●) and reacted (Δ) Pt/GaAs system (Ref. 4).

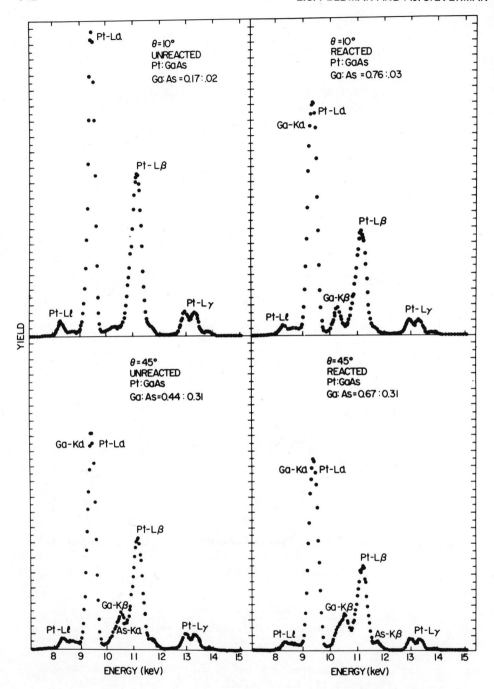

Fig. 5 The X-ray spectra from the Pt:GaAs system at angles of 10° and 45° in the reacted and unreacted films.

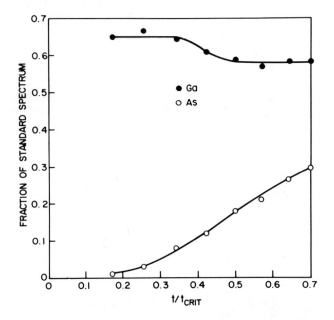

Fig. 6   Yield of Ga and As K-X-rays as a function of angle for the reacted Pt-GaAs system.

Ga yield will always be significant as the X-ray production cross section is a maximum in the Ga layer at the surface.

The formal solutions as described by Eq. (3) have been carried out to extract the profile for this particular case from the integral data. The results are shown in Fig. 7 and are consistent with the qualitative description given above. The general features of this analysis agree with more detailed analysis of the backscattered spectrum and the results of an Auger analysis of similar films.[5] However, it must be pointed out that certain small admixtures of the material would be difficult to detect. For example, 10 atomic percent As in the Pt-Ga layer should be detectable, while 10 percent Ga in the Pt-As layer would be difficult.

CONCLUSION

We have shown that ion-induced X-rays can be used to give details on the elemental depth profile of thin films. The

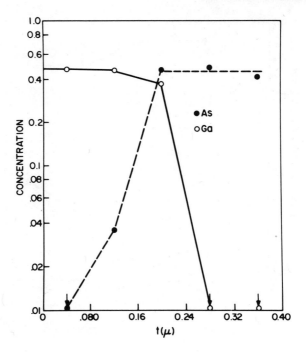

Fig. 7   Differential profile of Ga and As in the reacted film.

analysis has been carried out quantitatively for the first time. It is clear, however, that only the simplest cases are amenable to this quantitative technique. Experience has shown that errors propagate in such a manner that false results can easily be obtained unless the analysis is done cautiously and the relative cross sections are well known.

Qualitatively this technique can be an extreme useful supplement to ion backscattering as described here. Of course, for absolute identification of impurities, X-ray analysis is unsurpassed. The techniques described here can be useful in controlling and understanding the depth probed in such analyses.

ACKNOWLEDGMENTS

We gratefully acknowledge the help of J. Poate, F. Reinhart, W. M. Augustyniak and L. Dodd.

REFERENCES

1. L. C. Feldman, J. M. Poate, F. Ermanis and B. Schwartz, Thin Solid Films 19, 81 (1973).

2. Wolfgang Pabst, Nuc. Inst. and Meth. 124, 143 (1973).
   W. Pabst, Nucl. Instr. and Meth., 120, 543 (1974).

3. F. Rheinhart, private communication.

4. A. K. Sinha and J. M. Poate, Appl. Phys. Letters 23 (1973).

5. V. Kumar, J. Phys. Chem. Solids 36, 535 (1975).

SENSITIVITY IN TRACE ELEMENT ANALYSIS OF THICK SAMPLES USING PROTON INDUCED X-RAYS

Finn Folkmann

GSI, Postfach 541, D-6100 Darmstadt, West Germany

The optimum sensitivity in trace element analysis with proton induced x-rays is found for thick targets and compared with the results for thin targets. The effect of the slowing down of incident protons (1-4 MeV) on the x-ray production yield and the background contribution is theoretically evaluated, both by numerical integration over the proton path and by arguments based on power expansions in the proton energy for the involved processes. The x-ray attenuation is included giving results between the thin and thick target values and also the background from complex matrices is simply related to the contributions from the main components.

1. DIRECT EFFECTS OF THE SLOWING DOWN OF THE PROTONS

The fundamental limitations for the sensitivity in analysis with ion-induced characteristic x-rays have previously been evaluated for thin targets from a study of the most important processes involved: the production of characteristic x-rays from potential trace elements and the background from a representative matrix due to bremsstrahlung from secondary electrons and directly from the incident protons[1,2]. To extrapolate these results to thick targets we consider the slowing down of the protons in the matrix, expressed by the stopping power dE/dt, and study the effect which the changing energy has on the different processes. The experimental setup is sketched in fig. 1 and we are summing the contributions from positions X as function of the penetrated target depth t or of the related proton energy $E_p(t)$. With an initial energy $E_0 = E_p(0)$ the proton energy $E = E_p(t)$ inside the target varies from E_0 to 0 and the yield per projectile is

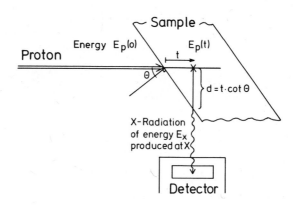

Fig.1 Schematic of integration procedure. The thick target yields are calculated with the production cross section at point X which via the distance t depends on the proton energy $E_p(t)$. The x-rays are attenuated over the distance d in the sample.

$$Y = \int f \cdot \frac{\sigma \cdot dt}{AM} = \int_{E_o}^{0} f \cdot \frac{\sigma(E)}{AM} \cdot \frac{dE}{\frac{dE}{dt}(E)} \qquad (1)$$

where the cross section for atoms of mass number A is σ, $M=1.66 \cdot 10^{-24}$ g is the mass unit and the thickness t is measured in mass per area (e.g. in g/cm^2). The factor f in eq.(1) is the detection efficiency for the x-rays produced at X. It includes the solid angle of the detector, detection efficiency and absorption of the x-rays on their way to the detector. In this chapter we will set f=constant as is the case when $\theta \sim 90°$. In the next chapter the effect of the attenuation of the x-rays in the sample is included through f's dependence on E_x and on $d=t \cdot \cot\theta$.

For the stopping power we will use the approximation

$$\frac{dE}{dt}(E) = -Z^{-0.33} (\frac{E}{1\text{MeV}})^{-0.7} \cdot 0.425 \text{ MeV} \cdot \text{cm}^2/\text{mg} \qquad (2)$$

which is in good agreement with the values of Northcliffe and Schilling[3] as shown in fig. 2 for the energy region of primary interest of x-ray analysis. Eq.(2) implies that the range of the protons is

$$R(E_o) = Z^{0.33} (\frac{E_o}{1\text{MeV}})^{1.7} \cdot 1.38 \text{ mg/cm}^2 \qquad (3)$$

and that the stopping power can be expressed as a simple power function of E/E_o

SENSITIVITY IN TRACE ELEMENT ANALYSIS

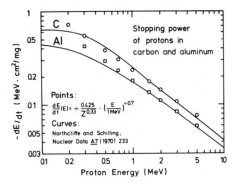

Fig.2 Proton stopping power in C and Al from ref. 3 and eq.(2).

$$\frac{dE}{dt}(E) = \left(\frac{E}{E_o}\right)^p \cdot \frac{dE}{dt}(E_o) \qquad p = -0.7 \qquad (4)$$

If also $\sigma(E)$ in eq.(1) can be expressed $\sigma(E) = (E/E_o)^u \cdot \sigma(E_o)$ the integral (1) is readily found to be $Y = Y_o \cdot (1-p)/(u+1-p)$, where Y_o is the yield assuming no dependence of $\sigma = \tilde{\sigma}(E_o)$ on E, i.e. the thin target yield multiplied up to the thickness $R(E_o)$. This estimate for the integral is made for the primary x-ray production and for the bremsstrahlung from secondary electrons and from the protons. The proton bremsstrahlung (B) is simply approximated

$$\sigma_B(E) = \left(\frac{E}{E_o}\right)^s \cdot \sigma_B(E_o) \qquad s = -0.8 \qquad (5)$$

whereas the corresponding expression for the secondary electron bremsstrahlung (SEB) has an exponent r depending strongly on the radiated energy E_r in units of $T_m = E_o \cdot 4m_e/M$, where m_e is the electron mass. T_m is the maximum energy transfered to a free electron by the protons of energy E_o and determines the rapid fall-off with higher E_r of the SEB spectrum

$$\sigma_{SEB}(E) = \left(\frac{E}{E_o}\right)^r \cdot \sigma_{SEB}(E_o) \qquad r\left(\frac{E_r}{T_m}\right) \text{ in fig. 3} \qquad (6)$$

r is shown in fig. 3 calculated for $E_o=2$ and 3 MeV on C and Al matrices averaged over an energy loss of 25 %. The rapid variation $r \sim 8$ for $E_r > 0.8 \cdot T_m$ is due to the primary strong decrease with E_r in this region and the shift of the complete SEB spectrum to smaller energies with decreasing E. The x-ray production cross section for different trace elements Z_T has also an expression

$$\sigma_x(E) = \left(\frac{E}{E_o}\right)^q \cdot \sigma_x(E_o) \qquad q(Z_T, E_o) \text{ in fig. 4} \qquad (7)$$

where the exponent q depends on Z_T^2/E_o or physically on the K binding energy $E_K(Z_T)$ ($\sim$ the energy of emitted K x-rays $E_x = Z_T^2 \cdot 0.01$ keV) in units of T_m. The range of q goes from 0, when $E_K(Z_T) \simeq T_m/4$, i.e. when the mean electron velocity equals the proton velocity, to 4 for the highest Z_T.

Integrating the approximations in eq.(4), (5), (6) and (7) with eq. (1) we get the ratio of the yield in a characteristic x-ray peak to the background (SEB or B) for a thick target related to the same value for a thin target with the same incident energy E_o

$$\frac{Y_x}{Y_{SEB}}(\text{thick}) = \frac{r+1-p}{q+1-p} \cdot \frac{Y_x}{Y_{SEB}}(\text{thin}) \simeq \frac{r+1.7}{q+1.7} \cdot \frac{Y_x}{Y_{SEB}}(\text{thin})$$

$$\frac{Y_x}{Y_B}(\text{thick}) = \frac{s+1-p}{q+1-p} \cdot \frac{Y_x}{Y_B}(\text{thin}) \simeq \frac{0.9}{q+1.7} \cdot \frac{Y_x}{Y_B}(\text{thin})$$
(8)

In the region $E_r = E_x \gtrsim T_m$ we may for a rough estimate set $q \sim 2-4$ and $r \sim 7$ so we get $Y_x/Y_{SEB}(\text{thick}) \sim 2 \cdot Y_x/Y_{SEB}(\text{thin})$ and $Y_x/Y_B(\text{thick}) \sim 0.2 \cdot Y_x/Y_B(\text{thin})$, i.e. that the analytical condition for a thick sample is a factor 2 better in the low energy region around T_m, but a factor 5 worse in the high energy part of the spectrum compared to a similar thin sample.

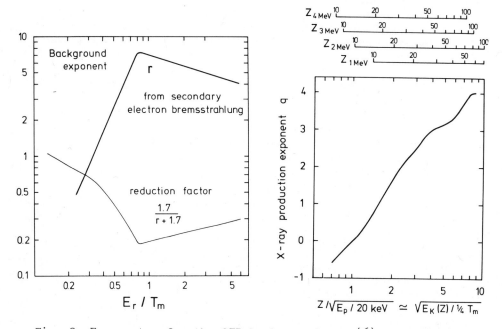

Fig. 3 Exponent r for the SEB background, eq.(6), as function of the radiation energy E_r in the spectrum. $T_m = E_o \cdot 4 m_e/M$. Calculation with the BEA procedure of ref. 1. The reduction factor occurs in eq.(13).

Fig. 4 Exponent q for the energy variation of the characteristic x-ray production from different trace elements (Z) with incident protons of energy E_p, eq.(7). Calculation from BEA theory[2].

Full calculations of the x-ray yields and SEB and B contributions have been made from eq.(1) with the procedures described in ref. 1 for the individual cross sections. With the background radiation taken within a width equal to the FWHM for a semiconductor detector, eq.(1) of ref. 2, we have in fig. 5 shown the concentration of trace elements (Z_T = 10 - 70) which gives a peak to background ratio of 1 for K x-rays from the trace elements. Results are stated for both thick and thin target for 2 and 3 MeV protons incident on carbon and aluminum matrices. Comparing the 3 MeV thick curves with the 2 MeV thin curves in fig. 5 it is seen that the overall sensitivities are similar. Hence the result of the calculations is that the optimum sensitivities for a thick target bombarded with protons of energy E_o is not much worse than those for a thin target, measured with a reduced proton energy $E_o' \simeq 0.8 \cdot E_o$.

The results stated here are derived from calculations of the background from only two processes (SEB and B) and both due to the theoretical nature of this study and due to the neglect of other important phenomena they are only the best possible limits to the capability of the analytical method. It has been found[1,2] that Compton scattering of γ-rays contributes to the background even higher than B for high values of E_o. E.g. for 3 MeV protons on Al this process dominates and it is thus not always possible to state that with a thick target one can raise the energy a little and thereby get a sensitivity as for a thin sample. The complications at high energies can make this procedure practically unacceptable, even if the higher energy can be delivered by the available accelerator. Also interference problems with peaks or tails of peaks for simultaneously occuring trace elements in the matrix have not been commented and in fig. 5 the peak to background ratio gives only part of the answer to the detection limit[1,2]. The total yield, which determines the statistical reliance with which a peak is observed is furthermore for a thick target reduced by the attenuation of the x-rays in the matrix.

2. X-RAY ABSORPTION

On its way out of the sample the x-radiation of energy E_x produced at the position X (fig. 1) is attenuated and only the fraction f can reach the detector

$$f = \frac{\Delta\Omega}{4\pi} \cdot e^{-\rho \cdot d} = \frac{\Delta\Omega}{4\pi} \cdot e^{-\rho(E_x) \cdot t \cdot \cot\theta} \qquad (9)$$

In this expression $\Delta\Omega$ is the solid angle of the detector (in steradian) and ρ the mass attenuation coefficient, in units of area per mass (e.g. cm^2/g), which can be found in the tabulation of Storm and Israel[4] for the matrix material of interest, and which is a function of the photon energy E_x, generally strongly decreasing for increasing E_x.

To perform the integration in eq.(1) including the effect of the absorption in a thick target f from eq.(9) is inserted and t is found as a function of E from eq.(3)

$$t(E) = R(E_o) \cdot (1-(\frac{E}{E_o})^{1.7}) \qquad (10)$$

with E as integration parameter. Also t can be used as variable with

$$E(t) = E_o(1- \frac{t}{R(E_o)})^{1/1.7} \qquad (11)$$

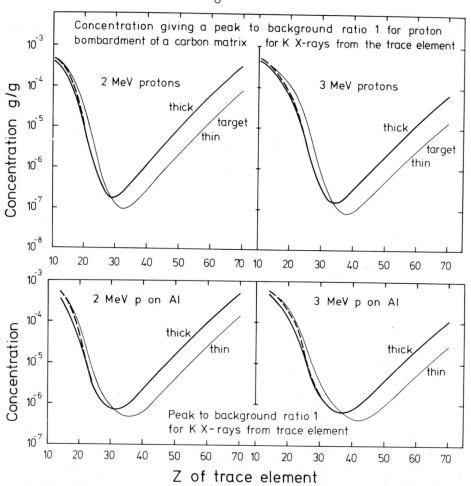

Fig. 5 Calculation of the minimum concentration of a trace element which can be detected if the characteristec K x-ray signal seen with a Si(Li) detector is higher than the bremsstrahlung background. Results are stated for thick targets (thick curve), for thin targets (thin curve) and for thick targets including x-ray absorption for $\theta=45°$ (dashed curve) with 2 and 3 MeV protons on C and Al matrices.

SENSITIVITY IN TRACE ELEMENT ANALYSIS

Using this procedure the absorption was included in the calculations presented in fig. 5 for the C and Al matrices observed under an angle $\theta = 45°$ and the result plotted as the thick dashed curve. This curve follows the thick target curve for high Z (small absorption) but gets close to the thin target curve at the lowest values of Z, which is easily understood as the low energy x-rays are only seen from the outermost layers of the sample and the peak to background ratio then is identical with the thin target value. The region for important absorption, i.e. where the dashed curve begins to deviate from the thick curve is characterized by the condition $\rho(E_x) \cdot R_{mean} \sim 1$, where R_{mean} is the mean depth from which the x-rays are produced and ρ depends on Z_T through E_x. This region is later outlined in table 1.

3. DEPTH INFORMATION

In x-ray analysis it is essential to know from which depth in the sample the x-rays from a certain trace element originate. To estimate the total observed yield we first introduce an effective range R_{eff}, over which a constant cross section $\sigma(E_o)$ can be integrated to give the observed yield. R_{eff} is useful for efficiency determination in applied work and for comparison with thin target yields and is found from eqs. (1), (4), (7), (9) and (11)

$$R_{eff}^{abs}(E_o, Z_T) = \int_o^{R(E_o)} \left(1 - \frac{t}{R(E_o)}\right)^{\frac{q}{1-p}} \cdot e^{-\rho(E_x(Z_T)) \cdot t \cdot \cot\theta} dt \quad (12)$$

Eq.(12) emphasizes the importance of the attenuation term, but as the absorption is a phenomenon which is well known from other x-ray technique we will first examine the ion-part of the integral and later determine in which cases the absorption is of importance. R_{eff} with no absorption is easily found using E as variable:

$$R_{eff}(E_o, Z_T) = \frac{1-p}{q+1-p} \cdot R(E_o) \approx \frac{1.7}{q+1.7} \cdot R(E_o) \quad (13)$$

For physical argumentation the similar mean range is also important. The mean distance along the projectile trajectory from which the x-rays are produced is with no absorption

$$R_{mean}(E_o, Z_T) = \frac{1-p}{q+2-2p} \cdot R(E_o) \approx \frac{1.7}{q+3.4} \cdot R(E_o) \quad (14)$$

With absorption we will find the mean distance from which the x-rays originate and for low Z_T it will be much smaller than R_{mean} in eq.(14). Multiplied with $\cos\theta$ it will give the mean depth perpendicular to the target surface. For $\theta \gtrsim 0°$ or $\theta \lesssim 90°$ one will emphasize the outermost layer of the sample, but especially for low photon energies and for $\theta > 0°$ one is very sensitive to the detailed shape of the surface, as we have in fig. 1 by the relation $d = t \cdot \cot\theta$ in eq.(9) assumed a strictly plane surface.

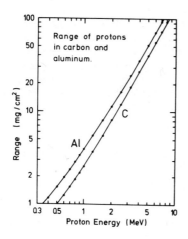

Fig.6 Proton ranges in C and Al given by Northcliffe and Schilling[3].

For typical interesting trace elements with $q \sim 2-4$ eq.(14) tells that the mean distance is 20 % to 30 % of $R(E_o)$. It is thus important for analytical estimates to know $R(E_o)$ which is shown in fig. 6 for C and Al from ref. 3 (or given by eq. (3)) and then reduce this value with a factor ~ 4. In table 1 we have for various proton energies E_o listed these values for C, Al and Ca matrices and have also given the value of Z_T for which $\rho(E_x(Z_T)) \cdot R_{mean}(E_o, Z_T) = 1$ with R_{mean} taken from eq.(14) and $q(Z_T, E_o)$ from fig. 4. This gives the region of Z_T for which the absorption of the x-rays becomes important and for lower trace element it has to be seriously taken into account.

		$R(E_o)/4 \sim R_{mean}$ (mg/cm^2)			Z_T for which $\rho(E_{K\alpha}) \cdot R_{mean} = 1$		
E_o	Matrix	C	Al	Ca	C	Al	Ca
1 MeV		.65	1.0	1.2	12	18	22
2 MeV		2.0	2.9	3.4	15	22	27
3 MeV		4.1	5.5	6.5	17	24	30
4 MeV		6.8	8.9	10.2	18	27	33

Table 1 For different proton energies E_o on a thick target is given a fourth of the range for different matrices ($\sim$ the mean depth from which the x-rays originate) and the Z_T, below which the trace element x-rays are strongly absorbed in the matrix.

4. COMPOSITE MATRIX YIELDS

If the matrix is not consisting of one element as has been assumed in the previous calculations but is a mixture of several major elements (Z_i, A_i) $i=1,N$, each one with the concentration W_i (by weight, normalized so $\sum W_i = 1$ with all summations in this chapter going from i=1 to N), one must for each separate process add the weighted yields

$$Y_{total} = \sum W_i Y_i = \sum f \frac{\sigma_i \Delta t}{A_i M} \cdot W_i \qquad (15)$$

which means that not the cross sections σ_i are directly important, but the ratios σ_i/A_i. However this is also the case in the calculations for fig. 5, where a peak to background ratio is used. In eq.(15) we have for simplicity assumed f independent of i (no absorption) and stated the yield for a thin target of thickness Δt (which also enters into the integral in eq.(1)). Eq.(15) applies e.g. to the proton bremsstrahlung yield

$$Y_B = \sum W_i Y_{Bi} \qquad (16)$$

and to the stopping power of the protons in the matrix

$$\frac{dE}{dt}(E) = \sum W_i \left.\frac{dE}{dt}\right|_{Zi}(E) \qquad (17)$$

with the individual dE/dt given by eq.(2), which means that for these quantities one gets a simple mean value between the results for the various matrix components. However for the most important background contribution from SEB we have to use eq.(15) separately for each of the three involved processes: the stopping power of the electrons, their bremsstrahlung and the production of secondary electrons. With the approximations of ref. 1 for the first two of these processes we then get for the total SEB yield

$$Y_{SEB} = \sum W_i \frac{<Z>}{Z_i} Y_{SEB\ i}$$

where
$$<Z> = \frac{\sum W_i Z_i^2/A_i}{\sum W_i Z_i/A_i} \simeq \sum W_i Z_i \qquad (18)$$

is really close to the mean value of Z - with the approximation of a constant Z_i/A_i ($\sim 1/2$). In the summation of eq.(18) there is the heaviest weight to the lightest elements of the matrix (which give the lowest yields in the interesting region $E_r \geq T_m$), but the result is still mostly some reasonable intermediate value bounded by the occuring separate element yields.

Only the summation of the yields from the background enters into the considerations of optimum sensitivity in ion-induced x-ray analysis as the x-ray yields from the trace elements are only influenced by the complex matrix by the absorption term f (for thick targets). The conclusion from eqs.(16) and (18) is that only the major elements in the matrix are important for the summation and that a reasonable mean-value will be obtained for the total background and thereby for the sensitivities.

The consequence of the different components of the matrix is more dramatic for the attenuation of the x-radiation (of energy E_x) in the sample, because although

$$\rho(E_x) = \sum W_i \rho_{Z_i}(E_x) \tag{19}$$

with ρ_{Z_i} from ref. 4, $\rho(E_x)$ enters into the absorption term $\exp(-\rho(E_x) \cdot t \cdot \cot\theta)$ in eq.(9). This means that even modest amounts of heavy elements can have an effect on the transmission of the x-rays due to the high absorption coefficients.

For the argumentation in this chapter we have assumed that the matrix was a homogeneous mixture of the elements. If grains of strongly deviating composition occur the results will be changed a little. However more serious is the problem of an inhomogeneous matrix for the absorption calculations. They are also especially sensitive to grains at the surface.

Although it is shown in this paper that the theoretical considerations determining the sensitivity for a thin target can be transfered to a thick target without much trouble or major changes of the results, one has to pay attention to several practical problems which get more pronounced with increasing thickness of the sample. The increased counting rate can not always be handled by the detector and the greater energy dissipation which causes heating and possibly damage to the sample may be a severe drawback considering the energy dependence of the stopping power in eq.(2). Also depth dependent distributions of trace elements in the matrix may be encountered, and the quantitative analysis of thick samples is more complicated than of thin ones. However, it is facilitated by a well defined straight line path of the incident protons, which makes the analysis more simple than for electron induced x-rays.

References

1. F.Folkmann, C.Gaarde, T.Huus and K.Kemp, Nucl.Instr. and Meth. 116, 487 (1974)
2. F.Folkmann, J.Phys.E: Sci.Instr. 8, 429 (1975)
3. L.C.Northcliffe and R.F.Schilling, Nucl. Data A7, 233 (1970)
4. E.Storm and H.I.Israel, Nucl. Data A7, 565 (1970)

SENSITIVITY IN TRACE ELEMENT ANALYSIS

DISCUSSION

Q: (B.R. Appleton) You have discussed the importance of knock-on electrons to the bremsstrahlung background. Is there a significant probability that the energetic knock-on electrons will excite characteristic X-rays in the target?

A: (F. Folkmann) There is a certain probability that high energy secondary electrons excite X-rays from trace elements. But their kinetic energy must be higher than the binding energy in the trace element, and there are normally not many electrons floating around with these energies (compared to the number of incident protons). Estimates of the effect have normally given less than a few percent. Also measurements trying to find the importance of this effect and the enhancements have been made with negative result on several samples. The enhancement is a similar secondary source for characteristic X-rays, where X-rays from a high Z-component in the matrix may excite X-rays from lower Z trace elements.

Comment: (W. Reuter) You mentioned that the detection sensitivity in IIXS (in ppm) could be better by as much as 3 orders of magnitude than the sensitivity achievable in electron probe microanalysis. This is not born out by our experimental results. We, (Lurio, Reuter, Ziegler) determined the detection sensitivity of As, Zn, Na and B implanted at a depth of a few hundred Å in silicon using both electron probe micro analysis and IIXS. Both techniques were applied under optimum conditions, i.e. the electron and proton energy was selected giving the best signal to noise ratio. Approximately the same total proton and electron charge was accumulated in both techniques. The result of this study was that at best an improvement of one order of magnitude can be expected in IIXS compared to electron probe microanalysis.

Q: (W.L. Brown) Could you comment on the possible value of synchrotron radiation as a tunable X-ray excitation source of X-ray fluorescence compared with ion induced X-rays?

A: (F. Folkmann) X-rays are traditionally produced by electrons and photons. For electrons physics is very similar to that of incident protons, but the bremsstrahlung background much more intense (3 orders of magnitude higher for 50 keV electrons). For incident photons, sensitivities in concentration can be obtained, which are a little worse than for ions on these targets, but not more than a factor 10. The best situation is to have a monochromatic beam incident on the target. It can be made from a broad band photon source (e.g. an X-ray tube)employing secondary targets as scatterers (reducing the intensity) or from radioactive sources. But whereas photon-excited X-rays penetrate - and thereby normally are applied to large quantities of material (>g, conventional

X-ray fluorescence), the force of ion induced X-ray analysis is, that small samples (less than 1 mg matrix) can be analyzed with good concentration sensitivity. Synchrotron radiation may therefore be used to generate X-rays from large specimens, and a tunable energy (at best within a narrow X-ray band) might emphasize best sensitivity for selected elements.

Q: (L.C. Feldman) Would you compare the sensitivity of ion induced X-ray techniques to X-ray fluorescence?

A: (F. Folkmann) The sensitivity expressed in terms of concentration is similar (ppm) for ion-induced γ-rays and X-ray fluorescence, but there are several important differences between the two related techniques. One is the general form of the spectra, where ion-induced X-rays have the highest intensity at low energies making attenuation by single absorbers very easy and the use of semiconductor detectors very attractive. The photon excited spectra have a high intensity at high energy near that of the incident photons due to Rayleigh and Compton scattering of the beam. Another difference is determined by the beam characteristics, where ions have a short range and can easily be focussed on small areas on the sample, whereas incident photons normally have a longer range and are bombarded on larger areas. This makes ion-induced X-rays best suited for analysis of small samples ($\lesssim$mg) and can make X-ray fluorescence favourable for the study of material, when larger quantities (>g) are available.

REVIEW OF TRACE ANALYSIS BY ION INDUCED X-RAYS

James F. Ziegler
IBM - Research
Yorktown Heights, New York 10598
USA

ABSTRACT

This review concentrates on the use of ion induced X-rays for trace analysis. The step by step procedure for absolute system calibration is outlined. The necessary data analysis for peak identification is reviewed. Several types of experimental chambers are noted, including one for non-vacuum analysis. The difficulties of using heavy ions for trace analysis are summarized.

Introduction

The use of ion induced X-rays for trace analysis has not been as widely accepted as other types of ion beam analysis because it more directly competes with long established analytic techniques: electron and photon induced X-rays. Recently, ion induced X-rays has been shown to be unusually sensitive and quantitative in trace analysis (1). Trace analysis can be defined as the detection of dilute impurities in a material in quantities in the range of .01% or less. The primary advantages ion-induced X-rays have are high cross-sections, high beam currents, and insensitivity to impurity particulate size (for surface impurities). The excitation cross-sections of H^+ beams (2 MeV) is slightly less than that for typical X-ray fluorescence, but the flux density of an ion beam is about 10^6 times that of photons from a radioactive source, and 10^2 that from standard high voltage X-ray tubes. The ion beam excitation cross-section is normally similar to that of electron induced X-rays, but the large bremsstrahlung background is greatly reduced. But one should not concentrate on relative merits as long as the techniques are comparable - the target studies available are so widespread that all techniques may be useful.

Ion Induced X-ray Calibration

The calibration of an ion induced X-ray analytic system is straightforward and can be done in a day or two. We shall show illustrations from the calibration of a 700 keV accelerator system (2) as used for pollution and contamination studies (surface impurities). The discussion of thick target analysis is more complex (4) and will not be included. There are four steps to the calibration: 1) determination of excitation cross-sections; 2) multiplying excitation by fluorescent yield to obtain the number of X-rays produced; 3) evaluation of absorption of X-rays by various windows between the target and the detector; and 4) evaluation of experimental efficiency (detector solid angle, detector efficiency, charge integration, etc). Once this is done, one constructs a graph which can be used to convert X-ray counts directly to surface impurity concentration (at/cm^2 or gm/cm^2).

(1) The purpose of determining the excitation cross-sections is not for direct calculation of yield, but for interpolation between calibration points. That is, we will describe how one might calibrate with Fe and Kr X-rays, and what we need is a way to interpolate between these experimentally determined values. For this purpose, the Binary Encounter Approximation (BEA) can be used (3). Its shape for K-shell ionization is quite good, and for L-shells the shape appears accurate to 30% (the absolute values are off up to factors of 3x for the heaviest element L-shell excitation). A simple polynomial expression, accurate to 1%, has been calculated (4) which allows generation of all necessary excitation cross-section curves.

(2) To determine the X-ray yield, one must multiply the excitation by the fluorescent yield (defined as the ratio of X-rays emitted divided by total excitation). There is no evidence that H^+ or He^+ ion excitation causes any deviation from standard fluorescence tables and these are found in many reports (5).

(3) The calculation of window absorption between the target and the actual detector (including the detector window) is also without problems. The various material thicknesses do not have to be known precisely as the calibration of step (4) will correct for small errors. Tables of absorption coefficients for many materials can be found in various sources (6).

Once the above three steps are taken a curve can be generated such as shown in Figure 1 for the K and L X-rays. The ordinate has arbitrary values and will be determined by the final calibration step.

(4) There are two simple ways to calibrate: to use standard X-ray sources mounted at the target positions, or to use targets

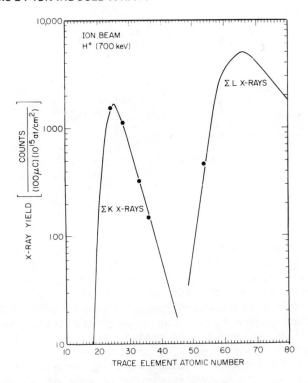

Figure 1. The curves represent the calculations of expected X-ray yield for the experimental arrangement of Figure 2. They include BEA cross-sections of target excitation by 700 keV protons, normal fluorescent yield, and absorption by the various foils between the target and the detector. The ordinate values of the plot were determined experimentally as shown by the data points and as described in the text. (From Ref. 2)

with known impurity concentrations. The latter technique allows a test of the X-ray excitation process as well as detection efficiency and is the better method. The targets can be made by dissolving metallic salts (e.g., $K_2Cr_2O_2$) in water in parts per billion quantities, then slowly drying droplets so that one obtains a residue on a target substrate. As long as the residue is smaller than the beam, one then can relate the detected X-rays to the metal atoms deposited. Standards can also be prepared by ion implantation if such facilities are available. These standards are more rugged and are not harmed by cleaning. Typical implant standards are Ne, Ar, Kr and Xe (which are easy to implant) and Cr, Mo, W to fill in the gaps. Implant doses need to be $\sim 5 \times 10^{15}$ at/cm^2 for accuracy. Once it is determined what the X-ray yield is for these standard targets, the curves generated in steps 1-3 can be normalized as shown in Figure 1. The curves allow interpolation between

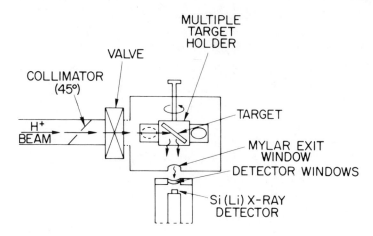

Figure 2. Experimental arrangement. The defocussed proton beam is defined by a collimator which is tilted 45°. The target chamber is insulated from the valve so the entire chamber acts as a Faraday Cup to measure the incident H^+ beam. The targets are arranged around a shaft which can be rotated to introduce each target in turn to the beam. The targets are tilted at 45° to the elliptical beam, so a circular spot is analyzed on the target. The exit window for the X-rays is 75 μm thick mylar film. There is a small air gap between the chamber exit window and the detector entrance window. The detector window is made of 40 μm Al and 25 μm mylar foils. (From Ref. 2)

the data points. The data shown is for 700 keV H^+ ions, using the scattering setup of Figure 2.

This low energy beam creates very little background, and its sensitivity is quite high. If we look at the water residue from 1 cc of water, Figure 3 shows the expected sensitivity for 10 minute exposures. This sensitivity converts to about 10^{-8} to 10^{-9} gm/cm^2 of impurities on the target.

X-Ray Analysis

The details of X-ray spectrum analysis are found in many texts, but the use of computers for general trace analysis is still developing. The advantages of computer techniques for trace analysis lie in the fact that all X-ray peaks have almost identical shapes. By computer fitting of the peaks to a Gaussian it is possible to determine peak energies to better than 10 eV. An example of such peak fitting is shown in Figure 4 (1). This final resolution is so accurate that all but a dozen of the primary X-rays can be identified. A second advantage of computer analysis is the removal of secondary X-ray peaks such as shown in Figure 5. The various L X-rays can be

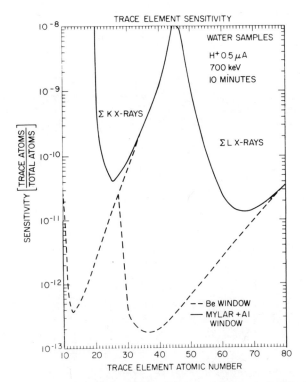

Figure 3. Trace element sensitivity of the residue from the evaporation of one cc of water is shown for a 10 minute analysis using a proton beam at 700 keV. The solid curves are theoretical calculations for the experimental setup of Figure 2, normalized by experimental measurement of targets with known trace atoms in the range of $10^{-8} - 10^{-9}$ in the above units. The dashed curves are calculated limits of sensitivity using 25 μm Be windows instead of the more absorbing windows of Figure 1. The quantitative trace element accuracy is about ± 40%. (From Ref. 2)

stripped from the spectrum once the large L_α or L_β peaks are identified. This is possible because the relative intensities between the various lines has been tabulated (7,8), although there is some disagreement in the values quoted.

Experimental Equipment

The only X-ray detector suitable for general purpose trace analysis is the lithium-drifted silicon detector (called the Si(Li) detector). Both the detector and the associated electronics are made by many manufacturers, and there do not appear to be any particularly important differences between them. The target chambers, however, can vary widely. Figure 2 shows a simple chamber with multiple target holders. It also shows one widely used condition:

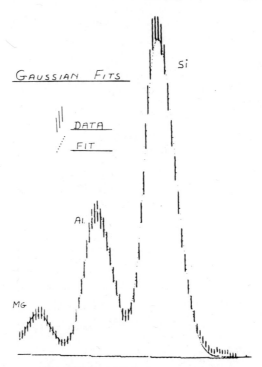

Figure 4. Computer fits of multiple Gaussians to X-ray data. Such fits can determine the X-ray centroid (and hence its energy) to better than 10 eV, allowing identification of almost all X-rays. (From Ref. 1)

A mylar window on the chamber so the detector is not directly coupled to the chamber. This eliminates ground loops, and more important isolates the detector from vibration of the chamber's vacuum pumps. The mylar attenuates X-rays somewhat in the energy range of 800-1400 eV, but this range is not of common interest in pollution and contamination studies.

A second chamber which allows the target to be outside the vacuum is shown in Figure 6. This remarkable chamber takes full advantage of the penetrating power of MeV energy protons. The beam is collimated, then leaves the vacuum through a thin nickel foil (~ 2 mg/cm^2). The chamber is filled with He at air pressure (the He inlet stream is usually directed at the foil to help cool it). The beam will travel over 10 cm in the He allowing the analysis of a wide variety of targets which would be difficult to analyze in a vacuum. The illustration shows the IIX analysis of used lubricating oil, but the chamber is equally adaptable to samples of flora or fauna which tend to disintegrate when subjected to both vacuum and intense radiation.

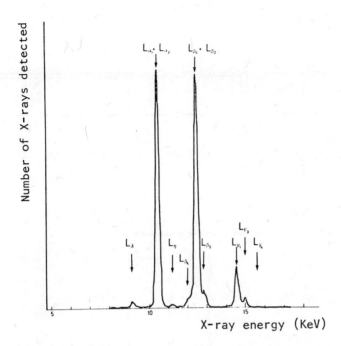

Figure 5. Typical spectrum of L X-rays from H^+ bombardment of Pb. Once the largest peaks are identified, the others may be stripped from the spectrum by a computer. (From Ref. 18)

Using Heavy Ions for Trace Sensitivity

The primary advantage of using heavy ions for analysis should be the possiblity of obtaining very high cross-sections at low projectile energies. In principle the process can be made so selective that one can stimulate X-rays from one impurity in a matrix, without producing any competing X-rays from the matrix (9,10). This phenomenon has created considerable interest, but has been found to be of little value in routine analysis (11). The main problems are:

1) Unexpected backgrounds and peaks appear for various ions on the same matrix. Figure 7 illustrates this for Ar and Kr ions onto silicon. The Kr ions give almost no background, while the Ar ions give a broad high background with an unexpected peak (the $Ar(K_\alpha)$).

2) Significant target damage. Figure 8 shows an electron-microscope picture of a Si surface before and after an Ar analysis at 2 MeV. The undulations are believed to be "blisters" similar to those found in fusion reactor studies.

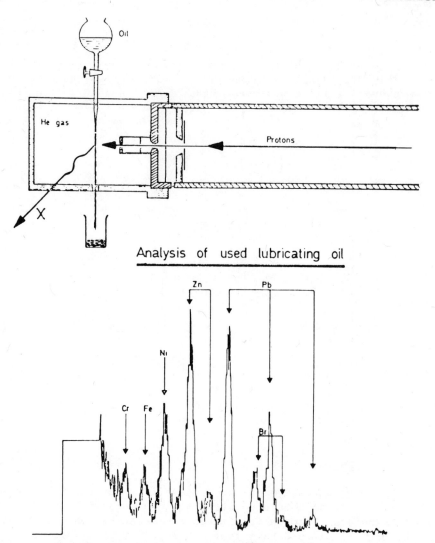

Figure 6. The upper figure shows a scattering chamber used for non-vacuum analysis of used lubricating oil. The beam is collimated, then goes through a thin Ni foil (2 mg/cm^2) into He at room pressure. The beam hits a stream of oil, exciting X-rays. A typical spectrum is shown below, with the major impurity peaks identified. (From Ref. 17)

3) The results are usually not quantitative. Because heavy ions can create cascade showers of substrate atoms, the substrate atoms may contribute significantly to impurity excitation process. If this occurs (and it is common), then the impurity yield will vary with impurity depth. This means impurities on the surface

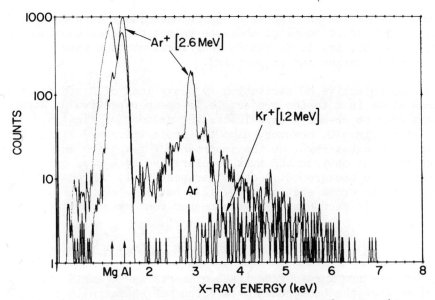

Figure 7. Superimposed are X-ray spectra from Ar^+ and Kr^+ beams bombarding Si. The Kr shows almost no background or peaks. The Ar shows considerable background, and an unexpected peak (the $Ar(K_\alpha)$). There is no way to predict beforehand what the spectrum will be like when energetic heavy ions hit thick substrates. The left-hand Mg and Al peaks are from filters to eliminate the strong Si(K) peak. (From Ref. 11)

(where there is no cascade) will yield fewer X-rays than those deeper in the substrate.

4) There are no rules for the selection of the right ion and energy to obtain high detection sensitivity for a given impurity (11). Usually a considerable number of trials must be undertaken to find high yield and low background.

Summary

The use of various ions for trace analysis is summarized in Table I (reproduced from Ref. 11). For general purpose trace analysis H^+ ions are the choice of most scientists. For thick targets there may be problems because of the deep penetration of these ions, but these can be untangled by detailed analysis (4). He ions are useful if one can simultaneously do nuclear backscattering (NBS). However at low energies (< 2 MeV) the He may not be totally ionized and the theoretical interpolation using the BEA theory may become inaccurate. Heavy ion excitation is useful only for specialized cases where considerable groundwork has been done on the impurity and matrix involved.

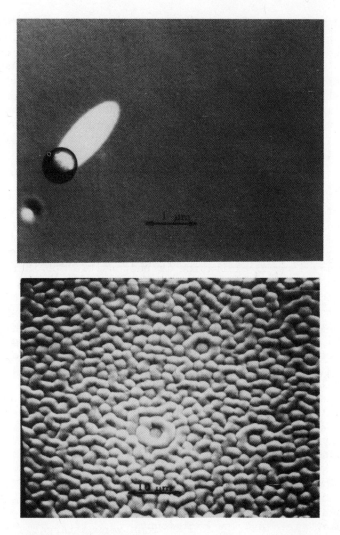

Figure 8. Electron micrographs of Si before bombardment (upper picture) and after bombardment by Ar^+ at 2 MeV ($\sim 10^{17}$ at/cm^2). The lower picture shows considerable target damage, a disadvantage of using heavy ions for analytic studies.

Table II summarizes some of the results of trace sensitivity from various authors (12) (converted to similar units). For very thin targets (where the beam loses almost no energy in the film) sensitivities of 10^{-9} gm/cm^2 appear attainable for experimental running times of 10 minutes, with easily fabricated substrates. For thick targets the sensitivity decreases to about 10^{-8} gm/cm^2. These values can be achieved with only a modest amount of equipment and time.

TABLE I

ION	ADVANTAGE	DISADVANTAGE
H^+:	Sensitivity Large σ Deep probe Quantitative	γ-rays may generate background Large absorption correction
He^+:	NBS Lower abs. correction	Less quantitative than H^+ Lower σ for large Z
Heavy Ions:	σ may be large at low E Shallow probe Selective excitation	Not quantitative Unexpected backgrounds Matrix cascade effects Target damage

TABLE II

Author	Ref	$S(g/cm^2)$	Z	Condition	Substrate
Cahill et al. (Exp.)	(1)	10^{-8}	10-83	16MeV He^+, 10 μC	Thin (0.6 mg/cm^2 mylar)
Folkmann (Theot.)	(13)	10^{-9}-10^{-10}	20-83	3MeV H^+, 100 μC	Thin (1 mg/cm^2 C)
Gilfrich et. al. (Exp.)	(14)	10^{-10}	19,35	5MeV H^+ 100 μC	Thin (20 μg/cm^2 C)
Johansson et. al. (Exp.)	(15)	10^{-11}	22,29	1.5MeV H^+, 9000 μC	Thin (C-Foil)
Duggan et. al. (Exp.)	(16)	10^{-12}(!?)	30	4.5MeV H^+, 1800 μC	Thin (20 μg/cm^2 C)
Ziegler et. al. (Exp.)	(11)	10^{-7}-10^{-8}	5,11,30,33	1.4MeV H^+, 200 μC	Thick (silicon)
Rickards et. al. (Exp.)	(2)	10^{-8}-10^{-9}	26,33,36, 54,80	0.7MeV H^+, 200 μC	Thick (Mylar)

References

1) T. A. Cahill, "Ion Excited X-ray Analysis of Environmental Samples," published in "New Uses of Ion Accelerators", J. F. Ziegler editor, Plenum Press (1975).

2) J. Rickards and J. F. Ziegler (to be published).

3) J. D. Garcia, R. J. Fortner, and T. M. Kavanagh, Rev. Mod. Phys. $\underline{45}$, 111 (1973).

4) W. Reuter, A. Lurio, F. Cardone, and J. F. Ziegler, Jour. Appl. Phys. $\underline{46}$, 3194 (1975).

5) W. Bambynek, B. Crasemann, R. W. Fink, H.-U. Freund, H. Mark, C. D. Swift, R. E. Price, and P. Venugopala Rao, Rev. Mod. Phys. $\underline{44}$, 716 (1972).

6) See for example: R. Theisen and D. Vollath, "Tables of X-ray Mass Attenuation Coefficients", Verlag Stahleisen MBH, Dusseldorf (1967); and B. L. Henke and E. S. Ebisn, Advances in X-ray Analysis $\underline{17}$, 50 (1974).

7) "X-ray Enission Line Wavelengths and Two-Theta Tables", ASTM Data Series DS37A, American Society for Testing and Materials, Philadelphia (1971).

8) E. Storm and H. I. Israel, Nucl. Data $\underline{A7}$, 565 (1970).

9) J. A. Cairns, D. F. Holloway, and R. S. Nelson, Rad. Eff. $\underline{7}$, 167 (1971).

10) J. A. Cairns, A. D. Marwick, and I. V. Mitchell, Thin Solid Films, $\underline{19}$, 91 (1973).

11) J. F. Ziegler and A. Lurio (to be published in Journal of Applied Physics).

12) H. Kamada, T. Tamura, and M. Terasawa, "Sixth International Conference on X-Ray Optics and Microanalysis", page 541, Univ. of Tokyo Press (1972).

12) W. Reuter (private communication).

13) F. Folkmann, C. Gaarde, T. Huers, and K. Kemp, Nucl. Inst. and Meth. $\underline{116}$, 487 (1974).

14) J. V. Gilfrich, P. G. Burkhalter, and L. S. Birks, Anal. Chem. $\underline{45}$, 2002 (1973).

15) T. B. Johansson, R. Akselsson, and S. A. E. Johansson, Nucl. Inst. and Meth. 84, 141 (1970).

16) J. L. Duggan, W. L. Beck, L. Albrecht, L. Munz, and J. D. Spaulding, "20th Conf. on Applic. of X-ray Analysis" (1971).

17) G. Deconninck, J. Radio-Anal. Chem. 17, 29 (1973).

18) G. Deconninck, G. De Mortier and F. Bodart, Atomic Energy Review, 13, 367 (1975).

DISCUSSION

Q: (G. Amsel) In a recent paper J.A. Cooper concluded that ion induced x-rays (2 to 4 keV protons, up to 30 MeV α's) have only disadvantages with respect to photon excitation for environmental sample analysis. Photon excitation requires much simpler and cheaper facilities. What is your opinion on this problem?

A: (J.F. Ziegler) The work of Cooper did not attempt to try for high sensitivity in the ion induced X-ray section of that paper. The portion on ion X-rays was done by Cahill who has since shown that the background present in the Cooper paper can be reduced by almost two orders of magnitude. For a more accurate presentation of the relative merits of the various X-ray techniques I refer you to the paper by Gilfrich (Gilfrich et al. Analytical Chemistry, 45, 2002 (1973).

Comment: (F. Folkmann) To the work by Cooper (Nucl. Instr. Meth. 106, 525 (1973)) I have a few comments. If you consider the spectra shown, it is clear that the best results are obtained for proton bombardment at lower energy (2 and 4 MeV) - except for one of the 2 MeV spectra which has a too high background. The conclusion of the paper reads that although photon excitation give results which are a little worse, it is much more practical or convenient to work with such a source than with an accelerator. Whether this last argument holds is not clear, and at least there is no contradiction to the statements on the sensitivity of ion-induced X-ray analysis given by Dr. Ziegler.

Comment: (W. Reuter) Comments to a question from the floor whether there is any advantage in the use of IIXS compared to photon excitation:
1. In X-ray fluorescence spectroscopy (XRF) thick, complex targets can be analyzed quantitatively, if standards similar in composition to the unknown are available. It is a tedious and often very difficult task to prepare such standards. In IIXS only pure element standards which are usually easily available are needed for the quantitative

determination of all constituents in the sample (W. Reuter, A. Lurio, F. Cardone and J.F. Ziegler, J. appl. Phys., July 1975). This is a very important advantage of IIXS over XRF.
2.) Detection sensitivities in a thick target are comparable in both techniques and are of the order of a few ppm.
3.) The detection sensitivity for thin targets is better by about one order of magnitude in IIXS, in part due to the shallower excitation depth with charged particles in comparison to typical photon ranges in XRF. This judgement is based on experimental data published by Gilfrich and Bircks, U.S. Naval Res. Lab.,who used a variety of X-ray techniques on identical samples.

Comment: (G. Deconninck) Comment to the problem whether high energy particles yield better sensitivity in X-ray analysis. We have experienced 40 to 100 MeV protons and α-particles. The sensitivity is worse than predicted because of nuclear satellites and multi-excitation problems. The best energy range is between 2 and 4 MeV for protons.

THE USE OF PROTON INDUCED X-RAYS TO MONITOR

THE NEAR SURFACE COMPOSITION OF CATALYSTS

J. A. Cairns*, A. Lurio, and J. F. Ziegler
IBM-Research, Yorktown Heights, New York, USA

D. F. Holloway and J. Cookson
A.E.R.E. Harwell, Didcot, Oxfordshire, England

ABSTRACT

This work describes how the technique of proton induced X-rays may be extended to analyse the elemental composition of catalysts in a quantitative manner with high sensitivity. It involves subjecting the catalyst, usually in the form of particles of less than 10 microns diameter and mounted on a thin carbon foil, to bombardment by energetic (few MeV) protons, and observing the characteristic X-rays from the constituent elements. The technique is shown to be complementary to X-ray photoelectron spectroscopy in that, although providing no chemical information, it generates relatively quickly a quantitative identification of the elements in the near surface region and thereby provides information which is of great value in catalyst research. For example, it can be used to differentiate between loss of catalytic metal and simple agglomeration and provides a very sensitive indication of the presence of surface foreign atoms which may cause poisoning effects. It is shown further that the forward scattered protons can be used to identify the light elements present in the catalyst, and to confirm the quantitative accuracy of the X-ray analysis by indicating the catalyst target thinness.

*Permanent Address: A.E.R.E. Harwell, Oxfordshire, England.

1. Introduction

This study seeks to demonstrate that there are special benefits to be gained from the application of the proton induced x-ray technique to the examination of the near-surface elemental composition of catalysts.

Before entering into detail, we may note some relevant background information. A catalyst is a substance which increases the rate at which a chemical reaction reaches equilibrium. When the catalyst constitutes an interface at which the reaction takes place, then it is said to be heterogeneous; an example is the use of a solid platinum catalyst for the oxidation of carbon monoxide gas to carbon dioxide. In fact, an enormous number of metals and metal compounds have been shown to be active catalysts for a huge range of reactions; hence heterogeneous catalysts are used widely in industry, e.g. in petroleum refining, drug manufacture etc.

It is well established that in heterogeneous catalysis, at least one of the reactants must be adsorbed chemically (i.e. chemisorbed) on to the catalyst surface. Hence it is often desirable that such a surface be as large as possible. This may be achieved by using a metal in a high surface area form, such as colloidal platinum, but it is more usual to disperse it over a high surface area support, thereby constituting a supported catalyst. The most widely used supports are alumina, silica and carbon. (Note the common feature here: these are all composed of relatively low atomic number elements. This can be used to advantage, as we shall see).

A further feature of relevance is that catalysts are often composed of a relatively small number of active metallic elements; for example, pure platinum on γ-alumina is an extremely versatile catalyst, but its range of application can be altered by the controlled addition of even one more element, such as ruthenium or rhodium. Similarly, a catalyst can be tailored for more specific application by the addition of trace elements, known as promoters. On the other hand, its performance can be altered adversely by the arrival on its surface of small amounts of undesirable impurities.

Hence there begins to emerge some of the specifications demanded of an analytical technique which may be used to monitor a catalyst preparation procedure, or to examine a catalyst after it has been used for some time. Such a technique must be sensitive to the presence of trace elements. It must be versatile, in that the range of elements of interest covers most of the solid elements in the Periodic Table. It must be a surface technique, but not be over-sensitive to the surface, because the elements of interest may be located within the micropores of a support. Finally, and perhaps most significantly, it must be quantitative, for then the precise role of specific elements in the catalyst kinetics may be elucidated.

At the present time, the two most widely used surface analytical tools in catalysis research are Auger Electron Spectroscopy and X-ray Photoelectron Spectroscopy (ESCA). Both of these, being surface techniques, tend to be rather insensitive to the presence of elements located within porous supports. In addition, it is not easy to render their information quantitative. However, they do constitute a most important means of monitoring actual surface composition, and can provide valuable chemical information, so they should be considered as complementary to the present technique.

We may now consider the special attributes of proton induced x-rays which render it appropriate for satisfying the analytical requirements detailed above. First, it is sensitive: concentrations down to 10^{-6}-10^{-7} by weight and absolute amounts in the region 10^{-9}-10^{-12} gm have been detected. [1] Next, it is versatile, in that it is sensitive to all elements of interest over the Periodic Table. The Si(Li) detector used in this work to detect the x-rays had a thin (0.008mm) beryllium window which tends to cut off the signals from low atomic number elements, but this is a positive advantage here, because these elements are generally of little interest in catalysis. However, if necessary, a forward scattering technique can be used to detect these light elements, as shown later.

In addition, although the spectrum will tend to be dominated by the surface elements, it includes also the near surface ones (i.e. within the first few microns). This has a most important consequence in catalysis: as well as monitoring the catalytic metal, the technique can detect any elemental changes in the composition of the support. Since the support plays a significant role in the mechanism of some important catalytic reactions, this is a most useful facility. Finally, the information can be rendered quantitative as follows: most catalysts are easily crushed to fine powders; if a sample is prepared with an average particle size of a few microns and dispersed on a thin carbonaceous film, through which a proton beam (of a few MeV energy) can pass, then the proton energy loss will be confined to within a few percent. [2]. This concept has been used in trace element analysis of pollution particulates, where it has been demonstrated that quantitative measurements can be made because the technique is insensitive to the particulate size found in their samples [3]. Hence we can obtain clean spectra, generated by protons of fairly well-defined energy. Now although in principle proton x-ray cross-section data could then be used to quantify the elemental composition, it is easier to take advantage of the fact that in catalysis the concentration of the main element of interest is always known. Hence to identify

the concentrations of other elements, we may simply make up a
solution of mixed salts of the main element plus the additional
elements of interest, add an appropriate amount of the crushed
support, and place a smear of the mixture on to a thin support
film as before. Then by measuring the ratio of the various x-rays
of interest, the spectra can be quantified, because we have thereby
eliminated complications associated with uncertainties in solid
angle of x-ray collection by the detector, detector efficiency
variations for different x-rays, proton ionisation cross-sections,
fluorescence yields etc.

The experimental arrangement will now be described, together
with practical illustrations of the application of the technique to
the study of catalysts.

2. Experimental

The results presented here were obtained on two facilities:
the IBM 3 MeV accelerator and the microprobe system of the Harwell
3 MeV IBIS accelerator although in the latter case the small beam
spot facility (down to 4μm) was not used; on both machines the
beam spot was of the order of 1mm diameter. The x-rays were
detected by means of a Si(Li) detector having a resolution at
5.9 KeV of $\sim$ 160 eV.

The powdered catalyst specimens were prepared simply as follows:
a small proportion of the catalyst was crushed manually, then
placed in a microniser air mill, which reduced its average particle
size to $\sim$ 2μm. A portion of this powder was then dispersed in a
suitable liquid (e.g. 2% polyvinyl alcohol) and one drop (0.01ml)
transferred to a 6μm aluminised mylar foil by means of a syringe.
The concentration of the dispersion was so arranged that the amount
of catalyst presented for examination was of the order of 10^{-5}gm,
i.e. 10^{-6} to 10^{-7} gm of catalytic metal (much greater than the
detection limit for the proton induced x-ray technique.) The time
required to accumulate a spectrum was typically 5 mins.

3. Results

As an example of a typical catalyst spectrum obtained by the
use of proton induced x-rays, fig 1 shows nickel oxide on alumina.
Apart from the expected Ni and Al peaks, there are many others.
Since the catalyst was unused, these additional elements were
either catalytic promoters, included by the manufacturer or
undesired contaminants. The background radiation in the lower energy
region of the spectrum is due mainly to bremsstrahlung, arising from
the relatively thick (6μm) mylar backing film on which the catalyst
was mounted.

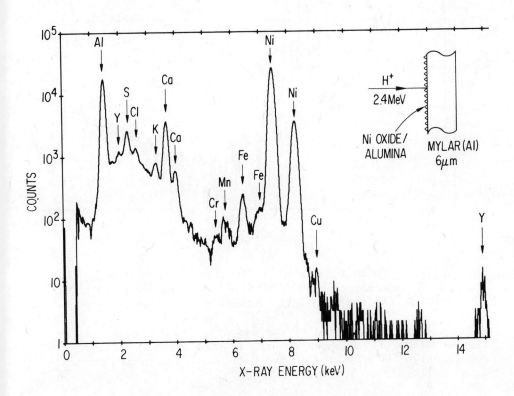

Fig 1   Proton-Induced X-ray Spectrum of Nickel Oxide/Alumina
         Catalyst (used as a hydrogenation catalyst).
         This, and all following spectra were recorded for the
         same proton dose.

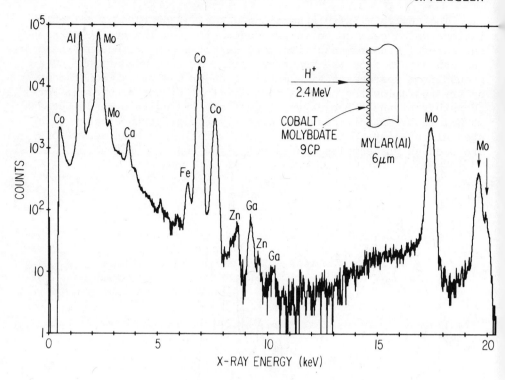

Fig 2   Proton Induced X-ray Spectrum of Cobalt Molybdate Catalyst. (This catalyst is used among other things for the removal of oxides of nitrogen).

As a further example of the simplicity and versatility of the technique, fig 2 shows the spectrum of another catalyst viz. cobalt molybdate obtained in the same way. Impurities present are zinc, gallium and calcium.

As mentioned in the introduction, the standard Si(Li) x-ray detector is relatively insensitive to the presence of x-rays from elements lighter than sodium, because of the absorption of such soft x-rays by its beryllium window. The point was made that this would not be a serious limitation, since such light elements are of little concern in catalysis. However, these light elements can be detected very easily without disturbing the specimen by using the technique of forward proton scattering, [3] [4] by which the proton beam, in scattering from the light nuclei, loses significant energy. This energy loss is a function of the mass of the nuclei, and so if the scattered protons impinge on a surface barrier detector, they are seen to have specific energies depending on the energy lost by collision with the various nuclei in the target.

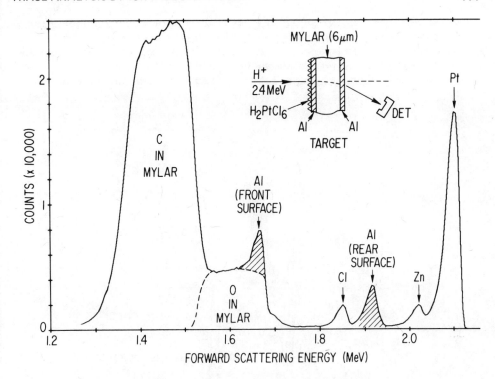

Fig 3   Forward Proton Scattering Spectrum from a drop of
Chloroplatinic Acid on 6μm Aluminised Mylar Foil.
The experimental arrangement is as shown in the
insert (Detector angle = 41°). The carbon and
oxygen of mylar are shown in concentration profile
because of the thickness of the film. The carbon
peak is not flat on top because the oxygen peak
partially overlaps it. The front and back surfaces
of the profiles are reversed from normal Rutherford
backscattering spectra because of kinematics. (Note
the two identified aluminium peaks). The sharpness
of the catalyst peaks indicates it is a thin layer
and therefore the ion induced x-ray cross sections
can be assumed to be constant through the layer.

An example of such a spectrum is shown in fig 3. The light elements in the foil are seen clearly. As others have noted, the sharpness of the peaks indicate that the layers being analysed are thin compared to the 6μm mylar support film. This gives added confidence that the sample being analysed will yield a quantitative measurement of the trace element concentrations.

It was also pointed out in the Introduction that Auger Electron Spectroscopy and Proton Induced X-Rays could be regarded as complementary techniques. This point may be appreciated further by reference to fig 4, which shows the changing Auger spectrum observed as a result of surface segregation effects occurring on a steel

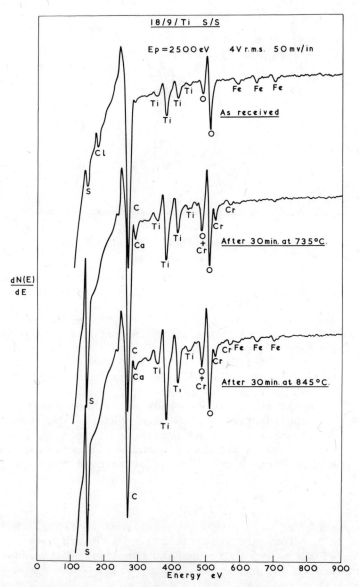

Fig 4   Auger Spectra from steel specimen after various heat treatments.

specimen at various temperatures. If these specimens are examined by the present technique, they give rise to identical spectra (fig 5), illustrating that the technique is insensitive to such surface changes. However, we see that the fig 5 spectrum indicates elements, particularly arsenic, niobium and molybdenum, which do not appear in the Auger spectrum. A total element analysis of the steel specimen was made by x-ray fluorescense and is shown in Table 1, where the two x-ray techniques are seen to yield similar results.

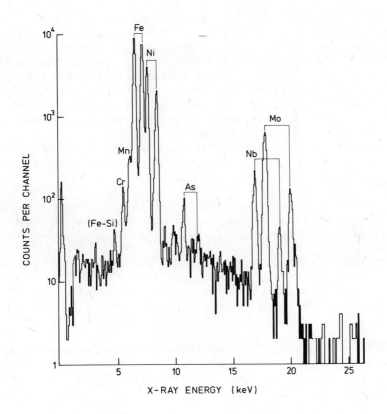

Fig 5   Proton-Induced X-ray spectrum of the same steel specimen as in fig 4. (The Ti X-rays have been cut off by an aluminium absorber).

TABLE 1

TOTAL ELEMENTAL ANALYSIS OF STEEL SPECIMEN (IN PERCENT)

Cr	Ni	Si	Mn	Ti	Mo	Co	Nb	As	Fe
17.8	9.54	0.69	0.69	0.45	0.27	0.10	0.06	0.01	BALANCE

Hence we conclude that (a), a combination of these two techniques permits a differentiation to be made between true surface elements and those in the near surface region; and (b) when using proton induced x-rays to examine catalysts, we obtain elemental information both from the main catalytic elements <u>and</u> from the surrounding substrate. This can be of great value in catalyst assessment. Its application will be discussed elsewhere.

4. Conclusions

This study has sought to demonstrate that the use of proton induced x-rays offers a rapid, sensitive and accurate means of evaluating the near-surface elemental composition of catalysts. It offers useful complementary characteristics to existing analytical techniques. For example, if a catalyst deteriorates in use, it may be due to the arrival of surface contaminant, or to the physical loss of catalytic metal; both of these situations can be monitored readily. Whilst Auger Electron Spectroscopy or X-Ray Photoelectron Spectroscopy could not unambiguously distinguish between surface agglomeration of catalytic metal and its actual loss from the surface, a combination of these techniques with proton induced x-rays could clarify the situation.

Furthermore, the technique is sensitive to most elements, and by using the specimen preparation procedures described here to obtain clean spectra, the information can be rendered quantitative.

By taking a simultaneous forward scattering spectrum one can confirm that the catalytic particulates are small enough to make the analysis quite quantitative and also information can be obtained about light elements (H through Cl) which cannot be seen by the x-ray detector.

Thus we conclude that catalysts, because of their combination of appropriate physical properties, constitute a class of materials which is particularly suited to examination by the technique of proton induced x-rays.

Acknowledgements

Grateful thanks is expressed to Dr. J.C. Rivière for providing the spectra shown in fig 4, and J. Keller and P.S. Drew for valuable experimental assistance.

References

(1) JOHANSSON, T.B. AKSELSSON, R. and JOHANSSON, S.A.E.
    Nucl. Instrum. Meth. $\underline{84}$, 141 (1970).

(2) NORTHCLIFFE, L.C. and SCHILLING, R.F.
    Nuclear Data Tables, $\underline{A7}$, 233 (1970).

(3) CAHILL, T.A. 'New Uses of Ion Accelerators',
    ZIEGLER, J.F. Ed., p.p 1-61 Plenum Press, New York (1975).

(4) JOLLY, R.K. and WHITE, H.B. Nucl. Instrum. Meth. $\underline{97}$, 103 (1971)

DISCUSSION

Q: (J. L'Ecuyer) How can you be certain that you are not loosing any element in the preparation or bombardment of the target?

A: (J.A. Cairns) By comparing the proton induced X-ray spectrum with a total elemental analysis of the catalyst before grinding, one may check that there have been no significant elemental losses. Similarly, by monitoring the proton-induced X-ray spectrum as a function of bombardment dose, it may be confirmed that there are no irradiation-induced compositional changes in the catalyst. In fact, typical doses are of the order of only a few microcoulombs.

Q: (B.R.Appleton) Is there a danger in grinding the catalyst that elements or compounds trapped say gases will be lost?

A: (J.A. Cairns) It is quite true that trapped gases could be released by the process of grinding the catalyst. However, the main purpose of this work has been to show that if the catalyst loses any of its original elements, or picks up impurity elements (such as lead) in use, then these changes can be monitored accurately.

APPLICATION OF PROTON-INDUCED X-RAY EMISSION

TO ELEMENTAL ANALYSIS OF OLIGO-ELEMENTS IN HUMAN LYMPHOCYTES

M.C. Bonnet[+], J.P. Thomas[°], H. Betuel[+], M. Fallavier[°]
and S. Marsaud[°]
+ - Centre de Transfusion Sanguine de LYON - FRANCE

° - Institut de Physique Nucléaire - VILLEURBANNE-FRANCE

SUMMARY

The proton-induced X-ray emission which represents a very sensitive multi-elemental method of analysis was used for the measurement of oligo-elements in human peripheral lymphocytes. The beam homogenized technique with an aluminium diffuser was employed. Quantitative analysis was obtained by two methods : external standards as thin evaporated layers of elemental Mn, Fe, Ni, Cu, Zn and secondly by internal standards with the use of Yttrium nitrate solution. In lymphocytes, besides the more common P, S, Cl, K, Ca elements, Mn, Fe, Ni and Zn were simultaneously determined. Though problems of contamination by the native medium and erythrocytes were encountered, the values of Fe and Zn were high and could be confidently assayed, in spite of a great dispersion.

INTRODUCTION

Great significance is attached to the dosage and analysis of oligo-elements in human blood on account of their role in biological processes. The proton-induced X-ray emission is specially suitable for this type of problem. It allows a multielemental analysis of great sensitivity due to the high values of the X-ray production cross section, though it is limited by a high background.

Different authors (1-4) have applied this technique to biological samples. The values obtained by Bearse et al. on a study on whole blood compare favourably with those given by atomic absorption spectrometry (5).

For leucocytes, data (6-8) are available concerning Zn, in particular ; however, to our knowledge no systematic study has been undertaken on lymphocytes. The aim of this paper is to present the results obtained in lymphocytes and to underline the problems raised at different levels of the analysis.

MATERIAL AND METHODS

1 - Irradiation-Detection System

Irradiations were done through the 4 MeV VDG accelerator of the Institute of Nuclear Physics, LYON, with a nominal energy of 2 MeV protons.

A constant homogeneous beam was achieved by interposing a diffuser rather than by defocusing the beam. With an aluminium sheet of 7.5×10^{-4} mm (E_i = 1, 975 KeV) placed at 1,125 mm from the targets, intensities around 10 nA could be maintained for periods of 3 to 4 hours. The upper limit is reached at 20 nA which entails target destruction. Two carbon diaphragms, 2 and 3 mm diameter and 50 mm apart, placed at the entry of the scattering chamber (10) prevent the diffusion of the beam on the lining of the target holder. The efficiency of the device is tested with the whole target-holder, including therefore the target frames, but if the targets are absent the background obtained is negligible. In the experimental conditions described, the detector Si(Li) (25 mm^2 - 3 mm) located at an angle of 90° to the incident beam has a resolution of 220 eV. Besides the beryllium window of the detector (25 µm) the X-ray detected must pass through the 9 µm Kapton window of the scattering chamber and through 10 µm of Aluminium absorber. The air thickness can be estimated at 9 mm. An opto-electronic preamplifier was used with a 6 µs shaping time constant ; signals were fed via an ADC to a HP 2116C computer. The different spectra are registered on magnetic tape, to be processed off-line.

2 - Samples and Standards

Pure suspensions of lymphocytes are prepared sterile in disposable plastic material. Healthy volunteers were bled on preservative free heparin (Roche), 2.5 mg for 20 ml of blood. After passage at 37° C on a column packed with nylon fiber which retains granulocytes and platelets, the filtrate is sedimented at 20° C in the ratio of 9:1 with a 5% saline solution of Dextran-500. Residual erythrocytes are lysed for 10min with 2.5 vol. of ammonium chloride solution (NH_4Cl 8.7 g/l). After centrifugation at 4° C at 250 g for 10 min, the pellet is washed **twice** in a saline solution (NaCl 9 g/l) and numbered. In the samples used the contamination in erythrocytes did

not exceed 3 per cent. The final lymphocyte pellet is suspended in
200 μl of saline and layered on a Millipore filter connected to a
vacuum pump. After drying, the targets are placed directly in the
scattering chamber. This mode of preparation avoids inconsistencies
due to the medium and yields homogeneous samples. Homogeneity is
checked both by microscope and by focusing the beam of protons so as
to make several impacts on the same target. Standard targets were
obtained by thermal evaporation on Millipore support, in elemental
form for Mn, Fe, Ni, Cu and in compound form for Zn (Zn F2). For
quantitative analysis with an internal standard, Yttrium was incorporated in the cells suspensions by addition of 20 μl of Yttrium
nitrate at 10^{-2} g/ml.

3 - Experimental Procedure

As regards quantitative analysis, the principle consisted in
obtaining spectra of simple elements from the standards and from
these reconstructing the spectra of the samples. The background inherent to the electronic bremsstrahlung in the target and the support has to be **subtracted** from or fitted to the spectrum of the
support itself. The operation is carried out by a computer program
reconstructing the spectrum out of these different components. Allowance is made to introduce the background shape calculated or experimental. The ratio of the peaks in the spectrum to those in the
standards is directly obtained with the reconstructed spectrum. With
this procedure, not only is the treatment automatized but it is also possible to choose the best possible background shape and to make
the requisite deconvolutions between peaks of adjoining elements.
Various authors have proposed (1, 11) an alternative to this use of
external standards. It consists in adding to the target investigated a known quantity of an heavy element not expected to be initially present. As regards our targets the validity of the method requires this element to be fixed in proportion to the number of cells
layered. The result is apparent on figure 1 which shows a good linearity with nevertheless an appreciable quantity bound to the support.
The general use of this method entails the setting up of an efficiency-curve of the detection system which must remained fixed.

RESULTS AND DISCUSSION

1 - Erythrocytes

The validity of the method was checked on red cells which have
been well studied (6, 9). The majority of detected elements is shown
in the spectrum of figure 2, typical of a normal individual. The
data obtained reveal some disparity which can be attributed to the
difficulties inherent to the mode of preparation. A first series of

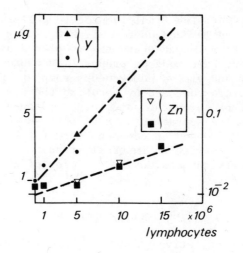

Fig. 1 - Linear fixation of Yttrium on the lymphocytes of two individuals (▲, •) and the respective Zn determination (▽, ■) from this internal standard.

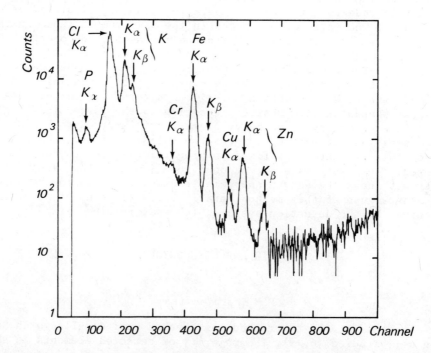

Fig. 2 - X-ray spectrum for human erythrocytes with a 10 μm Al absorber.

results established from a high number of cells (between 40 and 70 x 10^6) give absolute values for Fe and Zn notably inferior to those published : for 1 x 10^6 erythrocytes the mean value for Fe is 0.10 µg ; for Zn it is 1.4 10^{-3} µg. The values for Fe range between 5 x 10^{-2} and 1.2 x 10^{-2} µg for 1 x 10^{+6} erythrocytes and for Zn between 6 x 10^{-4} and 1.8 x 10^{-4} µg. The reason for these divergent results is not clear. A tentative explanation is the variability brought by an important layer of cells on the targets : when the number of cells increase absorption of X-rays is no longer negligible. It may also be due to our technique, it has not been compared with the internal standard method. However the mean ratio Fe:Zn is 72±15 which compares favourably with reported results standing between 60 and 100. This mode of preparation has its limitations for a low number of cells of the order of 10 x 10^6, because it is difficult to avoid the destruction of red cells which lose their hemoglobin, leading thereby to a patent decrease in Fe. In this situation, for Zn which is located in the cell membrane, the results are more valid : 1 to 1.5 x 10^{-3} µg for 1 x 10^6 cells and close to data previously reported (6). A confirmation of this type of error is brought when the amount of Fe is assessed in the low number of erythrocytes which contaminate lymphocyte suspensions.

2 - Lymphocytes

The different causes of error contributed by the support, the suspension medium and the contamination by erythrocytes have been examined. Among the elements investigated, the only one which can be determined in the Millipore filter is Cu. Even values inferior to 10^{-3} µg/cm^2 interfere significantly on the spectrum of the cells and invalidate the determination of this element in the cell suspensions. The saline suspension medium for the cells was filtered on Millipore and analyzed by the same procedure. The values for Fe and Zn which do not exceed respectively 10^{-2} µg/cm^2 and 5 x 10^{-3} µg/cm^2 represent a limitation to the sensitivity of the method.

Considering the content of Fe in erythrocytes a correction must be introduced, since this element may reach 30% of the amount under analysis. Such a correction is negligible for Zn.

Apart from Fe, Zn and the more current elements : P, S, Cl, K, Ca, the presence of Mn, Ni and Cu is also apparent in the spectrum on figure 3. The presence of an element at a high percentage, Fe with respect to Mn, or the presence of an element in the support (Cu) impair the quantitative analysis of these elements : Mn$\leqslant 10^{-3}$ µg, Ni$\leqslant 10^{-4}$ µg, Cu$\leqslant 10^{-4}$ µg for 1 million lymphocytes.

To determinate the accuracy of the method for Fe and Zn, a more discriminative test is shown on figure 4. Increasing amounts of lymphocytes from the same donor were measured. The dose-response curve

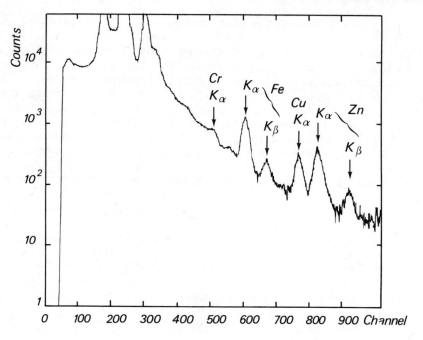

Fig. 3 - A typical proton-induced X-ray spectrum for human peripheral lymphocytes.

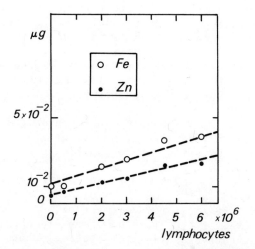

Fig. 4 - Simultaneous determination of Fe and Zn in the lymphocytes of the same individual using external standards.

is linear, within the experimental uncertainties ; no response is obtained for a number of lymphocytes equal or inferior to 5×10^6 /cm^2 which may represent the lower limit of experimental sensitivity of this method. The contamination in Fe and Zn brought by the medium is represented on ordinate zero and must be substracted for each sample. A high number of cells superior to 30 millions, besides the large amount of blood it implies, leads to erroneous results. An average of 15×10^6 cells/cm^2 seem to represent an optimal amount. The determination of Fe and Zn in lymphocytes was performed on 10 random blood donors. The average value for Fe is 9.9×10^{-3} µg for 1×10^6 lymphocytes with a dispersion of 70 % and for Zn 4.7×10^{-3} µg for 1×10^6 lymphocytes with a dispersion of 60 %. The ratio Fe/Zn attains 2.4 (with a less important dispersion of 25 %), both elements being determined simultaneously.

Considering the different causes of error : comparison with standards, statistic calculations, subtraction of background, the accuracy of an individual measurement is close to 50 % of the content measured. Experimental precision is essentially controlled by the accuracy in the standards and their resistance to the intensity of the beam. In this respect it appeared that plastic backings are more suitable than Millipore. The accuracy for the determination of the thickness of these standards is close to 10 %.

At the outcome of this first investigation, it is clear that the detection should be improved. No basic modification is, for the moment, apparent in the choice of backing imposed by the nature of the material to be analyzed. Thinner filters of Millipore are more frail, easily soiled by settlings and are often more contamined especially by Cr. The decrease in background due to electronic bremsstrahlung in the target is not susceptible of improvement except by back angle detection (10). To maximize the rate of counts, the geometry of the system of detection could be appreciably changed as well as the choice of the absorbers essentially for reducing P, S, Cl, K, Ca. Finally a more precise study on the fixation conditions of Yttrium or other heavy elements (11) should be undertaken in view of the results already obtained. As regards the preparation of the samples themselves, the use of filters is imperative since it is compulsory to separate the cells from their native medium which is a major source of contamination.

CONCLUSION

The results of this study show the exploitation offered in the analysis of lymphocytes by the technique of proton-induced X-ray emission. As regards the amount detected of 10^{-9} or 10^{-10} g for Fe and Zn the accuracy of the determination can be better than 50 % provided that some improvements are brought at the level of detection. Considering the major role played by these cells in immunological processes and in the defense of the body these results are encouraging and deserve to be confronted with other methods.

REFERENCES

1 - VALKOVIC V., LIEBERT R.B., ZABEL T., LARSON H.T, MILJANIC D., WHEELER R.M. and PHILLIPS G.C.
   Nucl. Instr. Meth., 114, 573, 1974

2 - YOUNG F.C., ROUSH M.L. and BERMAN P.G
   Int. J. appl. Rad. Isot., 24, 153, 1973

3 - WALTER R.L., WILLIS R.D., GUTKNECHT W.F. and JOYCE J.M
   Analyt. Chem., 46, 843, 1974

4 - KUBO H.
   Nucl. Instr. Meth., 121, 541, 1974

5 - BEARSE R.C., CLOSE D.A., MALANIFY J.J. and UMBARGER C.J
   Analyt. Chem. 46, 499, 1974

6 - VALLEE B.L and GIBSON J.G, 2 nd
   J. Biol. Chem., 176, 445, 1948

7 - DENNES E., TUPPER R. and WORMALL A.
   Biochem. J., 78, 578, 1961

8 - FRISCHAUF H., ALTMANN H. and STEHLIK G.
   Proc. 9th Congr. europ. Soc. Heamat., Lisbon 1963 (S.Karger, Basel/New York, 1963), 234.

9 - BOWEN H.J.M.
   Trace elements in Biochemistry (Academic Press) London/New York, 1966.

10 - THOMAS J.P.
    Thesis, LYON, 1974

11 - ISHII K., MORITA S., TAWARA H., CHU T.C., KAJI H. and SHIOKAWA T.
    Nucl. Instr. Meth., 126, 75, 1975.

DISCUSSION

Q: (M.A. Chaudhri) What were the special advantages of using this technique as compared to the standard techniques of neutron activation analysis and even charged particle activation analysis? These techniques are no more complicated or time consuming and could have provided even better sensitivities in some cases.

A: (J.P. Thomas) The proton induced X-ray emission method was chosen because the oligo elements contained in lymphocytes are not known. The advantages offered by our technique are: the multielementarity associated with a high sensitivity for all elements with $Z \geq 12$, it is also a non-destructive analysis.

Neutron activation analysis and charged particle activation analysis often require a long period of irradiation followed by chemical separation in order to obtain a good sensitivity especially as regards Fe and Zn. Moreover neutron activation analysis may be hindered by the amount of Cl contained in the cell suspensions.

ELEMENTAL ANALYSIS OF BIOLOGICAL SAMPLES USING DEUTERON INDUCED
X-RAYS AND CHARGED PARTICLES

L Amtén, L Glantz, B Morenius, J Pihl and B Sundqvist

Tandem Accelerator Laboratory

Box 533, S-751 21 Uppsala, Sweden

ABSTRACT

A method for elemental analysis in biological samples is presented. Samples were exposed to 4 MeV deuterons and both charged particle and X-ray spectra were recorded. The method was tested on a biopsi of mouseliver- and a blood-serum-sample and was found promising for rapid analysis of samples associated with small amounts of material (µg).

1 INTRODUCTION

In recent years much work has been devoted to elemental analysis using charged particle beams. The method of using proton induced X-rays [1] has turned out to be a powerful tool in several cases. However information on light elements is hard to extract with that method. Therefore the combined analysis of nuclear scattering and charged particle induced X-ray production is an attractive method to determine the relative abundances of a large number of elements in a sample in one measurement. The main idea is to cover the low-Z elements with a nuclear scattering technique thus complementing the now rather well established method for analysis of high-Z elements (Z > 14) using charged-particle induced characteristic X-rays. The possibility to use the elastic scattering of protons and alphas has been investigated. The general experience is, however, that it is difficult to combine these methods with X-ray analysis. To get enough elemental separation in the particle spectra the energy of the bombarding particle has to be so high that the "bremsstrahlung" background in the X-ray spectra becomes a serious problem.

An alternative method, using (d,p) and (d,α) reactions in the low Z-elements carbon, nitrogen and oxygen, was therefore studied. Due to the positive Q-values of these reactions (e.g. 8.6 and 13.6 MeV respectively for nitrogen) rather low deuteron energy can be used. This has the advantage of a low "bremsstrahlung" background in the X-ray spectra.

2   EXPERIMENTAL ARRANGEMENTS

Deuterons of 4 MeV from the Uppsala EN-tandem van de Graaff accelerator were used in the experiment. The beam was defined by two collimators ($\emptyset$ = 2 and 1 mm) followed by an anticollimator ($\emptyset$ = 1.5 mm). About 30 nA beam current was used in the experiment. A Si(Li) X-ray detector was placed at 90° relative to the direction of the incoming beam and outside the chamber behind a 100 μm window of mylar at a distance of 15 cm from the target. The detector itself has a 7.6 μm window of beryllium. A Si(Sb) particle detector (area 450 mm^2, thickness 1 mm) was placed at 165° relative to the direction of the beam and at a distance of 13 cm from the target. To prevent backscattering from the Faraday cup, it was placed at a large distance from the detectors and a collimator was placed between the cup and the target.

The detector signals were handled with standard electronics. Particle and X-ray spectra were recorded at the same time with a PDP-15 system.

3   TARGET PREPARATION

In the experiment two different samples were used namely biopsi of mouseliver and blood-serum. The targets, which should be less than a few hundred μg/cm^2 thick and reasonably homogenous, were preparated in the following way. Backings of about 80 μg/cm^2 were prepared of polyethylene using a wellknown technique [2] and the liquid samples were dropped onto glass slides covered with the backing. The glass slides with backing and sample were then freeze dried. After this treatment the targets were floated off onto target frames in water. In the actual cases the amount of sample in the targets was of the order of μg.

4   RESULTS AND DISCUSSION

The spectra recorded for blood-serum and biopsi of mouseliver are shown in fig 1.

In table I the main peaks in the particle spectra are identified. Particle spectra were used to get measures of the abundances of

ELEMENTAL ANALYSIS WITH DEUTERON INDUCED X-RAYS

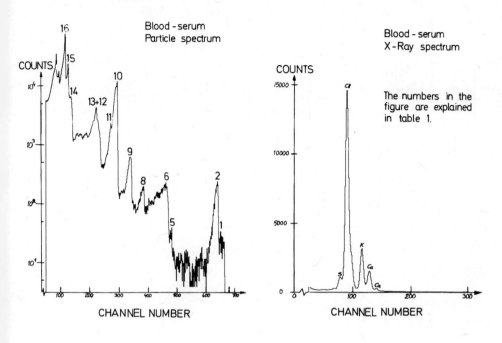

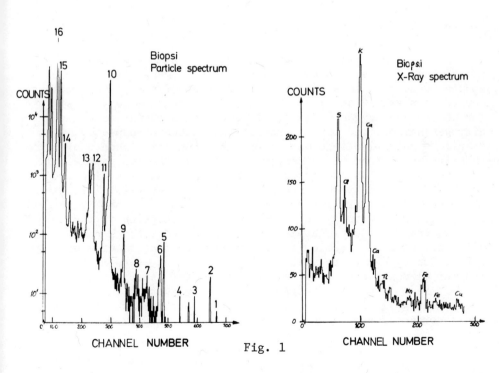

Fig. 1

Peak no	Reaction	Peak no	Reaction
1	$^{14}N(d,\alpha)^{12}C$	9	$^{14}N(d,p)^{15}N$ (1st and 2nd exc st)
2	$^{14}N(d,p)^{15}N$	10	$^{12}C(d,p)^{13}C$
3	$^{32}S(d,p)^{33}S$	11	$^{16}O(d,p)^{17}O$
4	$^{32}S(d,p)^{33}S$ (1st exc st)	12	$^{16}O(d,\alpha)^{14}N$
5	$^{13}C(d,p)^{14}C$	13	$^{16}O(d,p)^{17}O$ (1st exc st)
6	$^{14}N(d,\alpha)^{12}C$ (1st exc st)	14	$^{16}O(d,d)^{16}O$
7	$^{32}S(d,\alpha)^{30}P$	15	$^{14}N(d,d)^{14}N$
8	$^{32}S(d,\alpha)^{30}P$ (1st exc st)	16	$^{12}C(d,d)^{12}C$

Table I Identification of the main peaks in fig 1.

carbon and oxygen relative to nitrogen. The number of counts in peaks 1 and 2 were used as a measure of nitrogen, the number of counts in peak 10 as a measure of carbon and finally the number of counts in peaks 12 and 13 as a measure of oxygen. The values given for carbon and oxygen in table III are ratios of the carbon and nitrogen numbers and oxygen and nitrogen numbers respectively and therefore only give information on the amount of carbon and oxygen relative to nitrogen in a comparison between the two samples. To get the relative abundances of these elements the reaction cross sections have to be taken into account in the analysis. This has not been done so far. The number of counts in the peaks in the X-ray spectra however were corrected for absorption in the windows and for the probability of ionization of the K-shell in different atoms. In table II the correction factors used are shown. Consequently the values in table III for sulphor and heavier elements also give information on the relative abundances of different elements detected in the X-ray spectra in one sample. The values given in table III for these peaks are integrated peak areas corrected according to table II and divided by the nitrogen number.

S	Cl	K	Ca	Ti	Mn	Fe	Cu
.77	1.00	1.29	1.21	1.00	0.70	0.62	0.42

Table II Correction factors for the peaks in the X-ray spectra.

Element \ Sample	Blood-serum	Biopsi
C	38.8±0.1	1270±10
N	1.00±0.02	1.00±0.10
O	23.1±0.1	45.3±0.3
S	0.2±0.01	2.2±0.1
Cl	9.3±0.2	0.29±0.06
K	1.01±0.03	1.8±0.2
Ca	0.52±0.02	1.5±0.1
Ti		0.09±0.01
Mn	0.004±0.001	
Fe	0.012±0.0004	0.75±0.02
Cu	0.010±0.002	0.19±0.05

Table III The preliminary analysis of this experiment. The first three measures are taken from the particle spectra and the others from the X-ray spectra. The uncertainties are due to counting statistics only.

The investigated method was compared with two accepted techniques for multi-elemental-analysis namely neutron activation and spark ion source mass spectroscopy. The same amounts of material used in the analysis described above i. e. liquid samples of 2-3 µl (2-3 mg) were also used with these two accepted methods. The general experience was that such quantities are too small for these techniques. However for the elements detected with all three methods the results agree within the estimated uncertainties of the different methods. For heavier elements neutron activation showed a somewhat higher sensitivity than the described method. As can be seen in the X-ray spectra in fig 1 the background is relatively high and sets a limit for the sensitivity to detection of heavy elements. It is not yet clear if this is due to Compton scattered γ-rays from nuclear reactions induced by the deuterons or if the beam collimators produce an X-ray background in this preliminary experimental set up. This will be further investigated using collimators made of a light element material like e.g. carbon.

In the future work on this method the intention is to calibrate the method using known chemical nitrogen compounds as targets. With such a procedure it will be possible to give the abundances of detected elements relative to nitrogen in an unknown sample. In conclusion the advantages of the presented method are considered to be:

a) Small amounts of the material to be analyzed are needed (∼ μg)

b) The targets are easily prepared with no complicated chemistry involved.

c) The analysis time is short (∼ 10 min)

d) It is a multi-element-analysis determining the content of elements relative to nitrogen which is of special interest in biological samples.

REFERENCES

1 T B Johansson, R Akselsson and S A E Johansson, Nucl Instr Meth 84 (1970) 141

2 S Matsuki, M Yasue and S Yamashita, Nucl Instr Meth 94 (1971) 338

DISCUSSION

Comment:(G. Amsel) I just want to point out that the statement of reduced sensitivity for low Z nuclei due to lower nuclear reaction cross sections is incorrect. Contrary to ion induced X-rays nuclear reactions yield practically backgroundfree detection. Typical sensitivities are much below a monolayer.

A: (L. Amtén) The cross section for nuclear reactions is of the order 100 less than those for the induced X-rays. Of course, this makes the particle spectra less sensitive. The spectra I have shown you today show that the particle spectra are not backgroundfree.

Q: (R.D. Vis) What sensitivity is obtainable for the lighter elements? The cross-sections are of course much lower if you use nuclear reactions instead of X-ray production.

A: (L. Amtén) The "accepted" methods havn't given any results for these low Z-elements, so I don't know how many ppm of O or N there really are in the sample. The cross section is however of the order 100 times smaller for nuclear reactions than for induced X-rays.

Q: (M.A. Chaudhri) Did you try to measure the induced activity in your foils after irradiations to explore the possibilities of using charged-particle activation for elemental concentration. Even with 4 MeV deuterons a no. of measurable activities are induced in different elements so that quantitative estimation of the concentrations of these elements may be quite possible.

A: (L. Amtén) As I said in my talk, this is the first and therefore very preliminary setup with deuteron beam. Such a measurement that you suggest, hasn't been done. However, it is my opinion it would'nt be successful. As I said we have tested the reported method against neutron activation and the experience was that is was too small amount of sample for this method. One advantage of the reported method is that short time is required. I suppose that adding your suggested measurement one will need a lot of time.

SUPPRESSION OF RADIOACTIVE BACKGROUND IN ION INDUCED X-RAY ANALYSIS

H. Sobiesiak, D. Heck, F. Käppeler

Institut für Angewandte Kernphysik

Kernforschungszentrum, 75 Karlsruhe, P.O.B.3640, Germany

ABSTRACT

This paper describes a method to reduce the decrease in sensitivity of ion induced x-ray analysis by the presence of radioactive background. Measurements were made using a pulsed proton beam of 5 nA average current with a pulse width of 10 nsec and a repetition rate of 250 kHz. With typically 30 nsec time resolution of the Si(Li)-detector a background suppression of about 40 : 1 was achieved. An improvement of this value is possible by using a smaller repetition rate and a narrower time window.

INTRODUCTION

The excellent sensitivity of ion-induced x-rays for the detection of trace elements can be decreased by the presence of interfering background radiation. Such background might arise from radioactive samples or from simultaneous ion implantation. In such cases use can be made of pulsed ion beams as they are available at many electrostatic accelerators. Then, the induced x-rays are time-correlated with the ion pulse and can be discriminated by an appropriate time window against most of the background radiation which is time independent. This possibility was demonstrated at the Karlsruhe 3 MV Van-de-Graaff accelerator where proton pulses of 10 nsec width and repetition rates down to 125 kHz are available.

EXPERIMENTAL SET-UP

The experimental set-up is schematically shown in Fig. 1 together with the block diagram of the electronics. In addition to the conventional arrangement of sample, detector and Faraday cup

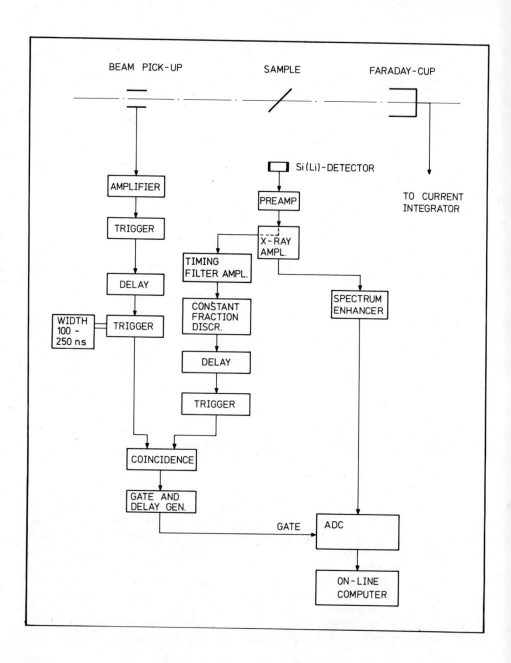

Fig. 1 Schematic view of the experimental arrangement and the electronic block diagram.

the beam passes through a pick-up electrode from which the signal for the time window is derived. After the beam pick-up electrode the proton beam passes through the sample where the x-rays are induced. It is stopped in a Faraday cup to monitor the integrated current.

The x-rays are detected by a Si(Li)-detector which is positioned perpendicular to the beam. After preamplification (Canberra 1708) and the first differentiation stage of the x-ray amplifier (Canberra 1713) the signal is split. The energy branch is fed through the spectrum enhancer (Canberra 1764) to the ADC (ND 2200). The time branch passes a timing filter amplifier (Ortec 454) and the time signal is derived from a constant fraction discriminator (Ortec 463). The time information of the beam pick-up is treated by a fast amplifier, two trigger stages and a suitable delay. Only if the beam pick-up signal and the x-ray detector signal have the correct time correlation which is checked by a fast coincidence unit, the gate of the ADC is opened. The width of the time window is adjusted by the variable width of the second trigger output, while its position is defined by delay lines.

MEASUREMENTS

To measure the timing behaviour and time resolution of the system the coincidence unit was replaced by a time-to-amplitude converter. The results are shown in Fig. 2. The targets were pure samples of Al, V, Fe, Cu and one sample of a Ca-S-compound. For higher energies the time resolution of the system is about 20 nsec FWHM. Besides the deterioration of the time resolution with decreasing energy also an energy dependent walk is observed, which could not be eliminated totally. The small peaks at the right end of the time distributions originate from γ-rays emitted after proton induced nuclear reactions and give the high energy limit of the walk.

If one is interested only in x-rays of energies larger than $\gtrsim 5$ keV a correct positioned time window of ~ 50 nsec is sufficient. (If also aluminum x-rays are to be observed, the window must be enlarged to about 200 nsec). Together with a pulse repetition rate of 125 kHz a suppression factor of 160 is obtained for the time independent background.

This is demonstrated in Fig. 3 which shows the energy spectra of a sample containing 2 µg Cu, 1 µg P, 1 µg S, 1 µg Br and 1.7 µg K.

To simulate the time independent background a 100 mCi γ-ray source of ^{137}Cs was positioned at a distance to produce a count rate of 10^3 counts/sec in the Si(Li) detector. The irradiated area of the sample amounted to 1.5 cm^2, at a beam current of 5 n Amp. and a total accumulated charge of 5 µCb. The pulse repetition rate for

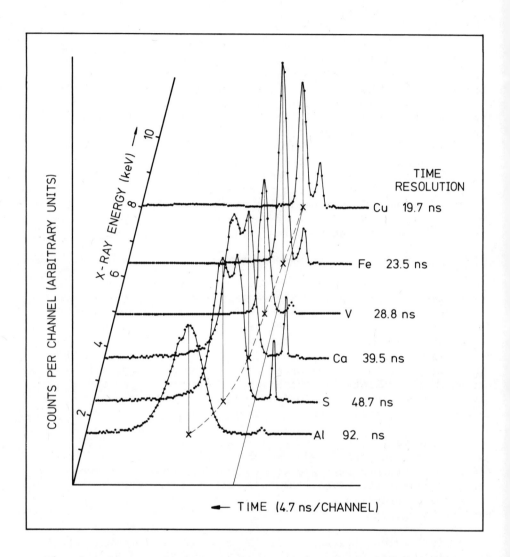

Fig. 2 Time resolution of the system as a function of x-ray energy.

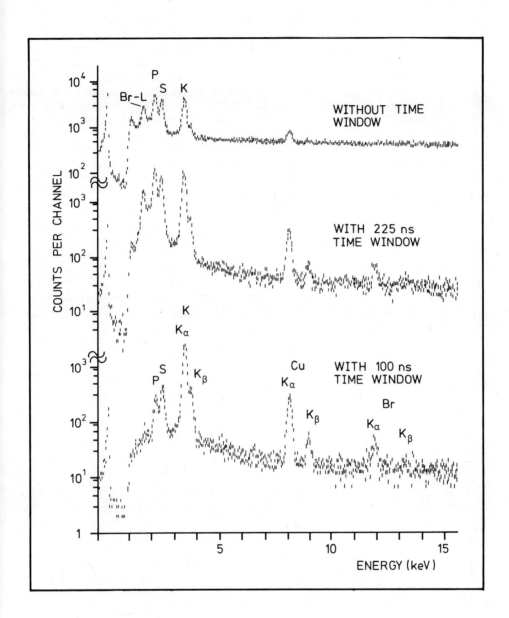

Fig. 3 X-ray spectrum of a test sample in the presence of a background count rate of 10^3 cps:
above: without time window
middle: 225 nsec time window
below: 100 nsec time window.

this measurements was 250 kHz. The spectrum at the top of Fig. 3 is taken without gating the ADC, thus allowing all detected events to be stored. The spectrum in the middle was accumulated using a time window of 225 nsec width which gives a cut-off well below the Bromium K line. Compared to the spectrum without time window now the Cu K_β and the Br K_α line are clearly visible. The better background suppression in the spectrum at the bottom is gained by a more narrow time window of 100 nsec width. Now also the Br K_β line is clearly visible. For the Cu K_α line a peak to background ratio of 0.35 without time window, 3.5 for the 225 nsec time window and 6.8 for the 100 nsec time window is derived. This means an enhancement in sensitivity by a factor of ~20.

CONCLUSIONS

It seems that an improvement of the time resolution and the elimination of the walk should be possible, resulting in considerably better values for the background suppression. Already now the gain in detection sensitivity is about two orders of magnitude at a comparably small expenditure of additional equipment. The method can be used also in conjunction with other x-ray detectors or even with other types of analytical methods such as the nuclear reaction technique.

Nuclear Reactions

DEPTH PROFILING OF HYDROGEN AND HELIUM ISOTOPES IN SOLIDS BY NUCLEAR REACTION ANALYSIS

J. Bøttiger, S.T. Picraux[+], and N. Rud

Institute of Physics, University of Aarhus

DK-8000 Aarhus C, Denmark

ABSTRACT

The paper reviews recent developments in the depth profiling of hydrogen and helium isotopes in solids by use of nuclear reactions. General considerations for depth profiling by the resonance method and by the energy-analysis method are summarized. The different nuclear reactions which have been used for profiling of H, D, T, ^{3}He and ^{4}He are intercompared with respect to depth resolution and sensitivity, and some of the limitations of the technique are discussed.

I. INTRODUCTION

Basic studies of the behaviour of the light atomic species, hydrogen and helium, in solids and the changes of material properties induced by such light impurities are being carried out in many laboratories with different experimental techniques. Important technological motivations for these studies stem from degradation of materials properties caused by H and He, for example in the fuel cladding of fast-breeder reactors, in the first wall of proposed controlled thermonuclear reactors, and in structural materials for hydrogen transport.

A powerful technique for the study of hydrogen and helium in the near-surface region of solids is analysis by ion-induced nuclear reactions. A particular advantage of this technique is its ability

[+] Visiting scientist supported in part by the United States Energy Research and Development Administration. Permanent address: Sandia Laboratories, Albuquerque, New Mexico.

to measure concentration – vs – depth profiles in the near-surface region ($\lesssim 10\mu m$) without layer removal techniques. This review summarizes important considerations in the use of nuclear reaction analysis for depth profiling and surveys recent developments in profiling hydrogen and helium isotopes. The discussion will be restricted to charged-particle, prompt-reaction analysis. Techniques using ion backscattering or requiring thin films will not be discussed. Detailed reviews on the application of nuclear reactions for near-surface analysis have been given previously [1-4].

II. GENERAL CONSIDERATIONS

1. Sensitivity

The sensitivity (minimum detectable concentration) for an isotope in the near-surface region is determined by the ratio of the signal from the nuclear reaction to the background observed in a given detector. An additional practical limitation is set by the total fluence of analyzing beam acceptable on the target before significant changes of the target have occurred. Also in some cases the analyzing times can be unacceptably long. In each specific case, one carefully has to consider the different reactions available to obtain the highest sensitivity (see Table I). The signal from the isotope can be maximized from a knowledge of the reaction cross section and detection efficiencies by proper choice of incident beam energy, detection system, detector angle and solid angle. However, consideration of the background often requires significant compromises.

In charged-particle detection, the background is primarily due to backscattered particles and particles coming from interfering nuclear reactions. In many cases the backscattered particles give an unacceptably high counting rate, due to the high cross section for scattering relative to nuclear reaction cross sections. In the case of resonance profiling (discussed below) the incident beam energy can in some cases be chosen at sufficiently low energies to be able to use a foil in front of the detector which stops the backscattered particles but allows transmission of the nuclear reaction product. For energy-analysis profiling techniques (see below) such foils usually introduce unacceptable energy straggling. For interfering reactions the energy of the reaction products relative to the desired signal can usually be changed by adjusting the incident energy or detection angle. Interferences due to high-energy light particles, as β or p, which penetrate the depleted region of a partially depleted surface barrier detector can be made to fall at different energies by changing the detector bias.

For γ and neutron detection the backscattered ions or charged particle reaction interferences present no problem since detection can be made outside of the vacuum chamber. However natural background

TABLE I. Summary of nuclear reaction methods that have been used for profiling hydrogen and helium.

Isotope Analyzed	Nuclear Reaction	Incident Beam Energy (MeV)	Energy Detected (MeV)	Method[a] (R. energy) (MeV)	Probing Depth[b,c] (Å)	Resolution[b,c] (Target) (Å)	Sensitivity (at. ppm)	Reference
H	$H(^{19}F,\alpha\gamma)^{16}O$	16–18	6.1, 6.9, 7.1 γ	R(16.4)	4,000	400 (SiO_2)	$\sim 10^3$ (d,e)	8
H	$H(^{11}B,\alpha)^8Be$	1.6–2.3	1–4 α	R(1.79)	4,000	600 (Si)	$\sim 10^3$ (d,e)	9
H	$H(^7Li,\gamma)^8Be$	2.7–7.0	14.7, 17.6 γ	R(3.07)	60,000	1000 (Fe, Cd)	~ 1 (d,e)	10
D	$D(^3He,p)^4He$	0.7–1.0	2–5 α	E-A	10,000	700 (ErD_2)	$\sim 10^3$ (e)	11
D	$D(d,p)T$	0.2	0.5–1.0 t or 2–3 p	E-A	30,000	1000 (Cu)	$\sim 10^4$	12
T	$T(p,n)^3He$	2.5	~ 1 n	E-A	100,000	9000 (Ti)	$\sim 10^3$	13
^3He	$^3He(d,p)^4He$	0.5	2–4 α	E-A	10,000	400 (Nb)	$. \sim 10^3$ (e)	14
^3He	$^3He(d,p)^4He$	0.4–3.0	12–14 p	R(0.43)	300,000	30000 (Nb)	~ 10	15
^4He	$^4He(^{10}B,n)^{13}N$	3.5–5.0	2–3.5 n	R(3.77)	10,000	600 (Al)	$\sim 10^4$	7

a) R = resonance method and E-A = energy-analysis method.
b) Value demonstrated or estimated from reference sited. Target material used to demonstrate method.
c) Probing depth and near surface resolution for perpendicular beam incidence, glancing incidence or analysis angles unusual can improve resolution by ≳ 2. Standard detector resolution assumed.
d) Depends on surface ^1H contamination level.
e) In practice sensitivity is limited by analyzing beam flux, due to sample heating, and total fluence, due to beam damage.

and detection efficiency as well as other γ or n interferences need to be considered. High beam currents coupled with target cooling can be an advantage in suppressing natural background.

2. Depth Profiling by the Resonance Method

The resonance method for depth profiling requires a sharp peak in the energy dependence of the nuclear-reaction cross section. In such measurements the incident-beam energy is changed in small increments, thereby changing the depth at which the resonance occurs. From the nuclear-reaction yield as a function of energy (depth), the profile is derived. A detailed discussion of the method is given in Ref. 5.

The depth resolution near the surface is determined by the resonance width of the nuclear reaction and the spread in energy of the incident beam. The energy straggling of the incident beam becomes the dominating factor in limiting the depth resolution at depths where straggling exceeds the resonance width. The maximum probing depth is usually limited by resonances or increased non-resonant yield at higher energies, unless the yields due to these can be subtracted from the data with sufficient accuracy. Otherwise the maximum probing depth will be set by the maximum allowable broadening in the depth resolution due to energy straggling of the beam, or by background interferences at higher beam energies.

An important limitation in the resonance method is the change in concentration, ΔC, that it is possible to measure relative to the total amount of impurity present. This is in contrast to the method of depth profiling by layer removal techniques, where it is generally possible to measure concentrations deeper into the sample orders of magnitude smaller than at the peak of the profile. The limitations in the case of resonance profiling arises whenever there is a non-zero value for the off-resonance yield, since to determine the impurity concentration with the resonance at a given depth some appreciable fraction of the signal should always come from the impurity at that depth. This limit on measuring the change in concentration does not depend on absolute concentration levels but on the relative areas under resonant and non-resonant parts of the cross section - vs - energy curve, and on the shape of the concentration profile. In addition, if the area under the resonance is too small then excessively long analysis times may be required.

3. Depth Profiling by the Energy-Analysis Method

The energy-analysis method of depth profiling can be applied to any reaction with a cross section that varies smoothly with energy provided a charged particle is emitted in the nuclear reaction. The depth information stems from an energy analysis of the emitted charged particles. A detailed discussion of the method is given in Refs. 1 and 6.

DEPTH PROFILING OF HYDROGEN AND HELIUM ISOTOPES

For perpendicular incidence an increment in depth Δx will correspond to a difference in energy ΔE observed for the emitted particle at angle θ with respect to the incident beam of

$$\Delta E = - [S_1 (\partial E_2/\partial E_1)(E_1, \theta) + S_2/\cos\theta]\Delta x \quad , \qquad (1)$$

where S_1 and S_2 are the average stopping powers and E_1 and E_2 are energies immediately before and after the nuclear reaction, for the incident and emitted particles, respectively. As seen from Eq. 1 the energy separation is dependent on both the reaction kinematics (1st term) and the rate of energy loss. Thus for a given energy resolution the depth resolution is determined by stopping powers, incident energy, detector angle, solid angle subtended by the detector and will vary with depth in the target.

The maximum probing depth will depend on the maximum combined energy straggling of incident and emitted particles that is acceptable, since this will eventually dominate in limiting depth resolution. Also the decrease in cross section of a reaction at sufficiently low energies together with upper limits on incident energy can limit the probing depth.

The change in concentration which can be measured for energy analysis profiling is normally determined only by counting statistics and sensitivity. As discussed in II.1 the backscattered particles are usually a major background limitation in this method. Unless electrostatic or magnetic separation of the emitted particles can be used, the beam currents must be kept low enough to prevent pulse pile-up and resolution degradation in the detector. In some cases this will imply unacceptably long analysis times.

III. EXAMPLES

In Table I are summarized several nuclear reactions which have been applied in different laboratories for profiling hydrogen and helium isotopes in solids. The quoted probing depths, resolutions and sensitivities are based on our interpretation of the papers of the references cited. In many cases depth resolution and sensitivity can be improved by special arrangements (optimized geometry such as glancing angle detection, coincidence techniques to get lower background, etc.). The depth resolution is given for perpendicular incidence. For both resonance and non-resonance reactions the depth resolution is always given close to the surface (no energy straggling). The sensitivities quoted are in some cases very uncertain. Frequently the practical sensitivity limit is set by the amount of analyzing beam acceptable on the target.

As seen from the table, the resonance nuclear reaction $^1H(^{19}F,\alpha\gamma)^{16}O$ has been used for depth profiling of ^1H where the emitted γ rays of 6.1 - 7.1 MeV are detected. The raw data for range-profile measurements for 12 keV H^+ implanted to a fluence of $4 \times 10^{16}/cm^2$

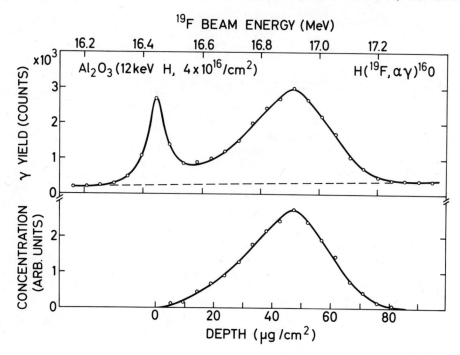

Fig. 1. Range profile of 12 keV H^+ implanted into Al_2O_3 at a fluence of $4 \times 10^{16}/cm^2$ measured by use of the nuclear reaction $^1H(^{19}F,\alpha\gamma)^{16}O$. Upper part of the figure is the raw experimental data and the lower part the extracted depth profile. The data are taken from Ref. 16.

into Al_2O_3 [16] are shown in the upper part of Fig. 1. The surface peak seen in the figure is presumably due to a hydrocarbon layer that is very often present on targets in accelerator laboratories with standard vacuum conditions. Unless special ultra-high vacuum and surface cleaning methods are used this surface contamination can be the primary limitation on sensitivity for near-surface profiling of ^1H. In this respect studies using D and T, where possible, can have definite advantages over working with the naturally abundant isotope. In the lower part of Fig. 1 is shown the depth profile obtained from the data of the upper part of the figure after subtraction of background and surface peak. A second resonance appears at a ^{19}F beam energy of 17.6 MeV and in practice, this second resonance often sets an upper limit to the profiling depth. Assuming the absence of a surface hydrocarbon layer, the sensitivity limit for the reaction is set by the natural γ background.

An example of the application of the energy-analysis method is shown in Fig. 2 for depth profiling by the reaction $^3He(d,p)^4He$.

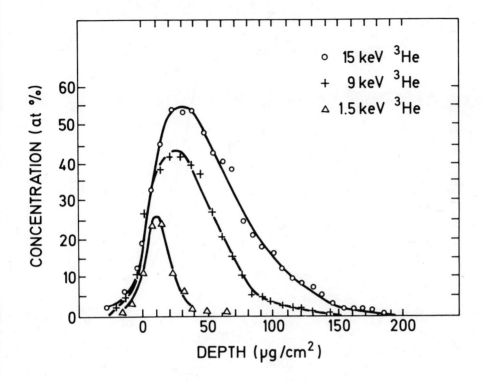

Fig. 2. Range profiles of 1.5 keV, 9 keV, and 15 keV ^{3}He implanted into niobium measured by use of the nuclear reaction ^{3}He(d,p)^{4}He. The data are taken from Ref. 14.

For profiling of ^{2}H, the reaction can be simply turned around, ^{2}H(^{3}He,p)^{4}He. Due to kinematic energy broadening (1st term of Eq. 1) the solid angle of the detector when using this reaction for profiling has to be kept small to obtain good depth resolution. This implies that in many cases a large fluence of analyzing beam has to be used. Another limitation in the use of the reaction is the backscattering of the incident beam into the surface barrier detector, since this limits the incident beam current and therefore the counting rates to low values.

In Fig. 3 is shown an example of profiling of ^{4}He by the resonance method using the reaction ^{4}He(^{10}B,n)^{13}N 7). In the work of Ref. 7 a liquid scintillation detector was placed at 60° to the incident beam. The detector had a solid angle of ~ 1 sr. and a detection efficiency of ~ 5%. Here the background is due to γ's from naturally abundant ^{40}K and ^{218}Th, and electronic discrimination between the neutrons and γ's is required.

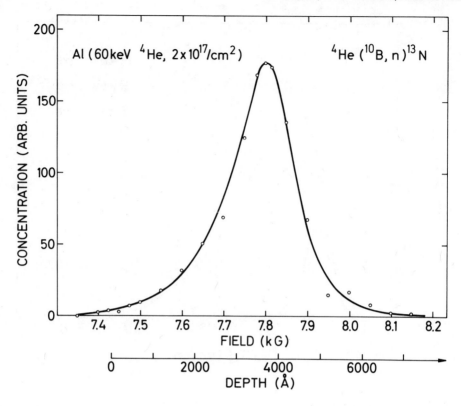

Fig. 3. Range profile of 60 keV ^{4}He implanted into Al at a fluence of $2 \times 10^{17}/cm^2$ measured by use of the nuclear reaction ^{4}He(^{10}B,n)^{13}N. The profile is taken from Ref. 7.

References

1. G. Amsel, J.P. Nadai, E. D'Artemare, D. David, E. Gigard, and J. Moulin, Nucl. Instr. and Methods 92, 481 (1971)
2. J.R. Bird, B.L. Campbell, and P.B. Price, Atomic Energy Review 12, 275 (1974)
3. E.A. Wolicki, New Uses of Low Energy Accelerators Ed. by J.F. Ziegler (Plenum Press, N.Y., 1975)
4. S.T. Picraux, G. Amsel and L.C. Feldman in Catania Working Data for Ion Beam Analysis (NSF and CNR, Catania, Italy, 1974)
5. K.L. Dunning, G.K. Hubler, J. Comas, W.H. Lucke, and H.L. Hughes, Thin Solid Films 19, 145 (1973)

6. A Turos, L. Wielunski, and A. Barcz, Nucl.Instr. and Methods 111, 605 (1973)
7. N.Rud, B. Bengtson, J. Bøttiger, and S.T. Picraux (to be published)
8. D.A. Leich and T.A. Tombrello, Nucl.Instr. and Methods 108, 67 (1973)
9. E. Ligeon and A. Guivarc'h, Rad.Effects 22, 101 (1974)
10. P.N. Adler, E.A. Kamykowski, and G.M. Padawer, "Hydrogen in metals", Ed. by I.M. Bernstein and A.W. Thompson, (American Society for Metals, Metals Park, Ohio, 1974) p. 623
11. R.A. Langley, S.T. Picraux, and F.L. Vook, J.Nucl Materials 53, 257 (1974)
12. P.B. Johnson, Nucl.Instr. and Methods 114, 467 (1974)
13. J.C. Davis and J.D. Anderson, J.Vac.Sci. Technol. 12, 358 (1975)
14. R. Behrisch, J. Bøttiger, W. Eckstein, U. Littmark, J. Roth, and B.M.U. Scherzer, Appl. Phys.Lett., august 1975
15. P.O. Pronko and J.G. Pronko, Phys.Rev. 9B, 2870 (1974)
16. J. Bøttiger, J.R. Leslie, and N. Rud, (to be published)

DISCUSSION

Comment: (R. Behrisch) The poor agreement between the measured and calculated mean range and straggling for protons in the low Z target material (Si) may be due to the approximation $M_1/M_2 \ll 1$ in calculation.

Q: (R. Behrisch) At this conference it has also been reported that depth profiles of these light ions ($M_2 \gtrsim 2$) in solids may as well be obtained by elastic proton backscattering. Can you please comment on how this method compares with the nuclear reaction method, especially in respect to sensitivity, depth resolution and total depths which can be probed.

A: (J. Bøttiger) In my opinion depth resolution by the elastic scattering technique is a factor of two poorer compared to $^3He(d,p)^4He$ using the energy analysis method. Sensitivity of elastic scattering is one to two orders of magnitude poorer in the case of elastic scattering depending on whether you use thin or thick targets as described by Blewer at this conference. On the other hand the elastic scattering techniques gives additional information concerning other isotopes and the analysis may be more straight forward.

Comment: (J. Biersack) $^3He(n,p)$ gives rather high senstivity for detecting He profiles due to the 5000 barn cross section. Also large detectors can be used without sacrificing too much resolution. Measuring time depends, however, very much on the type of reaction used.

ACHIEVABLE DEPTH RESOLUTION IN PROFILING LIGHT ATOMS

BY NUCLEAR REACTIONS

W.Eckstein, R.Behrisch and J.Roth

Max-Planck-Institut für Plasmaphysik, EURATOM-Association, 8046 Garching bei München, Germany

ABSTRACT

A comparison between calculated and experimental determined depth resolutions for near surface regions is given using the ^{3}He(d,α)H reaction. A depth resolution of 10 to 20 Å is achievable.

INTRODUCTION

The depth profiles of implanted light atoms in metals are of interest in connection with plasma-wall-interaction (1), drive-in targets (2) and diffusion (3). There are some methods available: Elastic backscattering (4), elastic recoil detection (5) and nuclear reactions (6,7,8). The second method is applicable only to thin foils and the achievable depth resolution is limited to a few hundred Å. Nuclear reactions give the best depth resolution for near surface regions. In this work a comparison between the theoretically achievable depth resolution and experimental results will be given. As an example the reaction ^{3}He(d,α)H is used to determine depth profiles of ^{3}He in Nb.

CALCULATIONS

The geometry of the experiments is given schematically in Fig.1, which gives all the abbreviations which will be used for the calculation. A deuteron with energy E_o penetrates the target and hits a ^{3}He atom in the depth x. The α-particle created in the reaction ^{3}He(d,α)H has an energy E_α in the laboratory system:

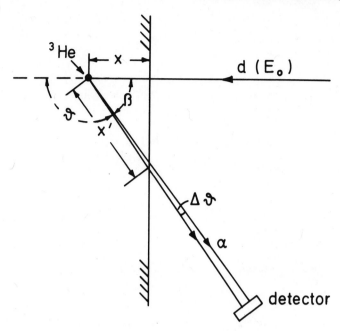

Figure 1. Schematic experimental setup

$$E_3 = \frac{M_4}{M_3 + M_4} Q + \frac{E_1}{(M_3 + M_4)^2} f(M_i, \vartheta, E_1, Q)$$

for the reaction $M_2(M_1, M_3)M_4$

$$f = 2 M_1 M_3 \cos^2\vartheta + (M_3 + M_4)(M_4 - M_1) + 2\cos\vartheta \, g^{1/2}(M_i, \vartheta, E_1, Q)$$

$$g = M_1^2 M_3^2 \cos^2\vartheta + M_1 M_3 (M_3 + M_4)(M_4 - M_1) + M_1 M_3 M_4 (M_3 + M_4)\frac{Q}{E_1}$$

dependent on the deuteron energy E_1, the Q-value of the reaction and the angle ϑ under which the α-particle is emitted. The path-length of the α-particle in the target is x'

$$x' = \frac{x}{\cos\vartheta} \quad , \quad \vartheta = \pi - \beta$$

for perpendicular incidence. The energy of the α-particle leaving the target is

$$E = E_3(E_1, \vartheta) - \int_0^{x/\cos\vartheta} S_\alpha(y) \, dy$$

For near surface reactions the stopping powers S_d and S_α of the incoming deuterons and the created α-particles can be put constant. For an energy loss of 10 % in the energy ranges regarded here, the stopping power changes only about 5 %, which is a lower limit of the accuracy for the stopping power data given in the literature. Energy and angular straggling is completely neglected. With these assumptions the last formula gives

$$\frac{dE}{dx} = \frac{\partial E_3}{\partial E_1}\frac{\partial E_1}{\partial x} - S_\alpha \frac{1}{\cos\vartheta}$$

and a depth resolution

$$\Delta x = \frac{\Delta E}{\frac{\partial E_3}{\partial E_1} S_d - S_\alpha/\cos\vartheta}$$

$$\frac{\partial E_3}{\partial E_1} = \frac{f(M_i, \vartheta, E_1, Q)}{(M_3 + M_4)^2} - \frac{M_1 M_3 M_4}{M_3 + M_4}\frac{Q}{E_1}g^{-\frac{1}{2}}(M_i, \vartheta, E_1, Q)\cos\vartheta$$

The uncertainty ΔE in the energy of the α-particles has two contributions: ΔE_g, originating from the finite acceptance angle $\Delta \vartheta$ of the detector 7

$$\Delta E_g = \frac{\partial E}{\partial \vartheta}\Delta\vartheta$$

$$= \left(\frac{\partial E_3}{\partial \vartheta} - x S_\alpha \frac{\sin\vartheta}{\cos^2\vartheta}\right)\Delta\vartheta$$

with

$$\frac{\partial E_3}{\partial \vartheta} = -\frac{2 E_1 \sin\vartheta}{(M_3 + M_4)^2}\left\{2 M_1 M_3 \cos\vartheta + g^{\frac{1}{2}}(M_i, \vartheta, E_1, Q) + M_1^2 M_3^2 g^{-\frac{1}{2}}(M_i, \vartheta, E_1, Q)\cos 2\vartheta\right\}$$

ΔE_d is the constant energy resolution of the detector. Then the total uncertainty ΔE is

$$\Delta E = (\Delta E_g^2 + \Delta E_d^2)^{\frac{1}{2}}$$

In order to give numerical results the following values have been used for the calculations: $Q = 18.341$ MeV, $E_0 = 500$ keV, $S_\alpha = 41$ eV/Å (for Nb), $\Delta\vartheta \approx 7 \times 10^{-3}$ radian (slightly dependent on β, spot size, target-detector distance and detector opening), $S_d = 16$ eV/Å.

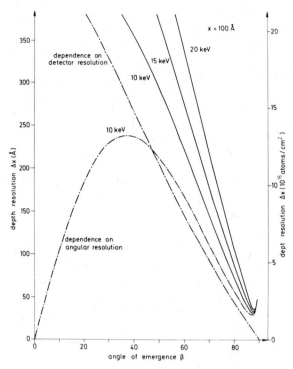

Figure 2. Depth resolution x as a function of angle of emergence ß at a depth of 100 Å.
— Total depth resolution for different energy resolutions of the detector.
—·— Depth resolution for a finite energy resolution of $\Delta E_d = 10$ keV of the detector alone resp. for a finite angular resolution ΔE_g alone.

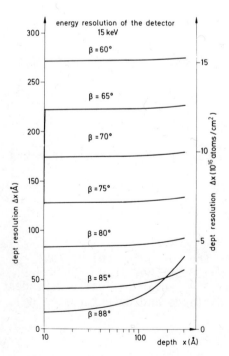

Figure 3. Depth resolution Δx as a function of depth x for an energy resolution of 15 keV of the detector. Parameter at the different curves is the emergence angle ß. Energy straggling which becomes the dominating factor at large depth is not included.

The results are shown in fig.2 and 3. Fig.2 gives the depth resolution at a depth of 100 Å as a function of the emergence angle β (perpendicular incidence) for three different detector resolutions between 10 and 20 keV. The broken lines give the dependence on detector resolution ΔE_d = 10 keV alone resp. on the angular resolution. The fig. 2 shows clearly that for small angles β the energy resolution of the detector is the limiting factor whereas for larger β the acceptance angle determines the depth resolution. The depth resolution is a strong function of β and an optimal value can be achieved at angles β very close to 90°. Only for a small depth x ($\approx$ 10 Å) both terms in ΔE_g are of the same order, for larger depths the second term is the dominant one, which gives a nearly linear relation between E_α and x for large β. The dependence of the depth resolution on the depth is shown in fig.3 with the angle β as a parameter and 15 keV detector resolution. For depths larger than 300 Å the straggling effects are dominating. At a depth of 10 Å one can reach 11 Å for β = 89°.

EXPERIMENTAL

Implantation of ^{3}He and probing with d are done by two different accelerators. Both beams have been directed through the same collimators of 1.1 mm diameter. The ^{3}He was implanted with energies of between 1.5 and 15 keV into a Nb single crystal and with currents of 0.1 - 2 uA and doses between 6 x 10^{16} and 1.5 x 10^{17} ^{3}He/cm^2. Subsequent probing was performed with a 1 MeV D_2^+ beam corresponding to 500 keV D^+ ions where the cross section is 66 mb/sr. A total dose of 10^{17} D^+/cm^2 was used for one analysis. The ^{4}He particles created in the nuclear reaction were detected at an angle β between 89° and 75° relative to the surface normal with a high resolution (15 keV) surface barrier detector. The acceptance solid angle of the detector was 3.9 x 10^{-5} sr. The Nb crystal had been annealed to 2100 °C for 10 min before the implantation.

EXPERIMENTAL RESULTS AND DISCUSSION

Fig.4 shows energy spectra of the α-particles for different observation angles β. The implantation of ^{3}He was done with 4 keV into a <100> direction of a Nb single crystal. With increasing β the spectrum is shifting to higher energies due to the kinetics of the nuclear reaction. The spectra become broader with increasing β due to the better depth resolution at larger β.

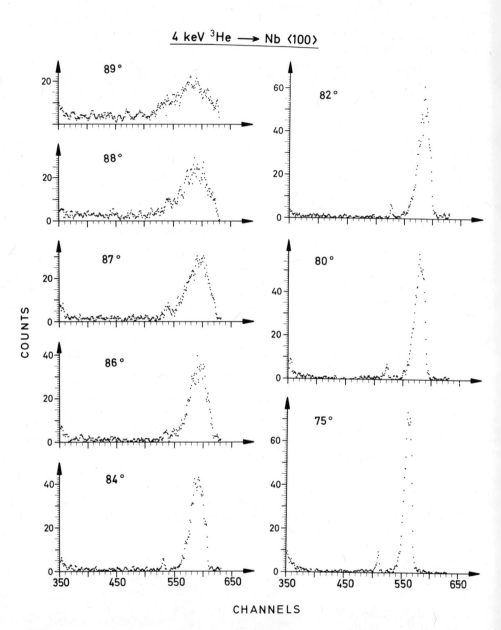

Figure 4. Energy spectra of the emergent α-particles for different angles β. ^{3}He was implanted with an energy of 4 keV in a ⟨100⟩ direction of a Nb single crystal. The dose was 1×10^{17} ^{3}He/cm^2.

In order to compare the calculated with the experimental depth resolution the following procedure has been used. The depth resolution at the surface is given experimentally by the sharp decrease of the spectrum at the high energy side. This sharp decrease should give an Gaussian distribution after differentiating this part of the spectrum. The fwhm of this Gaussian distribution is then the energy resp. the depth resolution. Practically it is not possible to construct the Gaussian distribution, but the following formula should give a good estimate for the depth resolution.

$$\Delta x_{exp} = \delta E \cdot n \cdot \cos\beta / S_\alpha$$

where δE is the energy interval corresponding to one channel and n is the number of channels in the linear high energy cutoff. The experimental results are compared with a calculation for a detector resolution of 15 keV in fig.5. There is

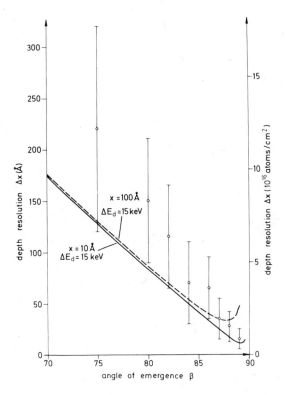

Figure 5. Comparison between experimental results for the depth resolution and the calculated curve (15 keV energy resolution of the detector). The solid line is calculated for a depth of 10 Å, the dashed line for a depth of 100 Å.

reasonable agreement for angles larger than 80°. The deviation for angles smaller than 80° is possibly due to a surface peak which appears already at 82°, however, it is well resolved at 89°. This peak corresponds to about one monolayer ^{3}He within about 12 Å. It should be pointed out that at grazing emergence one gets good depth resolution at the surface but the whole distribution gets so broad that the low energy tail of the spectrum moves to the proton peak. The energy losses become so large that there is no longer a linear relationship between the depth x and the energy E. The reasons for that are the energy dependent stopping power $S_\alpha(E)$, the energy straggling and the small angle scattering of the emergent α-particles.

Finally we have checked the depth resolution by implanting ^{3}He in Nb with two different energies, 15 keV in a ⟨100⟩ direction and 1.5 keV in a random direction. Both depth profiles are clearly separated as can be seen in fig.6.

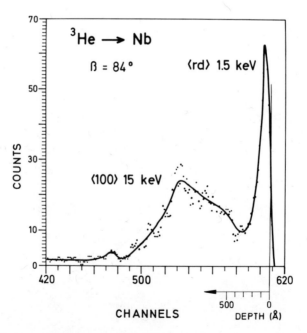

Figure 6. An α-energy spectrum for an emergence angle β = 84°. Implantation of ^{3}He with 15 keV in a ⟨100⟩ direction and with 1.5 keV in a random direction of a Nb single crystal. The doses were 1.5×10^{17} resp. 6×10^{16} ^{3}He/cm^2.

ACKNOWLEDGEMENTS

The authors wish to thank Dr.H.Vernickel for his constant interest in this work. The technical assistance of R.Heilmeier, H.Schmidl, S.Schrapel, and H.Wacker in planning and performing these measurements is gratefully acknowledged.

REFERENCES

1. R.Behrisch, B.B.Kadomtsev, Plasmaphysics and Controlled Nuclear Fusion Research (Proc. 5th Conf., Tokyo, 1974), IAEA, Vienna, 1975.
2. J.H.Ormond, Can.J.Phys. 52, p 1971 (1974) and J.C.Davis, J.D.Anderson (UCRL-75757), 21st Nat.Symp.Am.Vac.Soc. Anaheim, Oct. 1974.
3. W.D.Wilson, C.L.Bisson, rad.eff.19, 53 (1973) and S.Blow, J.Brit.Nucl.Energy Soc.11, 371 (1972).
4. R.S.Blewer, J.Nucl.Mat.53, 268 (1974) and Int.Conf. on Application of Ion Beams to Metals 1973 (Plenum Press 1974, 557).
5. B.L.Cohen, C.L.Fink, and J.D.Degnau, J.Appl.Phys. $\underline{43}$ (1972),19
6. G.Amsel, J.P.Nadai, E.d' Arpemare, B.David, E.Girard and J. Moulin Nucl.Instr.and Meth.92,481 (1971).
7. A.Turos, L.Wielunski and A.Barcz, Nucl.Instr.and Meth.III 605, (1973).
8. R.Behrisch, J.Bøttiger, W.Eckstein, U.Littmark, J.Roth, and B.M.U.Scherzer, Appl.Phys.Lett.27, 199 (1975).

DISCUSSION

Q: (J.S. Williams) When the detector is inclined at angles >85° to the surface normal, uncertainties in angular position of the detector (particularly at 89°) can introduce extremely large errors (∼50 %) in the determination of a depth scale.

A: (W. Eckstein) We agree that the uncertainty in angular position of the detector introduces a large error in the depth scale particularly at grazing angles to the surface. In our measurements the detector position was firstly determined optically and secondly with the beam by determining the cut off position at 90°. The uncertainty was about 0.1 degree.

Q: (W.C. Turkenburg) Contamination on the surface influences the depth-resolution, especially near to the surface. How do you measure the contamination? Do you see beam-induced build up of contamination ? What is the thickness of the contamination layer, normally, during your experiments?

A: (W. Eckstein) The vacuum in the target chamber was generally $\leq 10^{-8}$ Torr. Additionally the target was surrounded by a copper shield cooled with liquid nitrogen. Before the measurements the Nb single crystal was annealed in situ to $\sim 2000°C$. Thus the surface contamination of oxygen was reduced to less than one monolayer as determined by double aligned Rutherford backscattering. No beam induced build up of a surface contamination could be observed.

DEPTH PROFILING OF DEUTERONS IN METALS AT LARGE IMPLANTATION
DEPTHS USING THE NUCLEAR REACTION TECHNIQUE[*]

M. Hufschmidt, W. Möller, V. Heintze, and D. Kamke

Institut für Experimentalphysik der Ruhr-
Universität, 463 Bochum, Fed. Rep. Germany

ABSTRACT

Implantation profiles of deuterons at large implantation depths (up to 4 μm) are measured using the D(d,p)T nuclear reaction technique. In order to obtain reliable results, the mean range of the probing deuterons has to be chosen large compared to the depth region under consideration. Monte Carlo calculations have been carried out in order to compute the depth resolution, which is deteriorated by several straggling effects. As an application, density profiles of 400 keV implanted deuterons have been measured with 2 MeV probing energy. The lateral spread of the implanted deuterons could be determined by comparing profiles, which were implanted at normal and oblique incidence.

1. INTRODUCTION

During the past years many methods have been developed in order to determine density profiles of implanted ions in solid matter. The most important non-destructive method has become the Rutherford backscattering technique /1/, which is applicable for light implanted ions in the case of thin foils only /2/. In the case of thick absorbers different methods have been applied: If the concentration of the foreign atoms is sufficiently high, depth profiles may be obtained by backscattering using the additivity of the stopping cross section /3/. Secondly, the 0.43 MeV resonance of the

---

[*] Research sponsored in part by the Bundesministerium für Forschung und Technologie, Bonn, West Germany

$^3He(d,p)^4He$ nuclear reaction has been used with a limited depth resolution due to the large half width of the resonance /4/. For the same ions the 'direct' nuclear reaction technique is applicable, where the depth profiles are calculated directly from the energy spectra of the outgoing particle /5-8/. At high implantation depths, the depth resolution of this method is found to decrease drastically, as published in an earlier paper /9/.

The present paper describes the depth profiling of deuterons by means of the D(d,p)T nuclear reaction technique, which was used successfully even at high implantation depths (some µm). We achieve a satisfactory depth resolution by choosing suitable experimental conditions, in particular a high energy of the probing deuteron.

In the following section the method will be explained in detail. Section 3. deals with resolution problems, in section 4. a short description of the apparatus is given, and finally several applications will be described.

2. METHOD

Fig. 1 shows a solid absorber, within which foreign atoms are deposited according to a yet unknown distribution function. In order to determine the density distribution of the foreign atoms, we bombard the absorber with suitable probing ions, which are able to initiate a nuclear reaction with the implanted atoms as target nuclides. In the case of the D(d,p)T reaction, both implanted atoms and probing ions are deuterons. The reaction product (proton) is observed by a particle detector.

Let the incoming probing ions have the energy E_{10}. On their

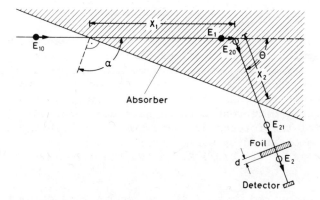

Fig. 1: Geometry of the nuclear reaction technique.

way x_1 to the reaction locus they lose some energy and may initiate a nuclear reaction with a foreign target atom with the energy E_1. The energy E_{20} of the reaction product depends on the energy E_1 and the deflection angle Θ. It is decreased by the energy loss on the way x_2 to the absorber surface, and on passage through a foil of thickness d, which is mounted in front of the detector in order to stop elastically scattered ions. Thus, the reaction product is detected with the energy E_2, which is a function of the reaction depth x_1, the energy E_{10} of the bombarding ion, the absorber angle α, the observation angle Θ, and the foil thickness d.

The latter four parameters are to be chosen suitably, so that the detection energy E_2 is a monotonic function of the reaction depth x_1, and the depth can be derived from the energy unambiguously. Furthermore, the derivative of the function $E_2(x_1)$ should be sufficiently large to gain an optimal depth resolution at given energy resolution. This is fulfilled for small observation angles Θ. Therefore, we chose an angle of $\Theta = 30°$ or $45°$ in our experiments.

From the energy spectrum of the reaction product particles we get not only the energy, but also the number of particles for each energy interval. The number dN_2 of reaction product particles coming from an absorber sheet dx_1 at x_1 is given by

$$(1) \qquad dN_2 = N_1 \cdot n(x_1) \cdot \left.\frac{d\sigma}{d\Omega}\right|_{Lab, E_1} \cdot \Delta\Omega \cdot dx_1$$

where N_1 is the number of the incoming probing particles, $n(x_1)$ the number of foreign atoms per unit volume, $d\sigma/d\Omega$ the differential cross section of the nuclear reaction, and $\Delta\Omega$ the solid angle of the detector. From eq. (1), the particle density distribution results as

$$(2) \qquad n(x_1) = \frac{1}{N_1} \cdot \left(\frac{d\sigma}{d\Omega} \cdot \Delta\Omega\right)^{-1} \cdot \frac{dN_2}{dx_1}$$

and, after introducing the channel contents dN_2/dE_2 of a multichannel analyzer:

$$(3) \qquad n(x_1) = \frac{1}{N_1} \cdot \left(\frac{d\sigma}{d\Omega} \cdot \Delta\Omega\right)^{-1} \cdot \left.\frac{dN_2}{dE_2} \cdot \frac{dE_2}{dx_1}\right|_{x_1}$$

The term $\left.dE_2/dx_1\right|_{x_1}$ is the derivative of the above discussed function $E_2(x_1)$.

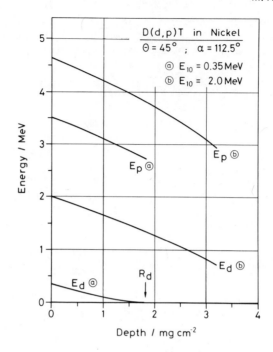

Fig. 2: Deuteron energy E_d at the reaction locus and energy E_p of the outcoming proton as function of the reaction depth at two different detection energies.

By means of eq. (3) we are able to calculate density distributions of implanted deuterons in Nickel from the proton energy spectra in an absolute scale. For that purpose, we have written a computer program. The energy loss functions in that program were given as polynomial fits to values from Northcliffe and Schilling /10/. Similarly, the cross section of the D(d,p)T nuclear reaction has been adjusted to experimental results of several authors.

3. PROBING ENERGY AND DEPTH RESOLUTION

In Fig. 2 the energy $E_1 = E_d$ of the incoming deuteron and the energy $E_2 = E_p$ of the outgoing proton is plotted versus the reaction depth in mg cm^{-2} (1 mg cm^{-2} corresponds to 1.12 μm Ni). We have calculated two situations: (a) The energy of the probing deuteron is 0.35 MeV: Such measurements could be done by using only one accelerator, the corresponding spectra could be measured during implantation. The observation depth is limited to the range of the deuterons, which is 1.82 mg cm^{-2}. (b) The energy of the probing deuterons is 2.0 MeV, the corresponding range 11.87 mg cm^{-2}. They

DEPTH PROFILING OF DEUTERONS IN METALS

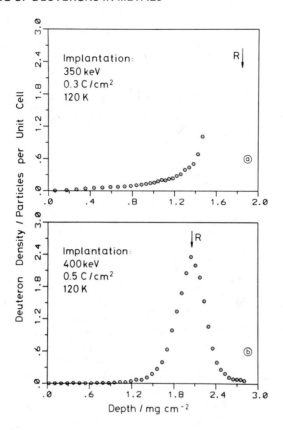

Fig. 3: Density profiles obtained from experiments with detection energies of 350 keV (a) and 2 MeV (b) (arrows: mean deuteron range from ref. /10/.

are used to detect deuterons, which have been implanted at an energy of 0.4 MeV. For that purpose, only the depth region from 0 to 3 mg cm^{-2} is of interest; at the end of that region, the probing deuterons still have an energy of 0.83 MeV.

Fig. 3 shows the density profiles, which have been computed from the experimentally measured proton energy spectra. The density is given in an absolute scale: 1 particle per (conventional) unit cell corresponds to 1 foreign atom per 4 nickel atoms or to $2.29 \cdot 10^{22}$ particles per cm^3. The absolute accuracy of these profiles is about 5% in the depth scale and 10% in the density scale. In the first case (Fig. 3a) the calculated density profile does not decrease for depths larger than the mean range, as must be expected. Instead, the calculated densities increase to infinity, when the depth approaches the mean range. Beyond the mean range no further evaluation is possible. The second part (Fig. 3b) shows

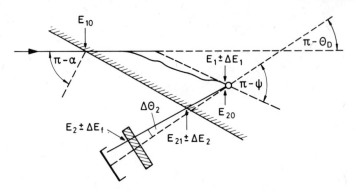

Fig. 4: Straggling effects deteriorating the depth resolution of the nuclear reaction technique.

a density profile in good agreement with the expectation: the density has its maximum near the mean range and decreases rapidly beyond it.

From that we can state, that the nuclear reaction technique operates successfully only if the range of the probing ions is large compared with the depth region of interest. This is due to straggling effects, which we did not take into account with our simple calculation of the density profile. These effects are illustrated in Fig. 4. After traversing the absorber to some extent, the energy and the direction of motion of the incoming deuterons are broadened considerably by multiple scattering with the absorber atoms. Also, the outgoing proton will undergo energy and angular straggling on its way to the absorber surface and through the stopper foil in front of the detector. The sum of these effects is, that nuclear reactions in a fixed absorber depth will cause an energy distribution of nonnegligible width of the detected protons instead of a sharp energy value.

In the case of low detection energies the above mentioned divergence of the calculated density profiles at the end of the mean range can now be explained: As the computations do not take into account the straggling effects, the energy of the incoming deuteron and therefore the cross section is set to be zero at the mean range of the incoming deuteron. On the other hand, due to the broadening of the proton energies, there will be observed some counts at the corresponding point of the proton energy spectrum. In order to determine the density from eq. (3), the number of counts is divided by the corresponding cross section. This quotient will increase to infinity when the depth exceeds the mean range.

In the case of higher detection energies, the straggling

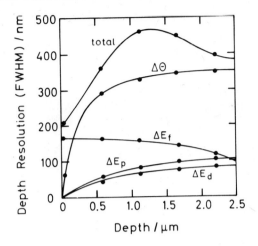

Fig. 5: Depth resolution due to multiple scattering and energy straggling: ΔE_d and $\Delta\Theta$ by energy- and angular straggling of the probing deuteron, ΔE_p and ΔE_f by energy straggling of the outcoming proton within absorber and foil.

effects merely deteriorate the depth resolution. In order to give a quantitative description, we carried out Monte Carlo calculations /11/ and determined the contributions of the different straggling effects to the depth resolution. The results are shown in Fig. 5: The main contribution at the absorber surface results from the energy straggling ΔE_f in the stopper foil, at larger depths the main contribution results from the angular straggling $\Delta\Theta$. The contribution of energy straggling of incoming and outgoing particle, ΔE_d and ΔE_p, may be neglected compared with the total depth resolution. In the region of interest, we can assume a depth resolution of 0.4 µm. This should be compared with the largest depth of 3.5 µm, which can be probed by 2 MeV deuterons. Removing the stopper foil, the depth resolution will be improved considerably near the surface. For the first surface layers it will be given by the detector resolution.

4. EXPERIMENTAL

In our experiments we used a nickel cylinder of 9 mm diameter and 5 mm length with its circular surface well cleaned and polished. It was mounted in a scattering chamber. By means of a heating and cooling facility, the absorber temperature could be controlled within -180°C and +1200°C with an accuracy of ±3°C under irradiation. During implantation the absorber temperature was usually held at -150°C.

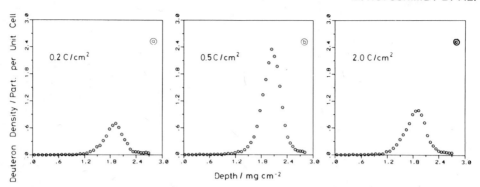

Fig. 6: Absolute density profiles of 400 keV deuterons in nickel at different implantation doses.

The scattering chamber was connected to a 400 keV SAMES electrostatic accelerator and to the 8 MeV Dynamitron Tandem accelerator as well. Both accelerators could be used one after the other; in general 400 keV deuterons from the SAMES accelerator were used for implantation, while the probing deuterons were taken from the 8 MeV tandem accelerator. As discussed above, it is very advantegeous to choose the energy of the probing deuterons large compared to the energy of the implanted deuterons.

The proton energy spectra were monitored by a silicon surface barrier detector connected to a multichannel analyzer. In front of the detector was mounted a thin nickel or aluminium foil, which stopped the elastically scattered deuterons from the absorber material in order to avoid high dose radiation damage of the detector. In recent experiments, we replaced the foil by an electrostatic field separator, which improves the depth resolution considerably.

5. RESULTS AND APPLICATIONS

Fig. 6 shows the deuteron density profiles obtained from the nuclear reaction method after implantations at a deuteron energy of 400 keV and different implantation doses of 0.2, 0.5 and 2.0 C/cm^2. At low doses (case (a) and (b)) we find by summing up the density profiles, that all implanted deuterons are still located in the observed absorber region (with an accuracy of 15%). At higher doses a considerable gas reemission is observed, which will be discussed in detail by a further contribution to this conference /12/.

As a further application we determined the deuteron depth profiles at different implantation energies (Fig. 7), once at normal ($\alpha = 180°$) and once at oblique ($\alpha = 112.5°$) incidence. In this case the stopper foil was removed (see section 4.) to obtain a depth

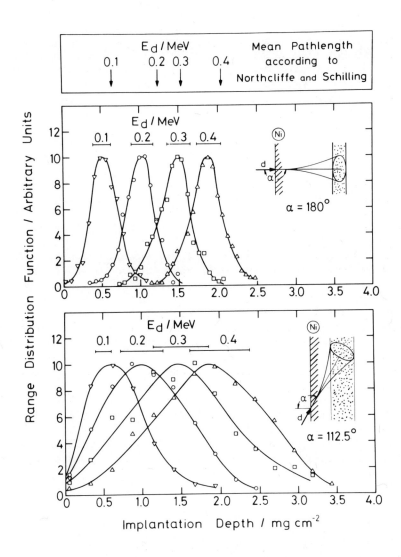

Fig. 7: Density profiles of deuterons in Ni at different energies after normal (a) and oblique (b) incidence; horizontal bars indicate the experimental depth resolution.

resolution as good as possible. The maxima of the depth distribution curves are positioned somewhat lower than the range values from ref. /10/ (which are indeed integrated pathlengths and expected to be larger than the mean projected ranges). Furthermore, the lateral spread of the implanted ions will introduce an additional broadening to the deposition zone, as indicated by the drawings of the implantation geometry inserted into Fig. 7. Thus, more strongly broadened profiles are obtained at oblique incidence compared with the normal incidence. The broadening is significantly higher than due to depth resolution change, which is indicated by the horizontal bars (taken from Fig. 5).

REFERENCES

/1/ D. Powers, W. K. Chu, P. D. Bourland, Phys. Rev. 165 (1968) 376
/2/ B. L. Cohen, C. L. Fink, J. H. Degnan, J. Appl. Phys. 43 (1972) 19
/3/ J. Roth, R. Behrisch, B. M. U. Scherzer, Appl. Phys. Lett. 25 (1974) 643
/4/ P. P. Pronko, J. G. Pronko, Phys. Rev. B 9 (1974) 2870
/5/ V. Heintze, D. Kamke, Appl. Phys. Lett. 10 (1967) 12
/6/ R. A. Langley, S. T. Picraux, F. L. Vook, J. Nucl. Mat. 53 (1974) 257
/7/ P. B. Johnson, Nucl. Instr. Meth. 114 (1974) 467
/8/ R. Behrisch, J. Bøttiger, W. Eckstein, U. Littmark, J. Roth, B. M. U. Scherzer, Appl. Phys. Lett. 27 (1975) 199
/9/ M. Hufschmidt, V. Heintze, W. Möller, D. Kamke, Nucl. Instr. Meth. 124 (1975) 573
/10/ L. C. Northcliffe, R. F. Schilling, Nucl. Data Tables A 7 (1970)
/11/ W. Möller, PhD Thesis, Bochum 1975 (to be published)
/12/ W. Möller, Th. Pfeiffer, D. Kamke, these conference proceedings.

DISCUSSION

Q: (B.M.U. Scherzer) Did you observe radiation enhanced diffusion due to the probing beam?

A: (M. Hufschmidt) No, we did not, even if we obtained several spectra from the same distribution each after another.

GAS REEMISSION AND BLISTER FORMATION ON NICKEL SURFACES

DURING HIGH ENERGY DEUTERON BOMBARDMENT

W.Möller, Th.Pfeiffer, and D.Kamke

Institut für Experimentalphysik der Ruhr -

Universität, 463 Bochum, Fed. Rep. Germany

Abstract

The formation of blisters on Ni-surfaces under deuteron bombardment (150 ... 400 keV) was investigated with scanning electron microscopy as well as nuclear reaction profiling technique. The dependence of blister morphology on incident particle energy, and of the critical dose on beam current and energy are reported. A model is developed to explain the experimental results.

INTRODUCTION

In earlier work on blistering of metals during bombardment with light ions /1-3/ the preferred method of investigation was scanning electron microscopy. It provided results with respect to quantities which admit direct observation, as the size or shape of the blisters or the dependence of the critical blistering dose on the irradiation parameters. The question of the blistering mechanism, however, led to the application of new techniques in the recent time, such as the ion backscattering method to detect the onset of blistering in connection with the observation of radiation damage /4/. Furthermore, the nuclear reaction technique has been applied to determine the depth distribution of the ions /5/ and to investigate the gas release during the implantation /6/.

The present paper describes measurements on the formation of blisters on nickel surfaces under deuteron bombardment in the energy range 150-400 keV, where the nuclear reaction technique /7/ as well as scanning electron microscopy are applied.

EXPERIMENTAL

Polycrystalline nickel absorbers (of cylindrical shape with 5 mm thickness and 9 mm diameter) were bombarded at normal incidence with deuterons from a 400-keV-SAMES accelerator. For the profile measurements (described in detail in a further paper at this conference /7/) a 2-MeV deuteron beam from the Bochum Dynamitron-Tandem accelerator was used. The beam current was monitored by direct current measurement on a rotating vane wheel in front of the absorber.

The temperature of the sample was controlled by means of a NiCr-constantane thermocouple in connection with electron-gun heating and liquid-nitrogen cooling facilities. Temperature adjustment was possible in the range between 80 K and 1500 K with a temperature instability of ±2 K maximum, even under irradiation conditions.

The implantations were carried out at 120 K to avoid gas re-emission by thermal diffusion. Critical doses of blistering were determined by visual observation during the irradiation. In order to mount the samples in the scanning electron microscope (SEM), the temperature was raised to room temperature and the absorber exposed to air, which sometimes caused a collapse of a blister dome due to hydrogen outdiffusion.

In some cases, blister covers had ruptured, allowing the determination of the cover thickness by means of the SEM. After lifting the exfoliated piece of skin carefully, the blister bottom plane could be inspected.

RESULTS AND DISCUSSION

Cover thickness and Critical Dose

Fig. 1(a-f) shows the dependence of the blister morphology on the incident deuteron energy. At energies larger than 300 keV one-dome blisters are formed (a,b) which cover the total area of irradiation. At lower energy, additional small blisters are observed (c,d), and finally (at 150 keV) the small blisters represent the only structure (e,f).

The cover thickness of the blisters is plotted in Fig. 2 versus the energy and compared with the mean pathlength of deuterons in nickel from Northcliffe and Schilling /8/, the mean projected range from depth profile measurements /7/, and the maximum of the damage distribution, which was computed with a Monte-Carlo

GAS REEMISSION AND BLISTER FORMATION ON NICKEL SURFACES

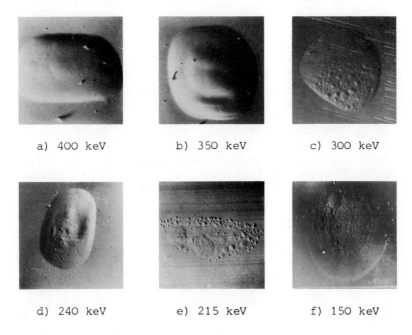

Fig. 1a-f: Morphology of the blisters at different implantation energies (scale 30:1)

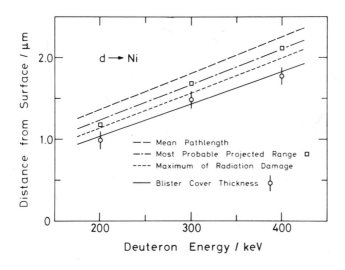

Fig. 2: Blister cover thickness compared with ion range and distance of damaged region from surface

program (taking into account primary events only). The diagram suggests that the early-stage processes of blister formation take place in a region of depth which is smaller than the mean ion range and which coincides with the depth of maximum radiation damage. This is in contrast to the observation of blistering at very low (few keV) energies/5/.

The critical dose of blistering was found to increase with increasing implantation energy. With increasing current density, the critical dose increases linearly at low current densities and shows a tendency toward a constant in the limit of high densities (Fig. 3).

In order to explain these experimental facts, it is necessary to develop a model of the blistering mechanism, which firstly depends on the kind of trapping the deuterons in the material after slowing down, and, secondly, on the amount of ions which are reemitted during the implantation period, i.e. the net deuteron concentration. In both questions, the nuclear reaction profiling technique proved to be a powerful experimental method.

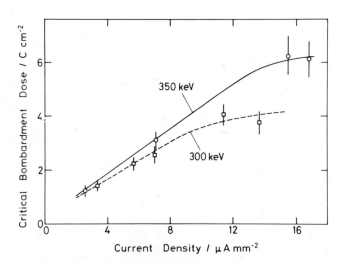

Fig. 3: Critical blistering dose as function of beam current density and energy

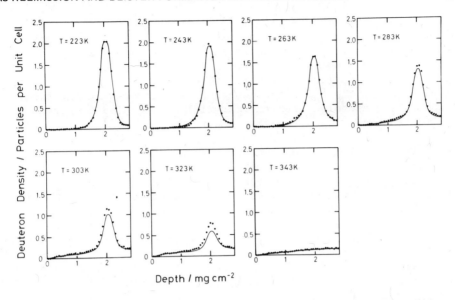

Fig. 4: Temperature variation of the deuteron depth profile

Bubble Formation

After an implantation of 0.5 C/cm^2 (below the critical dose) at deuteron energy of 400 keV and a temperature of 120 K, the absorber temperature was raised step-by-step (10 K each) allowing for an isochronal-annealing experiment, that was kept track of by simultaneous

Fig. 5: Bottom plane of a ruptured blister with grain boundaries and bubble structures (arrow)

depth profiling. The result is shown in Fig. 4: One observes that the half-width of the deuteron deposition peaks remains constant during the annealing process, which is in contrast to normal gas diffusion. Therefore, in a model calculation /9/ the total concentration is composed of one localized part and another freely diffusing fraction, assuming that the implanted deuterons are trapped in the form of gas bubbles and dissolve into the material at elevated temperature. The solid lines of Fig. 4 are the results of this model.

The picture of hydrogen trapping within gas filled bubbles is suggested also by the extremely low solubility in nickel at low temperature (due to the positive solution enthalpy), which is in the order of 10^{-5} at% even at pressures of about 10^4 at and the temperature of 120 K. Thus, if there exist any voids in the metal produced by radiation damage, they will be filled by deuterons immediately. A further evidence is given by the SEM observation of the bottom plane of a ruptured blister (Fig. 5). Besides the grain boundary structures small spots of about 4 μm diameter are visible which are believed to be relicts from the early-stage bubbles.

Gas Reemission

The nuclear reaction method was used furthermore to investigate the gas reemission during implantation. The result is shown in Fig. 6 in the case of a 400-keV deuteron bombardment at 120 K and a current density of 3 μA/mm^2. At low dose, the incident particles are fully trapped; above 0.5 C/cm^2 a radiation-enhanced outdiffusion takes place, which should lead to a constant asymptotic value (100% reemission) in the limit of high doses. Instead, one observes that the trapped dose strongly falls off at about 1.5 C/cm^2. In this case the critical dose was found at 1.3 C/cm^2. So the conclusion is that with the formation of the blister a strong reemission of particles starts. An estimation reveals that this is due to thermal diffusion across the cover of the large-dome blister (see Fig. 1) whose temperature is elevated by several 100 K with respect to the absorber body by the incident beam power.

Blistering Model

To develop a simple model of the blistering mechanism, we start with the assumption that the blistering is initialized by gas-filled bubbles, as suggested by the above experimental results. Small bubbles which may be identified with gas-filled voids will coalesce and form larger ones. The mobility is smaller for larger bubbles, which leads to an increase of the volume with irradiation time. Assuming a spherical-segment shape of the larger bubbles (Fig. 7), with a bottom plane diameter r and a height h, their volume is

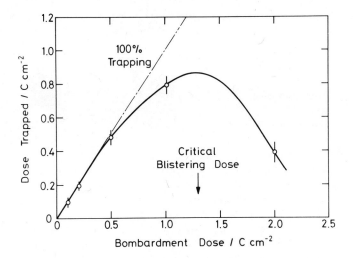

Fig. 6: Trapping curve of 400-keV deuterons in nickel (current density 3 µA/mm^2)

(1) $\qquad V = \dfrac{\pi}{3} h^2 (3\rho - h)$,

and the bending radius of the top surface

(2) $\qquad \rho = \dfrac{h}{2} + \dfrac{r^2}{2h}$.

The equilibrium condition is given by the gas pressure p and Young's modulus σ of the cover membrane of thickness d:

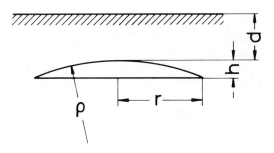

Fig. 7: Idealized shape of an early-stage bubble

$$(3) \quad \sigma = \frac{p\rho}{2d} = \frac{p}{4dh}(h^2+r^2)$$

Blisters originate in the plastic deformation of the cover; therefore, eq. (3) describes the blistering condition as well, if σ is replaced by the yield stress Y and p_c and ρ_c, respectively, denote the correspondent critical values /10/:

$$(4) \quad Y = \frac{p_c \rho_c}{2d}$$

At equal pressure in all bubbles, the largest one will therefore deform the surface first, and introduce a field of stress into its surrounding area. This affected area will be larger at higher cover thicknesses, i.e. higher deuteron energies. If another subcritical bubble is found in this neighbourhood, the additional stress will cause simultaneous blistering, and so on. Thus, the energy dependence of the blister morphology (Fig. 1) can be explained.

We now assumed that the volume of these bubbles increases as function of the irradiation time only, whereas the current density determines their number per unit area. This may be understood by the consideration that during the beginning of the irradiation more nucleation centers are formed at higher current density. By this assumption the dependence of the critical dose of blistering on the deuteron current density is explained as follows: At constant dose and at low current density the trapped gas is distributed into comparatively few bubbles, which will fulfil the critical condition of eq. (4) earlier than in the case of a larger number of bubbles.

In order to obtain a more quantitative description, we derived the gas pressure p from the amount of trapped deuterons, taking into account gas reemission. As further simplifications, the gas reemission and the growth rate of the bubbles were chosen independent on the incident energy, and the growth was assumed to take place mainly by lateral expansion (at constant h). In this way, we were able to obtain a fit to the experimental data of the critical bombardment dose, which is shown in Fig. 8. The cover thickness d has been replaced by the incident energy, according to the relation displayed in Fig. 2.

The simple model yields a satisfactory result to explain the dependence of the critical blistering dose on the incoming particle energy and current density. Both are found to be nearly linear in the parameter range under consideration. The saturation of the critical dose at high current density (see Fig. 3) may be due to the coalescence of two subcritical bubbles, which is not covered by this model.

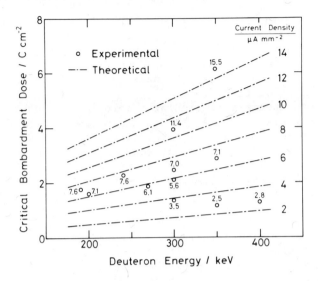

Fig. 8: Critical blistering dose as function of deuteron energy and current density: Experimental data and model calculation

REFERENCES

/1/ S.K.Das, M.Kaminsky, J.Appl.Phys. 44(1973)25
/2/ S.K.Erents, G.M.McCracken, Rad.Eff. 18(1973)191
/3/ H.Verbeek, W.Eckstein, Proc.Int.Conf. on the Application of Ion Beams to Metals, Albuquerque 1973; Plenum Press
/4/ J.Roth, R.Behrisch, B.M.U.Scherzer, J.Nucl.Mat. 53(1974)147
/5/ R.Behrisch, J.Bøttiger, W.Eckstein, U.Littmark, J.Roth, B.M.U.Scherzer, Appl.Phys.Lett. 27(1975)199
/6/ R.Behrisch, J.Bøttiger, W.Eckstein, J.Roth, B.M.U.Scherzer, J.Nucl.Mat. 56(1975)365
/7/ M.Hufschmidt, W.Möller, V.Heintze, D.Kamke, these conference proceedings
/8/ L.C.Northcliffe, R.F.Schilling, Nucl.Data Tables A7(1970)223
/9/ W.Möller, PhD Thesis, Bochum 1975 (to be published)
/10/ W.Primak, J.Luthra, J.Appl.Phys. 37(1966)2287

DISCUSSION

Q: (S. Das) 1. Your observation on dependence of critical dose for blister formation on incident ions flux (dose rate) is just the opposite to that observed by Verbeek and Eckstein two years ago. Can you comment on this?
2. Why do you assume a lenticular bubble in the bulk? I would expect the very small bubbles before coalescence to be nearly spherical as observed by transmission electron microscopy.
3. Lastly I want to comment that your blister skin thickness result for deuteron bombardment of nickel are very similar to our results for helium blisters in aluminum reported at the conference at Warwick last week.

A: (W. Möller) 1. The observations of Verbeek and Eckstein were carried out at a considerably lower energy (15 keV H^+ on Mo). At that low energies, probably different mechanisms take place.
2. We assume in our model that the blistering onset is due to the plastic deformation of the skin of one comparatively large bubble rather than coalescence of many small bubbles. The latter may be valid only in the case of helium, where the TEM observations have been carried out.
3. I think that at sufficiently high energy the cover thickness of the blisters will always be somewhat smaller than the mean ion range according to the damage distribution function. I don't know whether there is a correlation between He and D bombardment of fcc metals.

Q: (J. Roth) Could you calculate the activation energy for gas solution into the Ni-lattice from the temperature dependence of the amount of D-atoms retained within the range distribution? If yes, which value did you obtain?

A: (W. Möller) Yes, we could; the activation energy was about 0.2 eV in quite good agreement with extrapolated data from the literature.

Q: (R. Behrisch) How did you measure the deckeldicke and how accurate was this measurement?

A: (W. Möller) We used a stereographic method by means of the SEM. The accuracy was estimated to be within $\pm$ 5 %.

UNFOLDING TECHNIQUES FOR THE DETERMINATION OF DISTRIBUTION

PROFILES FROM RESONANCE REACTION GAMMA-RAY YIELDS

D. J. Land, D. G. Simons, J. G. Brennan* and M. D. Brown

Naval Surface Weapons Center

White Oak, Silver Spring, Maryland 20910

ABSTRACT

Analytical techniques for unfolding the concentration profile for a monoenergetic implant from the experimentally measured yield of a nuclear resonance reaction are discussed. The techniques include a least-squares fit on the profile assuming the particular analytic form of a split-Gaussian curve and a least-squares fit on the yield whose moments are related to the moments of the profile. Some mathematical comments on this general problem are made.

I. INTRODUCTION

Distribution profiles for 800 keV nitrogen ions implanted in a series of target materials from carbon (Z=6) to molybdenum (Z=42) have been determined by measuring the gamma-ray yield from the $^{14}N(p,\gamma)^{15}O$ resonance reaction at 1061 keV as a function of proton energy. This technique, while limited in applicability to those atoms whose nuclei have a convenient resonance reaction, is a fairly accurate and basically nondestructive method for profiling atomic distributions imbedded in a host material.

In this paper we discuss the problem of unfolding the concentration profile corresponding to a monoenergetic implant from the experimentally measured resonance yield. In particular we present a new approach which is proposed as a very fast computational

*Also Catholic University of America, Washington, D. C.

technique by which moments of the concentration profile can be estimated to good accuracy from appropriately determined moments of the resonance yield. The experiment which is discussed in this paper was performed on the 2.5 MV Van de Graaff accelerator at the Naval Surface Weapons Center. The experimental details and results are described elsewhere in this conference (1).

II. YIELD FROM RESONANCE REACTIONS

In this section we discuss the relation between the yield from a resonance reaction and the distribution of imbedded particles. Although we shall speak throughout this paper in terms of a (p,γ) reaction, it is clear that the analysis is of general validity.

The gamma-ray yield $Y(E_B)$ from an arbitrary distribution $n(R)$ of imbedded nuclei as a function of incident proton energy E_B is given by

$$Y(E_B) = \int dR\, n(R) \int dE\, \sigma_R(E) \int dE'\, I(E',E_B)\, F(E',E,R), \qquad (1)$$

where

$I(E',E_B)$ = the fluence of the initial proton beam at energy E' centered at energy E_B,

$F(E',E,R)$ = the distribution in the host material of protons having initial energy E' and energy E at the depth R,

$\sigma_R(E)$ = the resonance reaction cross section.

Equation (1) expresses the broadening effects upon the gamma-ray yield of the finite width of the incident proton beam, the effects of the straggling in energy of the protons as they transit the host material, the finite width of the resonance cross section, and finally the width of the distribution itself. If the first three processes were not present, i.e., they were infinitely sharp and expressed by δ-functions, the gamma-ray yield would profile the distribution $n(R)$ exactly.

It is of interest to write Eq. (1) in a more general form. We introduce the relative energy $\Delta = E_B - E_R$ (E_R is the resonance energy) and make the variable change under the first integral from R to $\Delta'(R)$, the average energy lost by the proton beam at the depth R. Then Eq. (1) is seen to be of the form of a Fredholm integral equation,

$$Y(\Delta) = \int d\Delta'\, n(\Delta')\, K(\Delta,\Delta') \quad, \qquad (2)$$

where the kernel $K(\Delta, \Delta')$ is a known function of Δ and Δ'.

The distribution function $F(E',E,R)$ is determined by the Vavilov solution to the particle transport equation (2). However, for energy losses greater than about ten times the maximum energy loss in a single collision $2 m_e v^2 \approx 2 \times 10^{-3} E_B$, the Vavilov solution can be approximated by a Gaussian distribution (3). Since $E_B \approx 1100$ keV in the present instance, this condition states that the Gaussian form may be used for energy losses greater than 20 keV. The distributions encountered in this investigation are sufficiently deep that this condition is satisfied.

The Gaussian distribution takes the form

$$F(E',E,R) = \frac{1}{\sqrt{2\pi\sigma_F^2}} \exp\left\{-\frac{[E - \varepsilon(E',R)]^2}{2\sigma_F^2}\right\}. \quad (3)$$

Here $\varepsilon(E',R)$ is the energy of a proton at the depth R in the target which has the initial energy E' at the surface and is found from the expression which relates range R to the stopping power $NS(E)$,

$$R = -\int_{E'}^{\varepsilon(E',R)} \frac{dE''}{NS(E'')}. \quad (4)$$

σ_F is the variance of the distribution which arises from the straggling in energy of the protons in the beam and was shown by Bohr (4) to have the form for high-energy (nonrelativistic) particles

$$\sigma_F^2 = N \Omega^2 R \quad (5a)$$

$$= 4\pi e^4 Z_1^2 Z_2 N R, \quad (5b)$$

where Z_1 is the effective charge of the incident particle ($Z_1 = 1$ for a proton at the energies of interest here), Z_2 is the atomic number of the target atom, and N is the density of the target atoms. In our numerical work to determine the distributions of the nitrogen atoms, we have used a set of calculated values for the proton straggling factor due to Chu and Mayer (5) and based on the work of Bonderup and Hvelplund (6). These can be written in terms of the Bohr expression by introducing a simple multiplicative correction. In the present analysis we shall simply write Eq. (5a), $\sigma_F^2 = N\Omega^2 R$, since it is the R dependence that is important.

To proceed with the evaluation of Eq. (1) we introduce two approximations:

1. the proton stopping power is expanded as a linear function of energy,

$$NS(E) = NS(E_B) + (E - E_B) \, dNS(E_B)/dE \; ;$$

2. the integral over E' in Eq. (1) is extended from $-\infty$ to $+\infty$. In addition we take the incident proton fluence $I(E',E_B)$ to have a Gaussian form with variance σ_B. Then, with the use of Eq. (3), it can be shown that Eq. (1) simplifies to

$$Y(E_B) = \int dR \, n(R) \int dE \, \sigma_R(E) \frac{\exp\left\{-\frac{[E - \varepsilon(E_B,R)]^2}{2(\sigma_B^2 + N\Omega^2 R)}\right\}}{\sqrt{2\pi(\sigma_B^2 + N\Omega^2 R)}} \qquad (6)$$

Equation (6) is the fundamental relation for our work. Clearly the use of the Gaussian form of Eq. (3) simplifies the analysis considerably.

III. ANALYTICAL TECHNIQUES

1. Least-Squares Fit on n(R)

The first method we consider for unfolding the concentration profile $n(R)$ from the measured yield $Y(E)$ is a least-squares fit on $n(R)$. A split-Gaussian curve, two half-Gaussian curves of different widths σ_- and σ_+ and joined at the peak R_p, was assumed for $n(R)$. This form is frequently used to describe the profile and has been discussed by Gibbons and Mylroie (7). The three parameters which describe this curve, R_p, σ_- and σ_+, were determined by performing a least-squares fit of the calculated gamma-ray yield $Y(E)$ given by Eq. (6) to the values of the measured yield Y_i. (It is understood that the various sources of background have been subtracted from the actual experimental counting rates to get the Y_i.) This procedure thus yields a specific distribution for the concentration profile whose moments are readily obtained.

In Fig. 1 we show the concentration profile obtained by the least-squares-fitting procedure for 800 keV nitrogen ions implanted in iron. Also shown for comparison are the gamma-ray yield calculated from this profile and the experimentally measured gamma-ray yield. The fit of the calculated yield curve to the experimental points is excellent. One can see clearly the effect of the proton straggling in the increased broadening of the profile at deeper

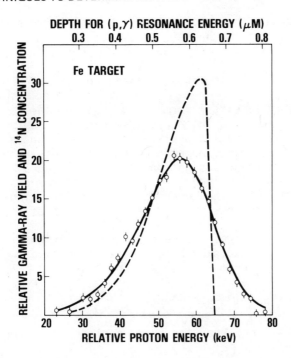

Figure 1. Calculated gamma-ray yield (solid curve) and concentration profile (dashed curve) for 800 keV $^{14}N^+$ ions implanted in iron. The circles show the experimental points with error bars.

layers of the target. This example shows the necessity for unfolding the profile from the resonance yield. One might also note the sharp falloff of the profile itself to the right of the peak. The split-Gaussian form used here to describe the profile cannot have a third moment larger in absolute value than the second moment. This profile has essentially reached this limiting case. However, although the specific form of this profile may not be uniquely determined, we believe that the values obtained for the moments are accurate. Discussions of the results of the other cases studied in this experiment are presented elsewhere in this conference (1).

The $^{14}N(p,\gamma)^{15}O$ resonance reaction used in the experimental work occurs at 1061 keV with a FWHM of 4.8 keV and a cross section at resonance of 0.37 mb (8). To represent it we have used a Breit-Wigner form and have attempted to carry out the integration in Eq. (6) over the entire curve by cutting it off at the energies at which the amplitude is 0.005 of its peak value. The values for the second moment of n(R) obtained by this procedure are not completely insensitive to the choice of cutoff. We will discuss the implications of this point in the next section.

2. Moments Analysis on Y(E)

The relatively simple form of Eq. (6) for the gamma-ray yield suggests another method which provides excellent estimates for the moments of a concentration profile formed by a monoenergetic implant from suitably determined moments of the yield distribution. This general approach sheds insight on the role played by the various physical processes involved in giving rise to the observed gamma-ray yield and also avoids the numerically lengthier calculations associated with the previously described least-squares-fit technique.

In the analysis we make several approximations:
1. The proton stopping power is taken to have a constant value NS. From Eq. (4) we find that

$$\varepsilon(E_B, R) = E_B - NSR .\tag{7}$$

2. The resonance reaction cross section is assumed symmetric, $\sigma_R(E) = \sigma_R(|E - E_R|)$.
3. The energy width of the incident proton beam is neglected.
4. All energy integrals are taken from $-\infty$ to $+\infty$.

With these approximations we may rewrite Eq. (6) for the gamma-ray yield as a function of the relative energy $\Delta = E_B - E_R$,

$$Y(\Delta) = \int_0^\infty dR\, n(R) \int_{-\infty}^\infty dE\, \sigma_R(E - E_R) \frac{\exp\left\{-\dfrac{\left[\Delta - (E - E_R + NSR)\right]^2}{2N\Omega^2 R}\right\}}{\sqrt{2\pi N\Omega^2 R}}$$

Notice that the exponential form in this expression is a normalized Gaussian distribution with respect to Δ; thus moments of this distribution with respect to Δ give rise simply to polynomials in R and $E - E_R$. We let $\overline{Y}$ be the first moment or centroid of $Y(\Delta)$ and $\overline{\Delta Y^n}$ be the n th moment evaluated with respect to $\overline{Y}$,

$$\overline{\Delta Y^n} \equiv \overline{(Y - \overline{Y})^n} = \int_{-\infty}^\infty d\Delta\, (\Delta - \overline{Y})^n Y(\Delta) ,$$

with corresponding definitions for $\sigma_R(E - E_R)$ and $n(R)$. (We define the variance $\sigma_R^2 \equiv \overline{\Delta \sigma_R^2}$, to establish consistent notation for the several variances we use.) One may readily find the following expressions for the first three moments of $Y(\Delta)$:

$$\overline{Y} = N S \overline{n}, \tag{8a}$$

$$\overline{\Delta Y^2} = \overline{(Y - \overline{Y})^2} = N^2 S^2 \overline{\Delta n^2} + N \Omega^2 \overline{n} + \sigma_R^2, \tag{8b}$$

$$\overline{\Delta Y^3} = \overline{(Y - \overline{Y})^3} = N^3 S^3 \overline{\Delta n^3} + 3 N S \overline{n} N \Omega^2 \overline{n} (\overline{\Delta n^2}/\overline{n}^2). \tag{8c}$$

The interpretation of these equations is obvious. The cross term that appears in the expression for the third moment arises from the asymmetry in R of the proton beam in the host material. These relations can be easily inverted to express the first three moments of the concentration profile in terms of the first three moments of the yield distribution.

To use these relations it is necessary to determine $\overline{Y}$, $\overline{\Delta Y^2}$ and $\overline{\Delta Y^3}$ from the experimental data. One might be tempted simply to compute these quantities directly from the data, e.g., by setting

$$\overline{Y} = (\Sigma_i \Delta_i Y_i) / (\Sigma_i Y_i),$$

where the summations extend over all the data points. However, the Y_i are known only in a statistical sense and, further, are frequently known only poorly near the lower and upper energy limits to which the moments are particularly sensitive. A better procedure is to smooth the experimental data by fitting the data to an appropriate analytic curve from which the required moments can be obtained. In the present problem we have accomplished this by performing a least-squares fit of the data to, again, a split-Gaussian.

One other quantity is also needed as input for this analysis, the variance associated with the resonance reaction cross section, σ_R. This cross section, normally represented by the Breit-Wigner form, describes the resonance closely in the vicinity of the peak, but it is not clear how well the actual experimental signal is represented in the wings, particularly in the presence of a large background. Furthermore, no moments exist for this analytic form. To determine the effective variance of this cross section, we have computed the gamma-ray yield from a thin layer of nitrogen nuclei and determined its variance from the point at which the yield falls to 0.61 of its peak value (i.e., we are implicitly assuming that the gamma-ray yield is Gaussianlike). From Eq. (8b) with $\overline{\Delta n^2} = 0$ we can determine σ_R^2. This procedure leads to the result $\sigma_R = 2.7$ keV for the $^{14}N(p,\gamma)^{15}O$ resonance reaction. This value might be compared with the FWHM for the reaction of 4.8 keV (8).

We summarize the numerical results of this method by showing for the three moments of the concentration profile, $\bar{n}$, $\overline{(\Delta n^2)}^{1/2}$ and $\overline{(\Delta n^3)}^{1/3}$ the percentage change of the value of each moment as obtained from the moments analysis on $Y(E)$ and from the least-squares fit on $n(R)$. This is done in Fig. 2 (closed circles) for the cases $Z_2=22$ to 32 and $Z_2=40$ to 42. We see that the agreement for the first and third moments is quite close. However the values for the second moment obtained by the least-squares fit on $n(R)$ is systematically lower by about 10% than the values obtained by the moments analysis on $Y(E)$. This difference arises from the different handling of the resonance cross section in the two approaches. In the least-squares fit on $n(R)$ we have integrated $\sigma_R(E)$ in Eq. (6) over the entire resonance curve. This gives rise to a large variance for $\sigma_R(E)$. The procedure employed for the moments analysis on $Y(E)$ described above gives rise to a smaller variance. Hence, by Eq.(8b) the profile width is smaller in the first case and larger in the second since $\overline{\Delta Y^2}$ remains roughly constant. The two methods of handling the resonance cross section were considered appropriate to the specific procedure to which each is applied. Basically this ambiguity arises because of our lack of knowledge of the exact shape of the resonance cross section.

The Pearson distribution (9) was also considered for use in the present context. This distribution has the following form appropriate to the present application:

$$f(x) = c(x) \big/ |b_0 + b_1 x + b_2 x^2|^p, \qquad p > 0,$$

where $c(x)$ is a slowly varying function that approaches different constant limiting values as $x \to \pm\infty$. The fact that it is a nonnegative, unimodal form and is a function of four parameters suggested that it could serve as a convenient form to obtain possibly a fourth moment for the profile. However, the least-squares fit for the yield curves indicated poor convergence and, when convergence was achieved, the numerical results for the higher moments show somewhat inconsistent agreement with our previous results. The difficulties may be attributed to the fact that this distribution tends to provide a somewhat better fit than the split-Gaussian to the yield curve in the wings. But this portion of the curve is not well determined by the data and can vary considerably from one element to another. Because the moments are sensitive to this region of the curve, the results are inconsistent. The ratios for the first three moments are also shown in Fig. 2 (open circles). One should note, however, that the first moments do show good agreement.

IV. CONCLUSIONS

The striking feature of the numerical results obtained in this study is the closeness to each other of the values of the first moment from all three approaches used. On the other hand, the

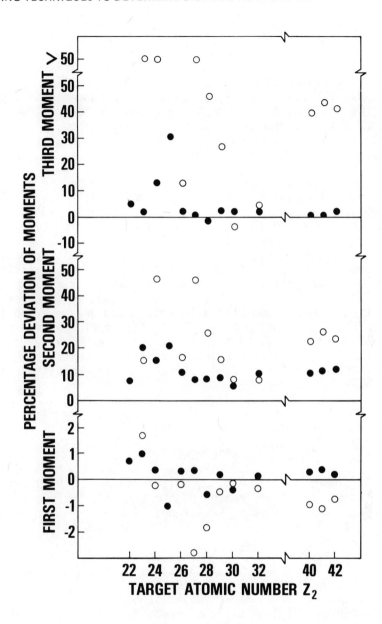

Figure 2. Percentage deviation of the first, second and third moments, where the solid circles and open circles represent the percentage deviation of the fits obtained by the split-Gaussian and Pearson distributions, respectively, on the yield $Y(E)$ with respect to the least-squares fit on the profile $n(R)$.

values of the second moment obtained from the moments analysis of the yield represented by the split-Gaussian form are consistently about 10% higher than those obtained by the least-squares fit to the profile. This difference, which arises from the difference in the representations used to describe the resonance reaction cross section, emphasizes the necessity of constructing a definitive description of the cross section. The third moments obtained from the same approaches are in agreement, with but two exceptions, to closer than 5%. The lack of agreement which was found with use of the Pearson distribution was commented on above.

The technique of unfolding the profile through the moments analysis on the yield is limited to 1) sufficiently deep implants and 2) monoenergetic implants. The first limitation, which arose through the appearance of a Gaussian distribution function for the particle beam in the host material, could in principle be relaxed somewhat by establishing an analytic form for this distribution valid nearer to the front surface, in the form of an expansion of the Vavilov distribution. The second limitation is real in that only concentration profiles which can be adequately described by the first few moments may be considered. However, when applicable, it is capable of providing an excellent description of the profile with a minimum of computational effort. The technique of unfolding through the least-squares fit on the profile is quite general in principle and is not restricted by either of the conditions outlined above.

REFERENCES

(1) D. G. Simons, D. J. Land, J. G. Brennan and M. D. Brown, to be published in the proceeding of the 2nd Int. Conf. on Ion Beam Surface Layer Analysis, Karlsruhe, 1975.
(2) P. V. Vavilov, Zh. Exper. Teor. Fiz. $\underline{32}$, 320,(1957). (Trans., JETP $\underline{5}$, 749 (1957)).
(3) S. M. Seltzer and M. J. Berger, Sec. 9, Natl. Acad. Sci. - Natl. Res. Council Publ. 1133, U. Fano, ed., 2nd printing, 1967.
(4) N. Bohr, Phil. Mag. $\underline{30}$, 581 (1915).
(5) W. K. Chu and J. W. Mayer (unpublished).
(6) E. Bonderup and P. Hvelplund, Phys. Rev. $\underline{A4}$, 562 (1971).
(7) J. F. Gibbons and S. Mylroie, Appl. Phys. Lett. $\underline{22}$, 563 (1973).
(8) P. M. Endt and C. Van der Leun, Nucl. Phys. $\underline{A214}$, 205 (1973).
(9) M. G. Kendall and A. Stuart, "The Advanced Theory of Statistics" (C. Griffin and Company Ltd., London, 1963).

DISCUSSION

Q: (W.K. Chu) Could you specify the direction on the systematic 10 % accuracy you have on the second moment calculation? Are you overestimating or underestimating the 2nd moment by up to 10 % in your analysis?

A: (D.J. Land) The 10 % difference is related to the two methods of analysing our data. The least-squares fit to the concentration profile, which we consider to be the more accurate analysis, gives a ΔR_p about 10 % lower than the least-squares fit to the yield distribution, the variation arising from the two differing treatments of the Breit-Wigner resonance curve.

Q: (A. Fontell) Can you give a rough estimate for the ratio of the number of implanted atoms to the number of target atoms in the maximum of the concentration profile because at least the form of the profile is easily affected by the stopping of nitrogen atoms already collected?

A: (D.J. Land) The ratio at the maximum is about 10 %. To check whether this level has affected the profile, we have made measurements on separately implanted iron and molybdenum targets for which the ratio at the maximum was 1 %. We found no significant difference for the first three moments in either case. Estimates of the effect of the nitrogen on the probing protons indicated an uncertainty in the projected range of no more than 1 %.

Z_2 DEPENDENCE OF THE ELECTRONIC STOPPING POWER OF 800 keV $^{14}N^+$ IONS IN TARGETS FROM CARBON THROUGH MOLYBDENUM

D. G. SIMONS, D. J. LAND, J. G. BRENNAN[*] and M. D. BROWN

Naval Surface Weapons Center

White Oak, Silver Spring, Maryland 20910

ABSTRACT

An oscillatory behavior of S_e as a function of Z_2 is inferred from concentration distributions of 800 keV $^{14}N^+$ ions implanted in metallic targets with atomic numbers from Z_2= 6 through 42. The distributions were measured using the $^{14}N(p,\gamma)^{15}O$ resonance reaction. Comparison of the first and second moments of these distributions with those predicted by the theory of Lindhard, Scharff, and Schiott (LSS theory) is made. By altering the electronic stopping power, S_e, in the LSS theory, the predicted projected range was adjusted to be equal to the first moment of the measured distribution.

INTRODUCTION

The accumulated results from experiments by several groups show that the electronic stopping power, S_e, exhibits a characteristic oscillatory behavior as a function of the charge of the incident particle, Z_1. This work was done for constant velocity ions in the range from lithium (Z_1= 3) to yttrium (Z_1= 39) incident on amorphous carbon targets (1, 2) and silicon targets in two different channeling directions. (3) The positions of the maxima and minima are roughly independent of target material. In addition, other experiments using the light ions of protons and ^{4}He show similar oscillations in S_e as a function of the target atomic number, Z_2, for a large range of target materials based on the results of several experiments. (4) It was the purpose of this investigation to extend the studies of the behavior S_e as a function of

Z_2 for heavy ions for a relatively wide range of targets. Some evidence for Z_2 oscillation of S_e for heavier incident ions has been presented by Hvelplund for 200 keV ^{16}O ions. (5)

The possibility of such oscillatory behavior has some interesting consequences as it relates to atomic structure in models which describe the electronic stopping. Models of electronic stopping developed by Lindhard et al (6) and by Firsov (7) predict a smooth behavior in the S_e as a function of both Z_1 and Z_2. However, extensions by several investigators (8) of the Firsov model which take into account the atomic structure of both incident and target atom show oscillations in S_e as a function of Z_1. Furthermore, calculations by Rousseau et al (9) show that S_e also has a similar oscillatory behavior as a function of Z_2 with incident ions of protons or ^4He.

EXPERIMENTAL METHOD

In the present experiment, we have determined the range and distribution profiles of 800 keV $^{14}N^+$ ions implanted in selected targets from Z_2= 6 to 42 using gamma-ray yield measurements from the $^{14}N(p,\gamma)^{15}O$ resonance reaction. From these distributions we have determined the values of the first three moments, quantities which are most readily interpretable in terms of theoretical analysis. Values of S_e were inferred from the transport theory of Lindhard, Scharff and Schiott (LSS) (10) by requiring the calculated projected range to be equal to the first moment of the measured distributions.

Both the ^{14}N implantation and the proton probing of the distribution using the (p, γ) resonance were carried out with the Naval Surface Weapons Center 2.5 MV Van de Graaff accelerator. In both phases contaminate build-up on the targets was minimized by placing a liquid nitrogen, cold finger directly in front of the target. Mercury pumps were used but on the basis of a proton backscattering measurement on an Fe target, we estimate that there was less than a 1% change in range due to Hg build-up. No measurements on carbon build-up were taken. An electron supressor set at -1000V was used to ensure reliable current integration.

The ^{14}N distributions were measured using the 1061.0 keV resonance in the $^{14}N(p,\gamma)^{15}O$ reaction which has a width of 4.8 keV and a cross-section at resonance of 0.37 mb. (11) During the proton probing the targets were mounted on a water-cooled holder to minimize target overheating and possible diffusion of the implanted ions. The gamma rays were detected by a 7.62 cm by 7.62 cm diameter NaI detector placed 3.0 cm from the target at a 0° angle. The gamma-ray signal was amplified and passed through a single channel analyzer to a scaler, gated by the current integrator. The single channel analyzer was set to accept only the highest energy gamma-ray, 8.28 MeV. (12)

Targets of two types were used. For most cases bulk samples were used provided their surface condition was sufficiently smooth. For those cases with large (p,γ) backgrounds from the host material, thin targets of at least 3μ thick were prepared by evaporating the target onto a Ta backing, which has a low (p,γ) background in the energy range of interest.

The gamma-ray yield measurements were accumulated from typically six individual sweeps through the energy range of interest. This method was used so that apparent changes in the ^{14}N distributions which might occur through target overheating would be detected. Throughout all of the measurements, no discernable difference was detected in the gamma-ray yield curves and hence in the concentration distributions for any of the targets. This stability also tends to rule out extensive contaminant build-up on target.

DATA PROCESSING

The gamma-ray yield distributions were obtained after subtracting time dependent backgrounds from natural sources in the detector environment and charge dependent backgrounds from (p,γ) reactions in the host material and off-resonance reaction from the implanted ^{14}N. After background subtraction the gamma-ray yield distributions were unfolded to obtain the ^{14}N concentration distribution. In order to do this several processes were included. These are: (1) The finite width of the incident proton beam, (2) the finite width of the reaction cross-section represented by the Breit-Wigner resonance cross-section and (3) the energy straggling of the proton beam as it penetrates the target. Comparative techniques by which this unfolding can be accomplished are discussed elsewhere in the conference.

RESULTS

Selected targets from carbon (Z_2= 6) through molybdenum (Z_2= 42), were implanted with 800 keV ^{14}N ions at fluences of the order of 10^{17} ions/cm^2. The resulting distributions were profiled utilizing the 1061.0 keV resonance of width 4.8 keV from the $^{14}N(p,\gamma)^{15}O$ reaction. Representative results of measurements for Z_2= 22 - 32 are given in Fig. 1 where the relative gamma-ray yield is shown as a function of the incident proton energy relative resonance energy. All of the distributions are normalized to the same area. The errors in the measured yield distributions are counting errors only. The solid curves shown in Fig. 1 are calculated yield curves

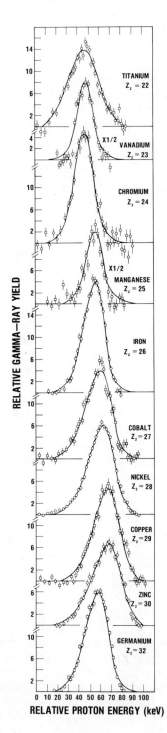

Figure 1. Gamma-ray yield distributions for 800 keV N^+ ions implanted in metallic targets for Z = 22 through 32 obtained from the $^{14}N(p,\gamma)^{15}O$ resonance reaction at 1061.0 keV. Error bars are counting errors only. The solid curves are least-square fit gamma-ray yield curves obtained from split-Gaussian concentration distributions. All curves are normalized to the same area.

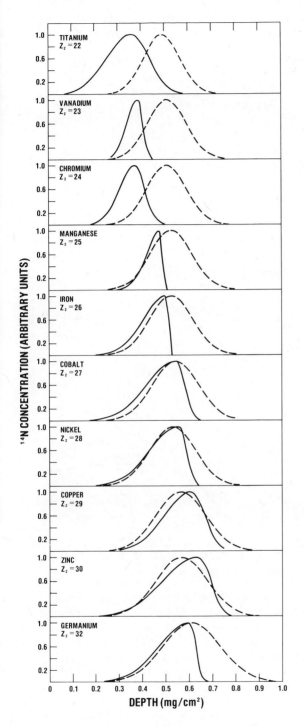

Figure 2. Concentration distributions of 800 keV $^{14}N^+$ ions implanted in metallic targets for Z = 22 through 32. The solid curves are split-Gaussian functions which result in the least-squares fit to the gamma-yield curves. The dashed curves are the distributions predicted by the LSS theory. All curves are normalized to the same peak height.

obtained from a least-squares fit to the data assuming a split-Gaussian concentration distribution for the implanted ions.

The concentration distributions obtained from the least-squares fit to the gamma-ray yield distributions for those distributions in Fig. 1 are shown in Fig. 2. The dashed curves are those predicted by the LSS theory in which a Gaussian distribution is assumed. All of the curves are normalized to the same peak height for ease in comparison of the distribution shapes. For all of the targets investigated the distributions are generally shallower and narrower than those predicted by theory. In addition, we note that the

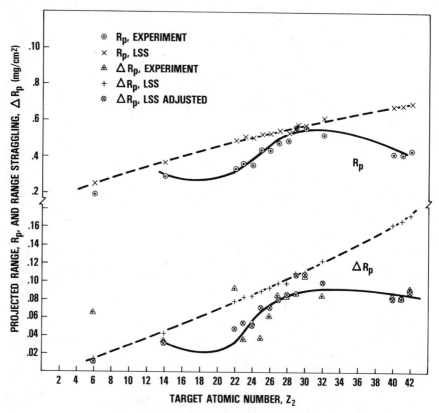

Figure 3. The projected range, R_p, and straggling, ΔR_p, for 800 keV $^{14}N^+$ ions implanted in targets of atomic numbers ranging from $Z_2 = 6$ through $Z_2 = 32$. The curves are drawn here to guide the eye with the solid curves representing the experimental results, and the dashed curves the LSS prediction. The experimental points are the first and second moments of the experimental ^{14}N concentration distributions. The LSS adjusted points are resulting straggling values when the adjusted electronic stopping power is used in the LSS theory. In this Figure the calculated first moments are equal to the experimental first moments.

concentration distributions are not symmetric but skewed such that, from the peak position, all distributions decrease more rapidly toward the deeper depths than towards the surface.

From the concentration distribution, we have obtained the first and second moments for direct comparison with the projected range and range straggling as obtained in the LSS theory. These results are shown as a function of Z_2 in Fig. 3. The deviations between the measured concentrations and the predicted LSS distributions as shown in Fig. 3 show that stopping powers greater than those used in the LSS theory are needed to predict the distribution more accurately. Since the electronic stopping power, S_e, is dominant over the nuclear stopping power, S_n, from the implant energy of 800 keV down to about 30 keV, an alteration of this quantity was used to bring the theoretical values into agreement with the measured values.

Thus we adjust values of the S_e such that the theoretically determined values of the projected ranges are equal to the experimental values of the first moments of the concentration distributions. These values of S_e are displayed in Fig. 4 as a function of Z_2. The changes in S_e are made by altering the LSS value of S_e by a multiplicative constant. We, therefore, assume a linear dependence of S_e on v. From Fig. 4 we see that increases in S_e by almost 100%

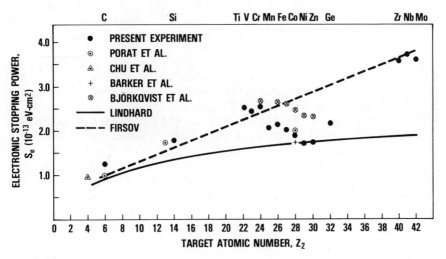

Figure 4. The experimental LSS and Firsov values for the electronic stopping power for 800 keV $^{14}N^+$ ions as a function of the target atomic number Z_2. The experimental values are obtained by altering the LSS value by a multiplicative constant such that the first moment from the LSS transport theory is in agreement with the experimental values of the concentration distribution.

are needed to account for the measured distribution, and S_e exhibits an oscillatory Z_2 dependence which is not accounted for by the Lindhard (6) or Firsov (7) pictures of electronic stopping.

We include in Fig. 4 values of S_e obtained by other investigators for 800 keV ^{14}N ions (13 - 16). Of particular interest are the results of Porat and Ramavataram (13) in which S_e was determined directly from the energy loss of ^{14}N through thin metallic foil We have good agreement with the 800 keV results shown here and with a measurement of 1200 keV ^{14}N on Ni. For this higher energy our results of 2.37×10^{-13}ev-cm^2 compare favorably with Porat's value of 2.46×10^{-13}ev-cm^2. The results of the experiments of Björkqvist and Domeij (14) and Barker and Phillips (15) do not compare quite so favorably with our results. In both cases, S_e was not measured directly but was inferred (as in the case here) from backscattering and resonance reaction measurements respectively.

The calculation of the profile parameter R_p from the adjusted values of S_e also results in adjusted values of the straggling (LSS adjusted). The values of ΔR_p (LSS adjusted) as shown in Fig. 3 follow the measured values much more closely than those obtained from the LSS theory.

Since these targets are implanted with relatively high fluences such that the highest atomic density is about ten percent of the host material, some concern was given to the possibility of a fluence effect on the measured concentration distributions. The implantation of heavy particles in a host material causes a change in the stopping power and straggling of that material from the addition of the particles themselves and possibly from the damage inflicted on the material by the implantation. This change is effective both during implantation and during proton profiling. We have made measurements of the ^{14}N concentration distributions in which implantation in Fe and Mo was carried out at a fluence of one-tenth that normally used in this experiment. For both targets the concentration profiles for the two cases of fluence differing by a factor of ten are not appreciably different. Thus, we conclude that the effects of fluence are not significant to the results of this experiment.

REFERENCES

* Also Catholic University of America, Washington, D. C. 20017
(1) J. H. Ormrod, J. R. Macdonald and H. E. Duckworth, Can. J. Phys. **43**, 275 (1965)
(2) P. Hvelpund and B. Fastrup, Phys. Rev. **165**, 408 (1968)
(3) F. H. Eisen, Can. J. Phys. **46**, 561 (1968)

(4) M. Bader, R. E. Pixley, F. S. Mozer and W. Wholung, Phys. Rev. 103, 32, (1965); D. Wayne Green, John N. Cooper, and James C. Harris, Phys. Rev. 98, 466 (1955); W. K. Lin, H. G. Olson and D. Powers, Phys. Rev. B, 8, 1881 (1973)
(5) P. Hvelplund, Kgl. Danske Videnskab. Selskab, Mat.-Fys. Medd. 38, No. 4 (1971)
(6) J. Lindhard and M. Scharff, Phys. Rev. 124, 128 (1961)
(7) O. B. Firsov, Zh. Eksperim. i Teor. Fiz. 36, 1517 (1959)
(8) I. M. Cheshire, G. Dearnaley and J. M. Poate, Proc. Roy. Soc. A311, 47 (1969); C. P. Bhalla and J. N. Bradford, Phys. Lett. 27A, 318 (1968); C. P. Bhalla, J. N. Bradford and G. Reese, Atomic Collision Phenomena in Solids, edited by D. W. Palmer, M. W. Thompson and P. D. Townsend (North-Holland, Amsterdam) P. 361; K. B. Winterbon, Can. J. Phys. 46, 2429 (1968); A. H. El-hoshy and J. F. Gibbons, Phys. Rev. 173, 454 (1968)
(9) C. C. Rousseau, W. K. Chu, and D. Powers, Phys. Rev. A 4, 1006, (1971)
(10) J. Lindhard, J. Scharff and H. E. Schiott, Kgl. Danske Videnskab. Selskab, Mat.-Fys. Medd. 33 No. 14 (1963); J. Lindhard, V. Nielson and M. Scharff, Kgl. Danske Videnskab. Selskab. Mat.-Fys. Medd 36, No. 10 (1968)
(11) P. M. Endt and C. Van der Leun, Nucl. Phys. A214, 205 (1973)
(12) A. E. Evans, B. Brown and J. B. Marion, Phys. Rev. 149, 863 (1966)
(13) D. I. Porat and K. Ramavataram, Proc. Phys. Soc. London, 78, 1135 (1961)
(14) W. K. Chu and D. Powers, Phys. Rev. 187, 478 (1969)
(15) P. H. Barker and W. R. Phillips, Proc. Phys. Soc. 86, 379 (1965)
(16) K. Björkqvist and B. Domeij, Radiation Effects 13, 191 (1972)

SENSITIVITY OF FLUORINE DETECTION IN DIFFERENT MATRICES
AND AT DIFFERENT DEPTHS THROUGH THE $^{19}F(p,\alpha\gamma)^{16}O$ REACTION

M. A. Chaudhri*, G. Burns, J. L. Rouse and B. M. Spicer

School of Physics, University of Melbourne

Parkville, 3052, Australia

ABSTRACT

Sensitivities for detecting fluorine distributed in elements with atomic numbers, 6, 13, 20, 28, 34, 42, 50, 60, 68, 79 and 92 through the $^{19}F(p,\alpha\gamma)^{16}O$ reaction has been calculated using experimentally measured excitation functions and the available energy-range data. Thick target yields of the prompt 6.13 MeV gamma as well as of the three gamma lines 6.13, 6.92 and 7.12 MeV combined have been plotted as a function of the incident proton energy of up to 4.16 MeV. From these yield curves the sensitivity of detecting fluorine in thick or thin samples and even in a layer of known thickness at a particular depth within a thick target can be directly read for known bombarding and detecting conditions. The curves should also be valid, to a certain approximation, for neighbouring elements and for mixtures or compounds with similar average atomic numbers. It has been shown that at incident proton energies of 1 and 4 MeV fluorine concentrations of a few parts per billion and as little as a fraction of one p.p.b. respectively can be detected in most matrices with 1 μa beam intensity and an hour of irradiation using this technique. It is also shown how the yield curves may be used in depth profile analysis of fluorine, especially at low concentrations where higher energies are needed and the resonances in the reaction are not sharp.

INTRODUCTION

Detection of fluorine with chemical techniques is not only

*Also University of Islamabad, Islamabad, Pakistan.

destructive but may also give unreliable results due to a number
of reasons like the loss or gain of fluorine in the sample during
chemical processing, incomplete extraction and/or ionization of
fluorine in solution when a solid sample is dissolved etc. The
x-ray fluorescence method for fluorine determination is difficult
due to rather low energies of the fluorescent x-rays and lacks
sensitivity too (limit of detection 0.2%) /1/. Activation
analysis with fast neutrons and bremsstrahlung may lead to the
tedious task of having to separate exponentials when other positron
emitting isotopes are also being produced along with ^{18}F. On the
other hand fluorine determination through the $^{19}F(p,\alpha\gamma)^{16}O$ reaction
is not only reliable and convenient but very sensitive too (limit
of detection being a fraction of one p.p.b.) due to the large
reaction cross section and easily observable prompt gammas with
almost no interference from other gammas. This technique is
quick, non-destructive and can also be applied to "bulk" samples
and to depth profile measurements.

In order to exploit the full potential of this reaction for
fluorine determination it may be of great help to know in advance
the expected sensitivities of detection in different elements/
matrices under various bombarding conditions. Such an information
would assist in selecting the optimum bombarding and detecting
conditions for a particular problem and would also indicate if the
problem could be solved with the existing experimental facilities.

In this paper we present calculated detection sensitivities
of fluorine distributed uniformly in a number of elements throughout the periodic table for incident proton energies from 300 keV
to 4.16 MeV.

THEORY AND METHOD

The gamma yield from a thick target under bombardment with
charged particles is given by /2/

$$P(\text{disintegrations/sec}) = \frac{\rho_t}{\rho} \frac{N_p N_A}{A_t} \int_{E_o}^{E} \sigma_E \left[\frac{dE}{d(\rho x)}\right]^{-1} dE \quad (1)$$

where
N_p = number of charged particles per second incident upon the target which is equal to $6.26\ I_{\mu a}\ \frac{10^{12}}{Z}$, $I_{\mu a}$ being the beam intensity in microamperes and Z the atomic number of projectiles

N_A = the Avagadro's number

A_t = atomic number of the target element within the sample undergoing nuclear transformation (fluorine in this case)

ρ_t = density of the target element

ρ = density of the element/compound containing the target element

σ_E = the reaction cross section at energy E

E_o = the minimum energy at which any appreciable amount of nuclear reaction takes place

$\frac{dE}{d(\rho x)}$ = stopping power in g/cm^2 of the matrix (element/compound containing the target element) for the incident charged particles.

Using the computer method developed by Chaudhri and Burns /2/ the values of P in eqn. (1) have been calculated for fluorine uniformly distributed in elements with atomic numbers 6, 13, 20, 28, 34, 50, 60, 68, 79 and 92 for incident proton energies from 300 keV to 4.16 MeV. The criterion for the selection of these elements has been that they are evenly distributed in the periodic table and that their stopping power values for protons of different energies are easily available from the tables /3/. The values of P calculated for these elements should also be approximately applicable to neighbouring elements as well as to mixtures and compounds with similar average atomic numbers as the stopping powers do not change drastically with a slight change in the atomic numbers. The excitation functions for the three gamma lines combined used in the calculations have been taken from references /4/ (from 300 keV to 2 MeV) and /5/ (from 1.9 to 4.16 MeV) after normalizing the results of the former authors with those of the latter ones around 2 MeV. The reaction cross section values at different energies for the 6.13 MeV gamma alone have been taken from /5/ which only goes from 1.9 to 4.16 MeV. No reliable cross section data for this line was available below 1.9 MeV. This prompt gamma has specially been dealt alone too because this line is sharp and well resolved while the other two gammas are Doppler-broadened. In cases where there are other interfering gamma energies around these fluorine lines or where the background is substantial it may be advantageous to concentrate on the 6.13 MeV gamma alone instead of summing all the three lines. The values of P for different elements are plotted as a function of the proton energy to give the yield curves.

RESULTS AND DISCUSSION

Thick-target yield/sensitivity curves for fluorine determination in different matrices with proton bombardment of up to 4.16 MeV are shown in figures 1(a) to 1(k). The upper curve is due to the three gamma lines combined while the lower one refers to the 6.13 MeV gamma only. The steps in the upper curve are due to sharp resonances at low energies. The ranges corresponding to the proton energies in different elements are also

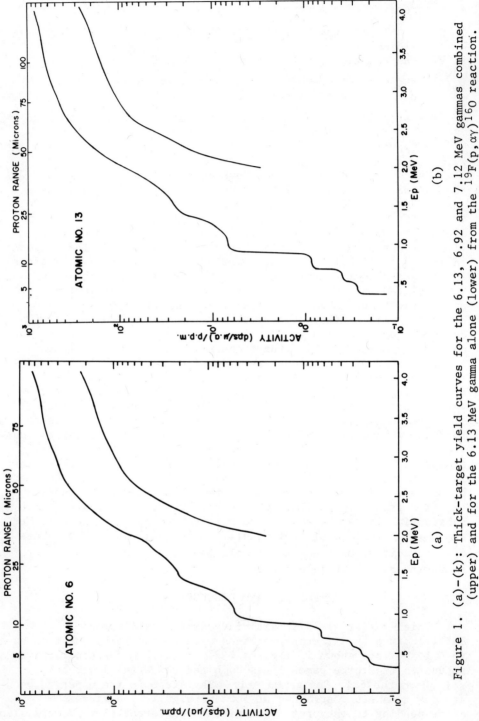

Figure 1. (a)-(k): Thick-target yield curves for the 6.13, 6.92 and 7.12 MeV gammas combined (upper) and for the 6.13 MeV gamma alone (lower) from the $^{19}F(p,\alpha\gamma)^{16}O$ reaction.

SENSITIVITY OF FLUORINE DETECTION IN DIFFERENT MATRICES

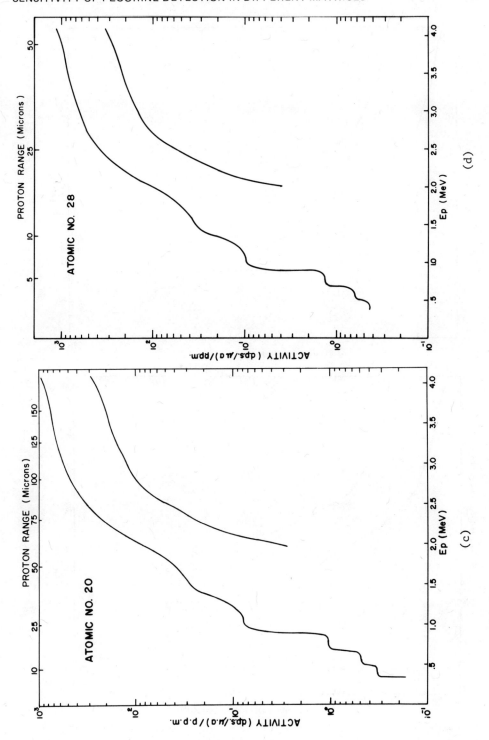

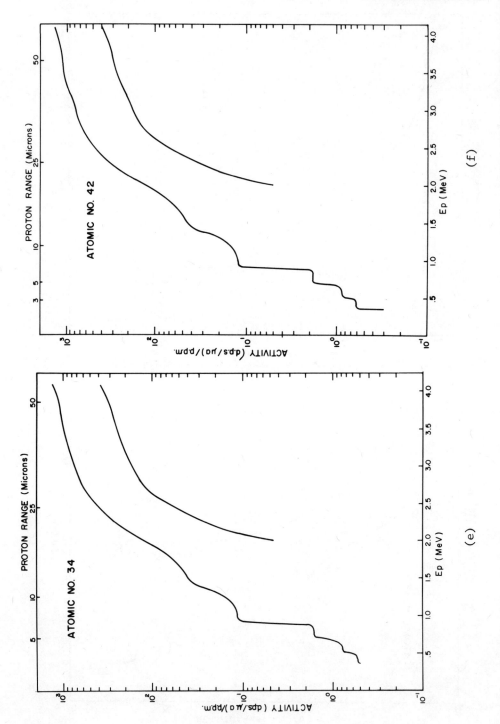

SENSITIVITY OF FLUORINE DETECTION IN DIFFERENT MATRICES 879

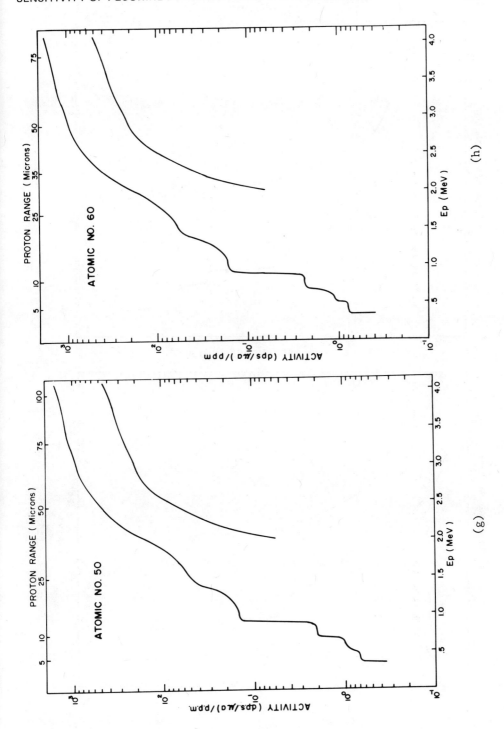

(h)

(g)

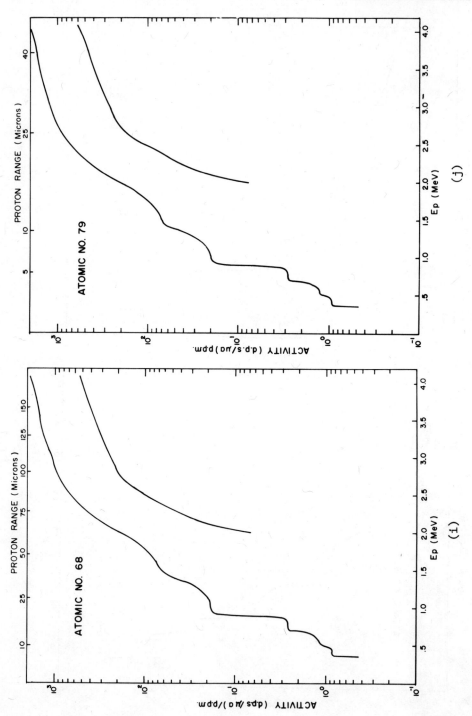

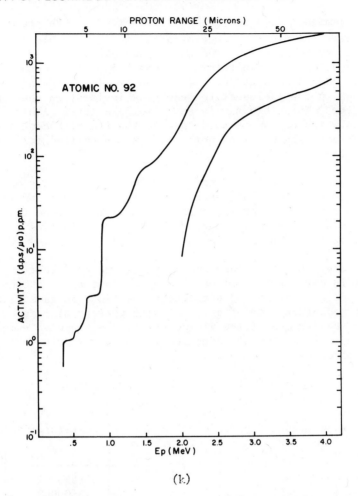

(k)

shown in the curves.

The ordinate in the curves shows the total number of gammas emitted in all directions per p.p.m. of fluorine distributed in thick samples of different elements under bombardment with 1 µa proton beam of energies ranging from 300 keV to 4.16 MeV. The number of actual gammas observed experimentally may, however, be lesser depending upon the solid angle and the efficiency of the detection system. This can be calculated or experimentally measured and should provide the normalizing factor for the particular set-up to be used with these sensitivity curves.

For thick samples the count rate expected from a particular concentration of fluorine under known bombarding conditions can be directly read from the curves. Conversely by observing the fluorine gammas from a sample under bombardment with protons one can estimate the fluorine concentration in that sample by making use of the yield curves. The curves are applicable in the case of thin samples of known thickness too. The expected number of gammas in such a case can be calculated by simply taking the difference in the yields at the energies incident on the sample and leaving it. This thin sample can also be a layer within a thick sample in which one wants to determine the fluorine concentration or to estimate the expected number of gammas emitted for a known concentration.

The yield curves show that for a thick sample many more gammas are produced at higher energies than at lower ones. Similarly more gammas are produced in heavier matrices than in lighter ones for the same concentration of fluorine and the same bombarding conditions.

In order to get an idea about the limits of detection of fluorine in different matrices we suppose that about 500 counts are needed to identify and quantitate the fluorine gamma lines for an irradiation time of one hour with protons of 1 and 4 MeV energy. Making use of the yield curves, the expected number of gammas and the limits of detection in various matrices are shown in the following table.

TABLE 1

at. no. of the matrix	1 MeV		4 MeV	
	gammas/h/p.p.m.	detection limit p.p.b.	Counts/h/p.p.b	detection limit p.p.b.
6	18000	28	2520000	.2
28	34000	15	3600000	.14
30	58000	9	5760000	.09
92	79000	6	7560000	.07

Depth profile analysis of fluorine in different samples is generally carried out by observing the shift in the strong and sharp resonance at 872 keV for the reaction $^{19}F(p,\alpha\gamma)^{16}O$ with the incident proton energy. However, if the fluorine concentration under investigation is rather low one has to use higher proton energies to obtain the desirable sensitivity. At these energies there are no sharp resonances and the above method is no longer applicable. In such cases depth profiling may be carried out using the yield curves which contain all the information about the variations in the gamma yield with the reaction cross section and the stopping power of the sample matrix for a uniform fluorine distribution. When such a curve is compared with an experimentally measured yield curve for the same sample containing non-uniformly distributed fluorine the difference in the yields at any one energy would be only due to the difference in fluorine concentration which could be studied at different depths. For depth profile analysis with this method one does not have to experimentally measure the entire yield curve. Only the range of interest may be covered by selecting suitable proton energies. This method can also be applied for depth profile analysis by using other gamma-producing nuclear reactions without any strong and sharp resonances.

REFERENCES

/1/ K. Norrish and J. T. Hutton, Geochim, Cosmochim, Acta 33 (1969) 431.
/2/ M. A. Chaudhri and G. Burns, to be published.
/3/ C. F. Williamson, J. P. Bougot and J. Piccard, Rapport CEA - R 3042 (1966).
/4/ H. B. Willard, J. K. Bair, J. D. Kington, T. M. Hahn, C. W. Synder and F. P. Green, Phys. Rev. 88 (1952) 849.
/5/ C. H. Osman, D. M. Rosalky and D. J. Baugh, Report ANU - p/450, Australian National University , Canberra.

ION BEAM ANALYSIS TECHNIQUES IN CORROSION SCIENCE

G. Dearnaley

AERE Harwell, England

ABSTRACT

Various methods of ion beam analysis which have proven of value in the study of corrosion in practical environments will be reviewed, with examples drawn from the nuclear energy industry and related fields. These practical applications contrast with the needs of laboratory studies of corrosion mechanisms. The techniques considered include ion-induced nuclear reactions, backscattering and X-ray emission, together with secondary ion mass spectrometry, and their relative merits will be discussed.

INTRODUCTION

Energetic ion beams, ranging from a few keV to a few MeV in energy, have proved to offer valuable methods of quantitative analysis of the surface layers of materials, as the many papers at this conference demonstrate. Probably the most costly surface phenomenon occurring is that of corrosion, which is responsible for an estimated annual expenditure in excess of 3 per cent of the GNP(1). Much of this enormous cost could be alleviated by a better understanding of corrosion mechanisms under practical conditions, since the behaviour of alloyed components in complex gaseous or aqueous environments is not deducible from laboratory experiments on pure materials. Such work can lead to a better choice of construction materials and of corrosion inhibitors. Corrosion is a particularly important subject in nuclear power reactors, where access is difficult, the consequences of failure are costly, and the choice of materials is restricted for a variety of nuclear and thermal (i.e. bulk) properties. Other challenging corrosion problems arise from the energy industry, e.g. in ensuring efficient fuel combustion, and in marine technology.

Practical corrosion films and scales may vary from thin tarnishing layers about 100 nm in thickness to heavy, layered deposits up to a millimetre thick. Although primarily oxides, these films contain a wide spectrum of impurities which play a crucial part in film growth, for example when corrosion inhibitors are used. Carbides, nitrides, fluorides, sulphides, and hydrogen are frequently present and it is desirable to know how these species are incorporated into the film and so influence its protective nature, stability, stresses and spalling. Corrosion films often provide an effective means of protecting a reactive metal (witness aluminium) or conveying desirable surface mechanical properties such as wear resistance, and there is a growing interest in tailoring the composition of corrosion films e.g. by ion implantation of the metal (2). Then it is clearly important to understand how and where the implanted species is incorporated into the oxide.

Laboratory studies of corrosion tend to differ in several respects which usually facilitate analysis. The materials employed are generally pure, or at least well-controlled in composition, the film thickness can be chosen to suit experimental requirements, environmental conditions are more reproducible, and specimens are of simple geometrical shapes. Oxidation can sometimes be carried out within the target chamber of an accelerator (3) and the kinetics of the reaction so observed. The aim of such work is to understand the basic ion transport mechanisms during oxide growth under varied conditions. Pre-eminent in this field has been the work of Georges Amsel and his group at the University of Paris, and their complementary use of nuclear reactions and ion backscattering, together with tracer species such as ^{18}O, has allowed the determination of transport processes in both thermal and anodic oxides of metals and semiconductors. Several excellent reviews of this work have now appeared (3,4).

The present review will therefore concentrate upon practical corrosion problems, which have done much to stimulate further developments of the technique of ion beam analysis. Ion microbeams have provided a new dimension in film analysis, and a wider variety of nuclear reactions is now employed for the detection of virtually all the light elements including hydrogen. Higher energies, to 5 and even 20 MeV, increase the scope for exploitation of nuclear resonances and ion penetration. In secondary ion mass spectrometry the imaging ion microprobe is a valuable tool for studying lateral composition variation, for example in intergranular corrosion.

PRACTICAL CORROSION PROBLEMS

The first problem encountered in the analysis of practical corrosion films is likely to be their complexity and the difficulty of achieving a reliable quantitative distribution of their constituents, particularly when these include light elements such as hydrogen, lithium or carbon. Such species are unsuited for techniques (e.g. EMMA) that rely upon X-ray emission, due to the softness of the radiation, while Auger techniques are sometimes vitiated by interference between electron spectra, e.g. from Fe and Li. Conventional techniques are often only qualitative unless laborious and questionable calibration procedures are undertaken, or they may require special preparation of the sample. Ion beam analysis provides the only practical alternative, with the following advantages:-

(i) depending upon the ion species and energy chosen the analysis may be made over depths ranging from about 100 nm to 10 μm, without destruction of the surface e.g. by sputtering;

(ii) the analysis is quantitative, usually with an accuracy better than $\pm$ 10 per cent;

(iii) the specimen surface need not be specially prepared and can be relatively rough, while the specimen size and shape can vary widely;

(iv) compacted spall particles can be examined, in quantities amounting to only a few milligrams;

(v) accelerator costs are typically $50 per hour, but many specimens can be loaded for sequential examination and bombardment times are short (1/4 - 1 hour). The technique does not require UHV and pump-down time is a few minutes. Thus the cost of this method of analysis is relatively low.

As a rather simple example of the techniques, I shall describe the analysis of thin Li-containing films on the inner wall of 316 stainless steel boiler tubing exposed to various alkaline solutions in a study of stress-corrosion cracking in power station boilers (5). The lithium distribution was readily obtained by use of the $^7Li(p,\alpha)$ reaction at 3 MeV, where there is a broad resonance. Lithium niobate was used as a comparison standard (Fig. 1), and the depth scale was obtained by calculation from the elemental stopping cross-sections. In a film of unknown composition this will require iteration as the various constituents and their concentrations are determined, using different reactions. Alpha back-scattering at 2.9 MeV (chosen to avoid prominent $^{16}O(\alpha,\alpha)$ scattering resonances) yielded the cation : oxygen stoichiometry, this time by comparison with a compacted sample of Cr_2O_3 powder (fig. 2). The sloping spectrum of particles scattered from transition metals contrasts with the almost flat spectrum from the homogeneous Cr_2O_3 and was found to be due to a marked non-uniformity in film thickness (fig. 3). Some ^4He ions penetrate thin oxide

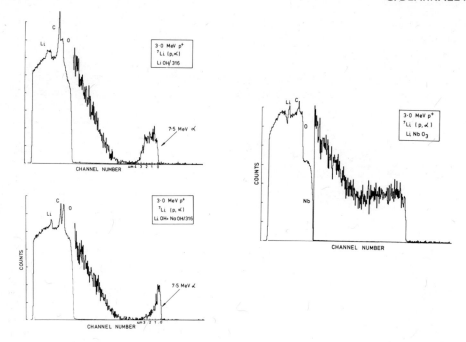

Fig. 1 Spectra of alpha-particles and protons from 3 MeV proton bombardment of two 316 stainless steel samples corroded in lithium-containing alkaline solutions, compared with (right) the corresponding spectrum from a LiNbO$_3$ standard specimen.

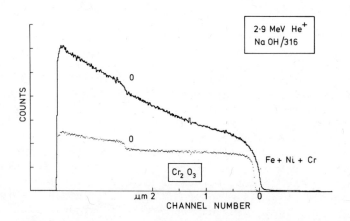

Fig. 2 Alpha backscattering spectrum at 2.9 MeV from a 316 stainless steel specimen corroded in NaOH solution, compared with that from a specimen of compacted Cr$_2$O$_3$ powder (lower curve); (from ref. 5).

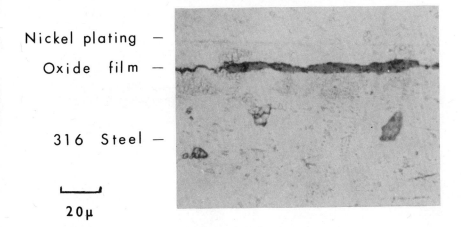

Fig. 3 Optical micrograph of a sectioned specimen of 316 stainless steel after corrosion in a mixture of LiOH and NaOH solutions, demonstrating marked non-uniformity in oxide thickness; (from ref. 5).

and are scattered from the steel substrate. Measurement of the mean oxide film thickness by 2 MeV proton backscattering provided results (Table 1) in close agreement with optical microscopy of section specimens (a more laborious and costly technique). This non-uniform oxide growth is a typical example of what can occur in practice: it is less common in the laboratory.

Table 1

Oxide Film Thicknesses

Sample	2 MeV p^+ (mean)	Optical	Remarks
LiOH/316	3.5 ± 0.2 μm	4-7 μm	Relatively uniform
NaOH/316	1.7 ± 0.1 μm	0.5-1 μm	Extremely variable in places
LiOH + NaOH/316	1.3 ± 0.1 μm	0.5-2 μm	Variable

Energy-dispersive X-ray analysis (EDAX) of sections of the thicker (3 μm) films revealed a depletion of Cr near the outer surface and an increased Fe concentration, not apparent in the backscattering spectra. The composition of the films derived from this combination of techniques was $LiNi Fe_4O_6$, and was found to be uniform throughout the layers except for small traces (100 - 1000 ppm) of Pb and Sn contaminants also present in the steel.

For these analyses, specimens were mounted on the periphery of a 12-sided wheel (fig. 4) which could be placed in a target chamber of the Harwell 6 MeV Van de Graaff. This wheel could be rotated or traversed horizontally by small stepping motors. Silicon surface barrier detectors were employed for observing reaction products, and standard nuclear electronics for pulse handling and data storage.

The example described shows the use of homogeneous calibration standards of known composition, and since scattering and reaction cross-sections are rarely known to better than 5% this is an important technique in quantitative work. We have found it advantageous to build up a collection of such standards, mostly simple and stable compounds of light elements. The standard selected should be comparable in stopping cross-section to the film being studied so that corrections for relative stopping are minimised. Powder specimens are naturally rough, and their surface topography

Fig. 4 Photograph of 12-sided wheel designed for mounting corrosion specimens for sequential examination. The two stepping motors provide remote control.

gives rise to a significant smearing-out (6) of the high-energy edge of a scattered spectrum. This is visible in fig. 2. On the other hand, the corrosion film under study may be equally rough, and an awareness of the 'shadowing' phenomenon is important in detailed comparisons. Single crystal or polished specimens give a more well-defined spectrum but as Abel et al. (4) have pointed out, ion channelling can give a significant dependence of yield upon angle of beam incidence. If single-crystal calibration specimens are chosen, for their purity and homogeneity, it is desirable to rotate them about an axis which does not coincide with the beam direction, so as to average the ion yield: it is not sufficient to choose some 'random' direction of orientation (4). Frequently the calibration specimens will be electrical insulators, and it may be difficult to make accurate current integration of the incident beam. A thin layer of evaporated Au can serve to dispel the surface charge and also to provide a convenient energy calibration of a scattered spectrum (fig. 5).

The next example is chosen to illustrate the complexity of some corrosion films, and the need for a wide variety of nuclear reaction processes in their analysis. Studies of corrosion in a magnesium alloy (Magnox A180) involved its exposure to nuclear reactor coolant gas (mainly CO_2) followed by aqueous corrosion in solutions containing alkalies and fluorine (7). The final specimens contained Mg, Al, Be, C, O, F, H and Na or K, and it was important to know how the impurities were taken up under different conditions. The reactions we used in this study included ^{4}He backscattering, $^{12}C(d,p)$, $^{16}O(d,p)$, $^{19}F(p,\alpha)$ and (to distinguish oxygen incorporated from reactor gas and from aqueous attack) $^{18}O(p,\alpha)$ combined with tracer experiments. Calibration specimens of MgO, $CaCO_3$, NaF and MgF_2 (covered with thin layers of Au) were found most useful, and fig. 5 shows comparative spectra of ^{4}He scattered from the Magnox and MgO. From these reaction spectra distribution profiles of ^{12}C, ^{19}F and ^{18}O were obtained (figures 6, 7 and 9 respectively). The ^{18}O tracer experiments provided valuable information concerning the transport of oxygen or water through the various duplex films. The corrosion data which has emerged from this study is to be published in detail (7). Having determined the distribution of all the other constituents of these films, the one lacking piece of data was the hydrogen content, e.g. present as $Mg(OH)_2$. Present work is being applied to the use of the $^{11}B(^1H, \alpha)2\alpha$ reaction for this purpose. Boron ion energies from the resonance peak at 1.8 MeV up to 4 MeV have been used to probe the hydrogen content both of the corrosion film and the metal beneath. Hydrogen, released from the chemical reaction is readily taken up by magnesium and similar metals (8) and yet there are few techniques capable of detecting it. Preliminary results have shown the presence of such hydrogen, and the technique promises to be of considerable value in the investigation of hydrogen embrittlement and stress-corrosion cracking.

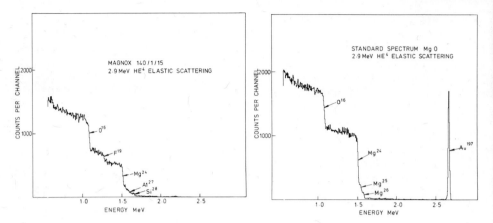

Fig. 5 Alpha backscattering spectrum at 2.9 MeV from a specimen of corroded Magnox alloy in gaseous and aqueous environments, compared with that from a pure MgO specimen coated thinly with Au (from ref. 7).

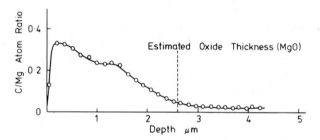

Fig. 6 ^{12}C distribution in a Magnox alloy specimen after corrosion in reactor gas. The mean oxide thickness is estimated from weight-gain measurements; (from ref. 7)

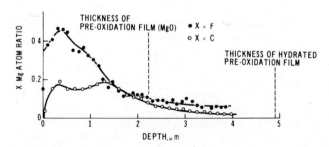

Fig. 7 ^{19}F distribution in Magnox alloy after corrosion in reactor gas and subsequently in an alkaline fluoride solution, with the ^{12}C distribution, both deduced from nuclear analysis; (from ref. 7)

ION BEAM ANALYSIS TECHNIQUES IN CORROSION SCIENCE

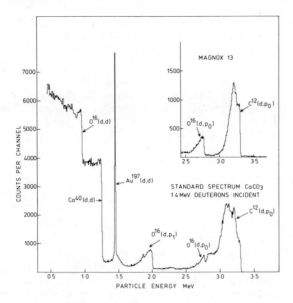

Fig. 8 Spectrum resulting from deuteron bombardment of a corroded Magnox alloy at 1.4 MeV, compared with that from a standard specimen of $CaCO_3$; (from ref. 7).

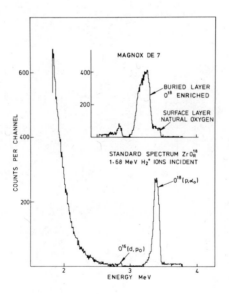

Fig. 9 Spectrum induced by proton bombardment of Magnox alloy corroded in aqueous alkali solutions, the second of which was enriched in $^{18}H_2O$. Below is shown the spectrum from a Zr sample anodized in $^{18}H_2O$; (from ref. 7).

An alternative reaction for hydrogen determination, namely the ^{7}Li(^{1}H,γ)^{8}Be reaction, has been earlier employed by Padawer et al. (9) in corrosion studies made on steel and aluminium alloys. In this case the 14 MeV and 17 MeV gamma rays are detected using a large NaI scintillation detector, and the sensitivity of the method is very high, typically a few ppm of hydrogen being detectable. The depth distribution is obtained by variation of the incident ion energy, and calibration is achieved using steel specimens of known hydrogen content. One difficulty not usually encountered with other species is the effect of beam heating in combination with the mobility of hydrogen: measurements must be repeated over a period of time in order to check whether the composition has remained constant.

In contrast with the Magnox example cited above, it may sometimes be necessary to obtain a qualitative comparison of complex corrosion films on practical components, without the justification for a complete analysis. Thus, in a study of the beneficial effects of ion implantation on the performance of burner tips through which fuel is injected into oil-fired power stations, it was possible to correlate the sulphur content of the film with its previous treatment, using the ^{32}S(d,p) reaction (10), figure 10.

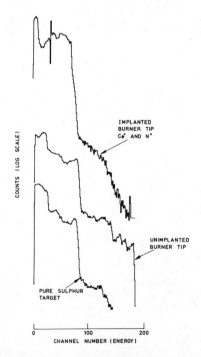

Fig. 10 Deuteron-induced spectrum from surfaces of burner tips for an oil-fired power station, after use, compared with that from a sulphur target. Qualitative similarities imply a suppression of sulphur uptake as a result of Ce$^+$ ion implantation; from ref.10.

The second major difficulty met in applying ion beam analysis to practical corrosion problems is that the film thickness may often exceed the few microns which MeV ions will penetrate. Corrosion scales between 50 μm and 1 mm thick are commonly found and there is considerable interest in understanding the layered structure of such scales and ion transport through them.

One established technique for this is secondary ion mass spectrometry (SIMS) in which deeper and deeper layers are exposed by sputtering during the course of the analysis. This method, as applied to thin films, is discussed by Hofer and Liebl (11) in the present conference, and it has been used by Stott et al. (12) for the investigation of oxides on Nimonic alloys. In thick films which are frequently inhomogeneous the technique is limited by cone formation resulting from non-uniform erosion (fig. 11). The film in this case was about 200 μm in thickness, and formed on stainless steel (13). Clearly this process will interfere with a depth analysis.

Fig. 11 Cone formation after sputtering away a 200 μm corrosion layer on oxidised 316 stainless steel, using secondary ion analysis; from ref. 13.

Fig. 12 Photograph of nuclear microprobe installed on the Harwell 3 MeV Van de Graaff accelerator, showing the 4-element magnetic quadrupole designed by Cookson et al. (14).

Because of these and other limitations of secondary ion analysis mentioned below, the development of ion microprobes of MeV energies has proved highly important for the analysis of thick films. Cookson et al. (14) at Harwell were the first to achieve beam diameters of only a few microns with useful current densities, by means of a magnetic focusing lens (fig. 12). If a taper section of a corroded specimen is prepared this focused beam can be scanned electrostatically across the exposed surface and prompt reaction products are measured in synchronism. Nuclear reactions, ion backscattering and ion-induced X-rays are measurable simultaneously, using appropriate detectors so that the depth distribution of light, medium and heavy elements can be carried out. Mayer and Turos (15) have pointed out how well these three techniques enable a high sensitivity to be achieved throughout the periodic table. The accompanying paper by Allen et al. (16) describes in detail the application of the nuclear microprobe technique to thick corrosion films on steels, up to 800 µm in thickness. With a taper section angle of $5°$, which is fairly easily obtainable, the thinnest films that can usefully be examined in this way are about 10µm in thickness: thus the microprobe complements the range of thicknesses suitable for normal-incidence examination. Figure 13 shows one example of the use of the nuclear microprobe in conjunction with ^{18}O tracer studies on thick films grown on boiler-tube steel, and the results have been important in distinguishing between two corrosion models.

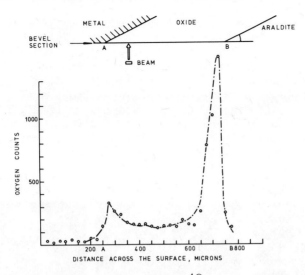

Fig. 13 Nuclear microprobe scan of ^{18}O in an oxide film grown on 9%-chrome steel. The tracer is taken up in two distinct layers of the oxide.

The ion microprobe mass analyser (IMMA) which makes use of secondary ion emission has a comparable or better spatial resolution than the nuclear microprobe, and can likewise be scanned across a taper section, but it suffers from the fact that the yield of a given species varies with its state of chemical combination. Poole and Jackson (13) report discrepancies due to this reason between electron-probe and ion microprobe analyses of corrosion films containing Fe and Cr. It may be that, with experience, these effects could provide useful information concerning chemical states of ions in oxides. The nuclear microprobe on the other hand is completely independent of chemical combination states.

With a reasonable variety of ion beams available (protons, deuterons, ^{3}He and ^{4}He) nuclear microprobe analysis can be extremely versatile, and some interesting possibilities are now being explored with ^{11}B microbeams for localized hydrogen determination and hydrogen depth distribution measurements. With high current densities, however, it seems certain that hydrogen mobility will pose problems, and measurement times will remain long.

The secondary ion mass spectrometer has undisputed value in corrosion science, for thin films and in its microprobe configuration for depth analyses. Its sensitivity for certain ion species is very high, and for this reason it was used by Antill et al.(17) for studying the location of yttrium implanted into a 25% Cr steel for corrosion inhibition. The yttrium was shown to remain close to the metal-oxide interface, where it acted perhaps by forming an impermeable perovskite $YCrO_3$ as a barrier to Cr^+ ion migration. The imaging ion microprobe (e.g. the CAMECA instrument) has important advantages for studying the two-dimensional distribution of chosen species, for example in the investigation of intergranular corrosion. Figure 14 shows ion images of O^+, Y^+ and Li^+ at grain boundaries observed by Quataert and Busse (18) in studies of corrosion by liquid Li of a tantalum heat pipe containing yttrium as a grain stabiliser.

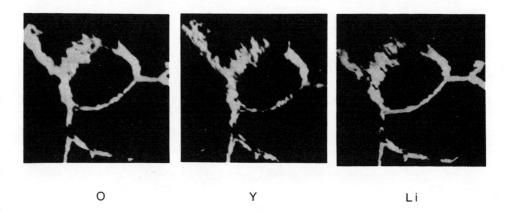

O Y Li

Fig. 14 Secondary ion images of O^+, Y^+ and Li^+ from intergranular corrosion of a tantalum alloy containing yttrium, exposed to lithium vapour (after Quataert and Busse, ref. 18).

CONCLUSIONS

Ion beam analysis utilising nuclear reactions, backscattering and ion-induced X-rays has been demonstrated to be a powerful quantitative tool in studying practical corrosion problems. Its use complements established techniques such as electron-probe microanalysis, EDAX, SIMS and scanning electron microscopy and, now that nuclear microprobes are available, it can be applied to films of almost any required thickness.

Experience in the use of standard specimens for quantitative analysis is growing, and higher energies (to 5 MeV and above) are enabling more resonance reactions to be exploited. More intense beams of B^+ and Li^+ are needed for hydrogen determination, but alterations in composition due to beam heating and other irradiation effects need to be investigated.

We may expect the interest of corrosion scientists in these still relatively novel analytical techniques to continue to grow, with an improving dialogue between them and the nuclear physicists. Ion beam techniques, from those of ion implantation, ion beam simulation of radiation damage, to ion beam analysis are now firmly established as important fields of materials science.

ACKNOWLEDGEMENTS

This work was carried out in collaboration with scientists at several laboratories, including CERL Leatherhead and CEGB Fawley Power Station. The author wishes to express his gratitude to them for allowing the use of material prior to publication elsewhere, and the discussion of work still in progress.

REFERENCES

1) Report of Committee on Corrosion and Protection, U.K. Dept. of Trade and Industry (HMSO, London, 1970); UN Seminar on Techno-economic Aspects and Results of Anti-Corrosion Measures in Engineering Industries, Geneva, January 1975 (to be published).
2) G. Dearnaley, in 'New Uses of Accelerators' ed. J.F. Ziegler (Plenum Press, New York, 1975) to be published.
3) G. Amsel et al., Nucl. Instr. & Methods 92, 481 (1971).
4) F. Abel, G. Amsel et al., J. Radioanal. Chem. 16, 567 (1973).
5) G. Dearnaley, R. Garnsey, N.E.W. Hartley, J.F. Turner and I.S. Woolsey, J. Vac. Sci. & Tech. 12, 449 (1975).
6) K. Schmid and H. Ryssel, Nucl. Instr. & Methods 119, 287 (1974).
7) B. Case, P. Bradford, G. Dearnaley, N.E.W. Hartley, J.F. Turner and I.S. Woolsey (to be published).
8) R.B. McLellan and C.G. Harkins, Mat. Sci. & Eng. 18, 5 (1975).
9) G.M. Padawer et al., Grumman Aerospace Corp. Report RE-464 (1973) and U.S. Patent 3,710,113.
10) D. Peplow, G. Dearnaley and N.E.W. Hartley (to be published).
11) W.O. Hofer and H. Liebl, this conference.
12) F.H. Stott, D.S. Lin and G.C. Wood, Corrosion Sci. 13, 449 (1973).
13) D.M. Poole and C.K. Jackson, AERE Report R7767 (1974).
14) J.A. Cookson, A.T.G. Ferguson and F.D. Pilling, J. Radioanal. Chem. 12, 39 (1972).
15) J.W. Mayer and A. Turos, Thin Solid Films, 19, 1, (1973).
16) C. Allen, N.E.W. Hartley and G. Dearnaley, this conference.
17) J.E. Antill, M.J. Bennett, G. Dearnaley and P.D. Goode (to be published).
18) D. Quataert and C.A. Busse, J. Nucl. Mat. 46, 329 (1973).

QUANTITATIVE MEASUREMENT OF LIGHT ELEMENT PROFILES IN THICK CORROSION FILMS ON STEELS, USING THE HARWELL NUCLEAR MICROBEAM

C.R. ALLEN, G. DEARNALEY and N.E.W. HARTLEY

Nuclear Physics Division, AERE Harwell, England

INTRODUCTION

The quantitative measurement of light element concentrations in materials is found difficult by using characteristic X-ray emission for Z numbers below < 15. Elastic scattering of nuclear beams is also only useful for heavy elements in light substrates. Nuclear reactions, however are very numerous for low Z number elements, and in many cases high energy reaction products are produced. This has led to their use in studies of light elements in many different materials [1,2].

Using a nuclear reaction technique, there is normally a limit to the depth that can be investigated. This limitation in depth varies with the reaction used, but is typically 5 μm in steels. The nuclear microbeam, attached to the 3 MeV IBIS Van de Graaff offers a means of extending the profiling depth of nuclear reactions, by producing a beam size of micron dimensions. This paper gives examples of the use of this facility for light element profiling in corrosion studies.

REACTIONS

Five different elements have been investigated; oxygen, lithium, hydrogen, nitrogen and carbon using the resonance reactions $^{18}O(p,\alpha)$, $^{16}O(d,p)$, $^{7}Li(p,\alpha)$, $^{1}H(^{11}B,\alpha)2\alpha$, $^{14}N(d,\alpha)$ and $^{12}C(d,p)$. The reaction cross-section data are available in ref. 2. In each case the light reaction product is emitted at an energy much higher than the incident beam energy. The yield from the reaction can therefore be easily resolved from elastically scattered counts off the target material.

To convert the reaction yield into elemental concentrations, a second specimen of known chemical composition is run as a standard, for the same beam dose. Fig. 1 shows the energy counts, collected at backward angles (135°) from different standards used in this study; ^{7}Li from single crystal, lithium niobate, ^{1}H from electrolytically-charged zirconium, ^{14}N or ^{28}Si from sintered silicon nitride, ^{16}O or ^{12}C from single crystal calcium carbonate. For microbeam analysis use of single crystal material is considered possible for standards since the beam is sharply convergent(3) at the target, and thus channelling effects are largely eliminated.

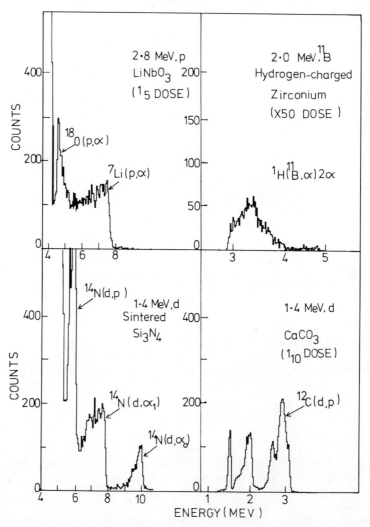

Fig. 1 Resonance Energy Spectra collected from four standard materials.

QUANTITATIVE MEASUREMENT OF LIGHT ELEMENT PROFILES

MICROBEAM TECHNIQUE

The 3 MeV Van de Graaff produces a variety of ion beams; protons, deuterons boron etc. each of which can be compressed using a set of four magnetic quadrupoles(3) into a micron-size beam spot, which represents a demagnified image of an objective slit placed close to the analysing magnet. The ultimate beam size produced by the quadrupole system is estimated as 4 x 4 μm square.

Although micron dimensions are available, beam current on the target is rather low for nuclear reaction studies where target currents in the 10 to 20 nAmp range are needed for good counting rates. A compromise between beam size and target current was made, and in these measurements 10 x 30 μm beam spots were used. This beam size was comparable to the specimens of interest. Fig. 2 shows

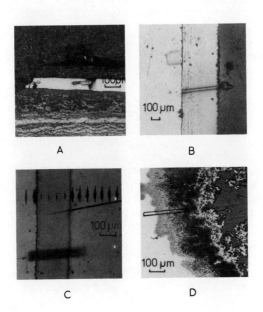

Fig. 2 Examples of corrosion products analysed on the microbeam; (A) 316 steel, almost completely oxidised after caustic corrosion, 65 μm metal remains. (B) High temperature gas corrosion of 9 Cr steel. (C) Nitrided 316, where the steel has been exposed to NH_3 gas at 1 atm, 600°C for 2.67 hrs. (D) Caustic attack on boiler tubing, where the solution contained LiOH.

optical micrographs of four corrosion films formed on steel. In each case, the specimens were mounted at an angle and bevelled to expose an elongated film section. Bevel ratios were normally close to 4:1. In the case of specimens where carbon was of interest care was taken in the preparation of the bevels to use non-carbon containing abrasives e.g. alumina Al_2O_3 grit. Specimen A (fig. 2) which was oxidised in strong caustic solutions has been almost completely oxidised except for a thin metallic strip in the centre, 65 µm wide. The beam staining during analysis shows the definition of the beam spot, and carbon and nitrogen levels were easily detected for concentration levels in the 0.1 atomic % range.

To investigate the variation of an element across a specimen over several hundred microns the beam was electrostatically scanned by a continuous 50 hertz saw-tooth voltage applied to deflector plates at the entrance to the target chamber. Nuclear particles were collected by a Au-barrier detector positioned with a solid-angle of 0.04 sr., at 135° to the beam. Detected particles were then time synchronised with the slow beam sweep, by using the amplified pulses to trigger a linear gate and integrator to sample the beam sweep voltage for each pulse detected. The sampled voltage then sorts the pulse into one of 32 memory stores in a 4096 channel multi-channel analyser. Each memory store comprises 128 energy bins, and the detected pulses are pulse-height analysed into appropriate energy position. Each store therefore relates to 1/32 of the complete beam sweep on the target. The technique[4] was used for both X-ray and nuclear particles emitted from the specimen.

The staining produced by the scanning analysis is shown in specimen B, C and D (fig. 2). Scan lengths of up to 800 µm have been used, and the staining show no appreciable beam broadening occurring towards the ends of the scan. The analysis technique described here is preferred to a static beam[5], since (a) it does not involve mechanical movement of the specimen during analysis (b) the high beam power incident on the specimen is effectively spread over larger areas (up to X80 an unscanned beam area). (c) accurate positioning of the beam on the specimen is achieved rapidly, by using characteristic X-rays emitted from the 32 positions on the scan.

MICROBEAM APPLICATIONS

A. ^{18}O tracer studies in a ferritic steel

The $^{18}O(p,\alpha)$ reaction allow experiments to be carried out in oxides to study the movement of oxygen during corrosion. The reaction has been used in many studies for investigating oxidation in thin oxides[7,8]. In a continuing series of experiments[6] enriched layers of ^{18}O have been deposited on steels of different types to monitor the oxygen movement under gas corrosion in the 300-600°C range. Over a period of weeks, corrosion layers of

QUANTITATIVE MEASUREMENT OF LIGHT ELEMENT PROFILES

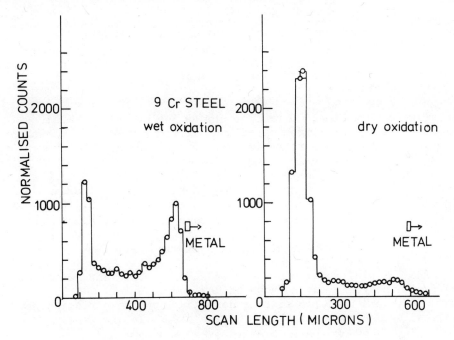

Fig. 3 ^{18}O diffusion profiles across thick oxides on a ferritic steel, where the outer oxide was grown in enriched CO_2^{18} gas.

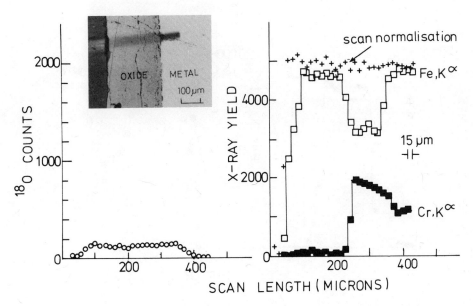

Fig. 4 ^{18}O and X-ray data across the same steel, oxidised in natural CO_2 gas.

40 to 90 µm can be produced, and towards the end of the corrosion test the corroding environment containing carbon dioxide is enriched with CO_2^{18}. The specimens are then bevelled (specimen B fig. 3) and ^{18}O analysis carried out at 840 keV with the microbeam. This beam energy corresponds to a peak in the yield from the $^{18}O(p,\alpha)$ reaction.

Fig. 3 shows the ^{18}O profile across two corroded specimens exposed for similar times but with a different corrosion atmosphere. As can be readily seen there is a sharp contrast in the movement of ^{18}O to the metal-oxide interface. The relative amounts of ^{18}O atom diffusion from the outer oxide can be found by comparing the enriched profile with a specimen having no enrichment, as shown in fig. 4. In this case only the naturally-occurring ^{18}O, (0.2 atomic %) contributes. The X-rays from the same area are also shown.

B. Lithium profiling in 9% Cr Steel

The $^{7}Li(p,\alpha)$ reaction is a particularly efficient reaction, the resonance yield having a broad maximum at 3 MeV. It has previously been used for lithium determinations on corrosion product produced by caustic attack on 316 steels[9]. In the previous study oxide thicknesses of up to 5 µm could be studied.

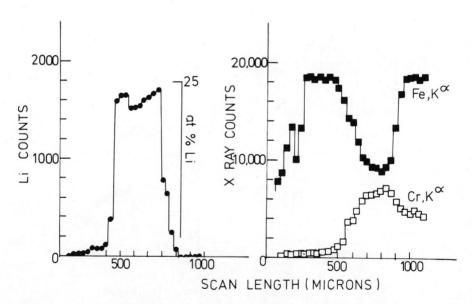

Fig. 5 Distribution of lithium in a thick oxide formed on 9 Cr steel, after caustic corrosion in NaOH/LiOH solution.

An equivalent measurement has been made on 9% Cr steel exposed to the same types of caustic solution containing LiOH where very thick and variable attack has occurred. An example of the resulting film on the steel is shown in fig. 2, specimen D, where the bevelled section shows a complex oxide growth composed on the outer edge of black crystallites.

The reaction is so efficient that 10 x 10 µm beam sizes could be used and the section of film analysed is marked in fig. 2. The lithium yield across this area is shown in fig. 5. where a broad lithium build-up was found in the chromium rich inner oxide. The atomic concentration of ^{7}Li in this region was determined by comparing the yield from the corrosion film, with that from a polished crystal of lithium niobate. Using proton stopping powers[12] for the $LiNbO_3$ and Fe_3O_4 the lithium concentration was found to be close to 25 atomic % in the inner oxide, falling in the outer Cr-depleted oxide to below 1 atomic %. Any background counts could be accurately evaluated from the steel, where lithium levels were below .01 at.%.

C. Hydrogen determination in metals

Hydrogen is thought to play a key part in the cracking of steels under caustic attack. It is however a difficult isotope to measure analytically. The microbeam offers a means of probing along cracks in steels, and preliminary measurements have been carried out for hydrogen determination using the $^1H(^{11}B,\alpha)2\alpha$ reaction. This reaction has been successfully used by

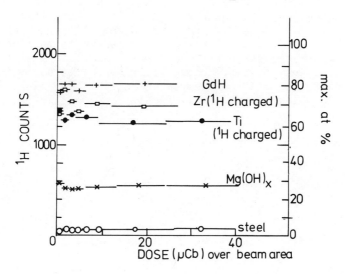

Fig. 6 Variation of hydrogen concentrations in six metals using the boron microbeam.

Ligeon et al.[10] for profiling shallow hydrogen profiles in semiconductors. The reaction is a reversal of the 170 keV resonance found with proton beams on boron.

A boron beam was successfully produced on the 3 MeV accelerator, and measurements were carried out with a 2 MeV beam on a number of hydrogen-containing specimens. The reaction produces a distribution of alpha energies, and so there is no way of determining the depth at which the resonance was excited. However a lower depth limit is set by the reaction, which has negligible yield below a boron energy of 1.6 MeV. This sets a depth limit of close to 1000Å in steels for a 2 MeV incident beam. There is also some yield from the surface so surface contamination problems were tested on each specimen. To minimise the beam staining (a hydrocarbon deposition) a liquid-nitrogen cooled trap was used inside the chamber.

Absolute hydrogen concentrations were calibrated to discs of zirconium and titanium cathodically charged with hydrogen[11]. Full hydrogen charging produces up to 68 atomic % hydrogen. The yield from the charged Zr disc is shown in the resonance spectra of fig. 1, where collection was made at $120°$ with a large detector covered with 2.5 μm mylar to exclude backscattered boron.

Fig. 6 show the hydrogen levels found in a number of metallic specimens, where the 90 x 30 μm beam spot was scanned over 600 μm. In each case the hydrogen counts per micro Coulomb of boron beam did not change with dose. This indicates that the hydrocarbon buildup on the specimen surface was not a dominant contribution to the hydrogen yield, for the levels of hydrogen investigated. Secondly the beam heating in the target was not appreciably effecting the hydrogen levels observed. This result is particularly encouraging for the boron microbeam, and points to its use in more detailed hydrogen studies. There will however be a serious limit to the sensitivity of the hydrogen detection. For these measurements levels below 0.1 at.% will require long machine runs on the microbeam.

D. Microhardness and nitrogen determination in steels

Nitriding of steels is a conventional technique for hardening surfaces, and so protecting the metal from corrosion. In profiling the nitrogen introduced into the steel the $^{14}N(d,\alpha)$ reactions have been used. Several $^{14}N(d,p)$ reactions are also excited by the deuteron beam, but these were excluded from the high energy part of the spectrum by using a thin depletion depth in the Au-barrier detector of 100 μm. This thickness of silicon fully stops the 10 MeV α_0 and 7.5 MeV α_1 while only partially detecting the (d,p) reactions. A standard of sintered nitride, Si_3N_4 was used as a calibration. The nitrogen yield is shown in fig. 1 for an incident

1.4 MeV deuteron beam. The thin detector also excludes energetic $^{28}Si(d,p)$ reactions which is also valuable in the study of steels where silicon trace concentrations are common.

Fig. 7 shows the nitrogen profile measured across a 316 steel nitrided to a depth of about 30 μm by heating in ammonia gas at 1 bar, 600°C for 2.2/3 hours. The heavily nitrided zone is visible on the optical micrograph as a dark vertical band (expanded to a level of 120 μm on the bevel section). The nitrogen concentrations were found to vary from 16 at.% at the outer surface to 10 at.% at the edge of the heavily nitrided zone. The level then fell rapidly.

The deuteron energy of 1.4 MeV is also useful for carbon investigations using the very efficient $^{12}C(d,p)$ reaction, which has a peak in the yield curve at 1.32 MeV. The proton is emitted at 3 MeV, and for this specimen where carbon levels were found to be low appreciable backgrounds were found from the nitrogen reactions. Accurate carbon determination was therefore difficult except in the surface region of the steel where a large carbon yield

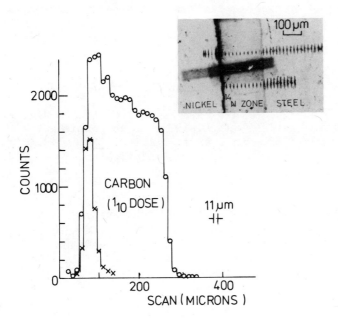

Fig. 7 Nitrogen and carbon profiles across an area of heavily nitrided steel.

was found, corresponding to a 1-2 at.% level (fig. 7).

The dominant feature across the nitrided specimen was thus the nitrogen diffusion profile. The nitrogen level was then correlated to microhardness measurements, shown by the stylus marks on the optical micrograph of fig. 7. As can be seen the length of the marks is reduced in the nitrided zone, and since the indentation length relates to the hardness of the surface, this parameter could be directly related to the nitrogen concentrations.

CONCLUSIONS

In each of the studies exampled in the previous sections, the nuclear microprobe has enabled specific elements to be investigated over tens of microns giving their atomic concentrations to a depth resolution of typically 2µm. Accurate knowledge of the position analysed can be pinpointed by scanning the beam across the specimen and simultaneously use the beam-induced X-rays to monitor the major chemical changes in the specimen e.g. metal-oxide interfaces, or potting interfaces. Additional checks are available using hydrocarbon staining, which is often produced on the specimen unless efficient cold traps are in use.

The sensitivity of the microbeam to an element is limited by the reactions used, and vary with detection angle and beam energy. In this study of oxygen, lithium, nitrogen and carbon we estimate levels in the 100 p.p.m. range as probably a fair sensitivity limit, without extended and costly machine runs. This figure does not apply for the hydrogen determination where difficulty would arise for levels below 1000 p.p.m.

REFERENCES

1. E.A. Wolicki, in 'New uses of Accelerators' ed. J.F. Ziegler (Plenum Press, New York, 1975) to be published.
2. G. Amsel et al. Nucl. Inst. & Meths. 92 (1971) 481.
3. J.A. Cookson et al. J. Radioanalyt. Chem. 12 (1972), 39.
4. J.A. Cookson (private communication).
5. T.B. Pierce in 'Characterisation of Solid Surfaces' ed. Kane and Larrabee (Plenum Press, 1974), 419.
6. C. Allen et al. Nature (to be published) 1975.
7. D. David et al. J. Electrochem. Soc. 122 (1975), 388.
8. J.M. Calvert et al. J. Radioanalyt. Chem. 12 (1972), 271.
9. G. Dearnaley et al. J. Vac. Sci. & Tech. 12 (1975), 449.
10. E. Ligeon, Radn. Effects 22 (1974), 105.
11. C.J. Smithels, Metals Reference Book Vol. II (Butterworths, 1961), 607.
12. L.C. Northcliffe and R.F. Schilling, Nucl. Data. Tables A7 (1970), 233.

DISCUSSION

Q: (J.W. McMillan) May I comment on the correlation of N% with micro-hardness. The correlation will only occur when the only element causing hardness is N; C could also cause hardening.

A: (G. Dearnaley) I quite agree. In our case C was absent and hence the good correlation with N concentration. Carbide formation is bound to influence the microhardness.

Q: (W. Brown) Would you comment on the relative value of mechanical scan of the target through the beam in comparison with electrostatic scanning used in the reported work?

A: (G. Dearnaley) In our present chamber the mechanical scanning is not sophisticated enough to be quite as good as electrostatic scanning. However, a new chamber has been designed with better mechanical manipulation, and this should be superior to electrostatic scanning over >100 µm.

Q: (W. Reuter) Could you comment on a practical lower limit of the beam size, fundamental limitation and the beam current obtainable under such condition. What is the anticipated detection sensitivity in ppm if one operates in this mode?

A: (G. Dearnaley) Practical lower limit to beam size is 5 µm, fundamental limit is estimated by Dr. J. Cookson to be approximately 1 µm. Beam current is $\sim 10^{-9}$ A at 5 µm dia, and would be $\sim 10^{-10}$ A at 1 µm. Detection sensitivity, say by ion-induced x-rays will depend on various parameters (Z_1, Z_2, E_1) and the time of observation. A typical sensitivity might be ~ 500 ppm.

NUCLEAR MICROPROBE ANALYSIS OF REACTOR MATERIALS

J. W. McMillan and T. B. Pierce

Applied Chemistry Division
U.K.A.E.A., A.E.R.E. Harwell
Didcot, Oxon, U.K.

SUMMARY

The Harwell nuclear microprobe is used extensively for the determination of carbon, nitrogen, and boron concentration profiles in reactor steels. The concentration ranges covered are from tens of ppm to a few percent. Spatial resolutions of as high as 10 μm are available and concentration profiles to depths of several millimetres can be measured. Several developments are in progress including the determination of variations in O/M ratios in ceramic nuclear fuels.

1. INTRODUCTION

During the last three years the Harwell nuclear microprobe has been used increasingly for the determination of elemental distributions in nuclear materials. Methods have been developed for the measurement of concentration profiles of carbon (1), nitrogen (2) and boron (3,4) in structural and cladding steels, and are now in regular use, the current sample throughput being approximately 300 per year. Present developments include the determination of oxygen to metal ratios in ceramic nuclear fuels, the variation of the stoichiometry of refractory materials such as boron carbide and the measurement of silicon and beryllium concentration profiles in steels.

The nuclear microprobe has proved emminently suitable for this type of analysis for a number of reasons. Prompt nuclear reaction methods are particularly favourable for light element analysis because of the high Q-values for many reactions, reasonably high nuclear cross-sections, the emission of particle groups and gamma-rays of characteristic energy, and the possible avoidance of matrix interference in medium and high Z number materials through coulomb barrier suppression of their reactions at low beam energies (5). Also, the analysis of beta/gamma active materials is feasible through the measurement of the emitted particle groups. These reactions when used with the small diameter charged particle beams available on the Harwell nuclear microprobe facility, provide excellent methods for the measurement of the distribution of light elements in materials, at spatial resolutions as high as 10 μm, over distances that vary from a few tens of micrometeres to several millimetres.

The apparatus, general techniques and application of specific methods to the analysis of light element concentration profiles in reactor materials, will be used to illustrate the power of nuclear microprobe analysis.

2. NUCLEAR MICROBEAM SCANNING FACILITY

The Harwell nuclear microprobe system has been described in detail previously (6), however, its essential features are shown diagrammatically in Figure 1.

Hydrogen and helium isotope beams are accelerated in the 3 MeV electrostatic generator IBIS. After magnetic analysis the beam shape is defined by a stop assembly, and the beam is focussed onto the target by a set of quadrupole magnets, the demagnification factor being 1:5.6. The relative movement of the beam and target can be obtained either by electrostatic deflection of the beam or by mechanical movement of the target using stepping motors. Particle, X-ray

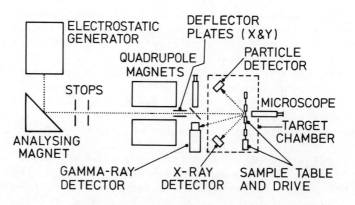

Figure 1 Diagram of nuclear microprobe system

and gamma-ray detectors are placed in or around the target chamber. Two microscopes allow the observation of the rear of a viewing quartz and the front face of thick targets. Equipment is also available for the measurement of beam current and integrated charge, for the electronic processing and recording of detector signals, and for the control of the scanning programme.

3. NUCLEAR MICROPROBE ANALYSIS

For the majority of the work described a fairly standard analysis procedure has been adopted (1,2).

Normally samples are cut to expose the cross-section of interest. The specimen is then mounted in Woods metal and polished to a fine finish with alumina or diamond abrasives. Calibration standards of known composition are similarly cut, mounted, and polished.

For linear profiles a narrow rectangular beam shape is preferred in order to obtain increased beam currents, while maintaining high linear spatial resolution. Beam currents of approximately 20 nanoamps are desirable to maintain a high speed of analysis.

To measure a profile the sample is aligned with the beam adjacent to its edge. The beam is then stepped discontinuously onto and across the specimen. The yield of charged particles or gamma-rays from the reaction of interest is recorded at each point while it is irradiated to a constant dose. Simultaneous measurement of X-rays allows a check on the movement of the beam onto and from the specimen. The particle or gamma-ray yield at each point is converted to an elemental or isotope concentration by calibration with standards of known composition. To compensate for short range variations in the concentration of light elements in standards, it is necessary to measure each at a minimum of ten points. When bulk diffusion behaviour is required it is essential to average several scans to smooth short range inconsistencies.

4. DETERMINATION OF CARBON PROFILES

The carburisation and decarburisation of steels in contact with liquid metals is of considerable importance in liquid metal cooled fast reactors. The carbon profiling method was developed specifically to study carbon diffusion in fast reactor steels, although the technique has applications in the study of materials from other reactor systems (1).

The method is based on the following reaction

$$^{12}C(d,p)^{13}C \quad (E_d = 1.3 \text{ MeV})$$

By measuring emitted protons, the carbon concentration profiles of both active and inactive specimens may be determined. The development of the method has already been described and its reliability proved by obtaining good agreement with chemically determined profiles (1). Analysis is rapid, point concentrations usually being measured in less than 30 seconds.

The profiles of carburised steels have been found to follow smooth error function like curves (1). Decarburisation has also been observed, but is more difficult to measure because of the low carbon concentration encountered, a typical curve for decarburised M316 steel is shown in Figure 2.

Measurement of carbon at low concentration is liable to interference from surface contamination, particularly carbon deposited by the beam from hydrocarbons in the accelerator vacuum. While the rate of carbon deposition can be controlled by placing a cold trap close to the sample surface it can rarely be eliminated completely. However the proton spectrum of deposited carbon differs greatly from that of carbon in a steel matrix as is indicated in Figure 3. Subdivision of the spectra into two energy bands, creates a lower band in which the matrix carbon protons predominate and an upper band in which the surface carbon protons predominate. The proportions of proton counts in each region can be measured for both types of target; for matrix carbon a 1% carbon in steel standard is used, and for the surface carbon a very low carbon content pure iron is used and a suitable thickness of carbon allowed to build-up on the surface from the beam. By using the same energy bands during the scanning of samples a simple correction can be made at each analysis point from the proportion of counts in the two energy regions. The correction method is only suitable for thin contamination films which do not appreciably reduce the energy of the incident deuterons before they enter the steel matrix, and is fairly sensitive to amplification changes. Even so, it is particularly useful at carbon concentrations of 1000 ppm and less as carbon from build-up is commonly equivalent to 100 ppm. The results of examining a specimen containing 19 ppm carbon (average chemical analysis) are given in Table 1, which shows the corrected and uncorrected carbon values. The mean corrected value of 22.7 ppm is in excellent agreement with the chemically determined value, and the standard deviation (σ) on the set of ten results is reasonably small and may in part be attributable to variations in the distribution of carbon in the specimen. The uncorrected results are highly erroneous. The very high value of the first uncorrected result reflects the fact that the beam dwelt on that point prior to commencing the analytical scan. The correction method allows profiles to be determined in the concentration range 0.001 - 5.0% carbon.

5. DETERMINATION OF NITROGEN PROFILES

Nitrogen can profoundly alter the properties of steels whether it is added deliberately or fortuitously. For instance nitriding

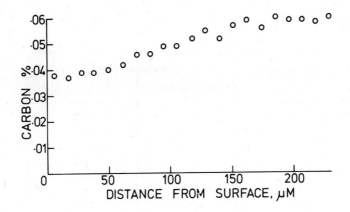

Figure 2 Carbon profile in decarburised M316 steel
(○, mean of 4 scans)

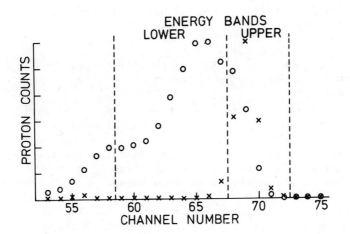

Figure 3 Proton spectra, ○ 1% carbon in steel, x,
thin carbon film on iron

Table 1

Nuclear microprobe analysis of low carbon (19 ppm) content iron

Point Number	Carbon ppm	
	Corrected	Uncorrected
1	16.6	414
2	26.1	220
3	26.1	183
4	19.4	161
5	22.5	160
6	13.7	188
7	31.3	204
8	22.7	193
9	17.1	180
10	31.1	173
Mean	22.7 σ = 6.0	208

strengthens 20/25 Ti stabilised austenitic steels. The distribution of nitrogen in materials can indicate the success of deliberate nitriding or the extent of unintentional nitrogen pick-up.

A nitrogen profiling method has been developed based on the following reactions (2).

$$^{14}N(d,p)^{15}N$$
$$^{14}N(d,\alpha)^{12}C \quad (E_d = 1.9 \text{ MeV})$$

Measurement of the emitted protons (p_o) and alpha particles (α_o) is again chosen to allow the determination of profiles in radioactive specimens. The development of the method has been described previously (2). It can be used successfully for the determination of nitrogen profiles in the concentration range 0.005 - 2.0%. The only important interference is from boron.

While carbon profiles in steels are usually smooth curves following an error function, some nitrogen profiles indicate the presence of precipitation fronts. Figure 4 shows the profile obtained for a specimen of partially nitrided 20/25 Ti steel. The plateau at approximately 0.45% nitrogen is caused by rapid preferential precipitation of titanium nitride which prevents the further movement of nitrogen into the specimen. The diffusion of nitrogen

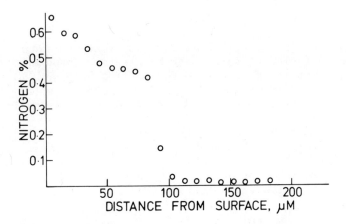

Figure 4 Nitrogen profile in nitrided 20/25/Ti steel (○, mean of 2 scans)

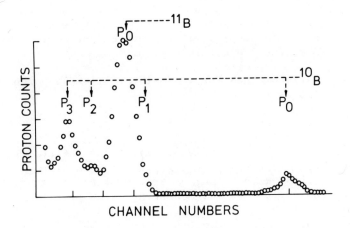

Figure 5 Proton spectrum, natural boron in steel

6. DETERMINATION OF BORON PROFILES

Because of the high thermal neutron cross-section of ^{10}B, boron is an important element in nuclear materials. Boron may be incorporated deliberately in materials for control purposes, or may be present fortuitously when problems of neutron economy may ensue or damage through the build-up of helium from the $^{10}B(n,\alpha)^7Li$ reaction.

Ideally, methods capable of measuring changes in boron distribution in control materials should be able to determine ^{10}B and ^{11}B simultaneously thus distinguishing between burn-up and redistribution. A nuclear microprobe method has been developed for this purpose and is based on the following reactions (3,4).

$$^{10}B(\alpha,p)^{13}C$$
$$^{11}B(\alpha,p)^{14}C \quad (E_\alpha = 2.67 \text{ MeV})$$

Measurement of emitted protons was chosen to allow the examination of irradiated specimens. The proton spectrum from a steel containing natural boron is shown in Figure 5. The p_o group from ^{10}B can be measured independently, but the ^{11}B p_o group is interfered with by the ^{10}B p_1 group and requires correction. Linear calibration curves have been obtained for ^{10}B and ^{11}B using enriched and natural boron containing steels. Elements most likely to cause interference are aluminium, sodium, and fluorine. The effectiveness of the method for the measurement of boron isotope ratios was tested by comparison with mass spectrometry. The results in Table 2 show reasonable agreement; the major error in the nuclear microprobe measurement of the enriched boron steel is in the ^{11}B determination.

An alternative method for the determination of boron profiles is by the following reaction.

$$^{10}B(d,p)^{11}B \quad (E_d = 1.3 \text{ MeV})$$

This reaction is relatively sensitive but suffers from nitrogen interference (2). However, for materials low in nitrogen and intentionally adulterated with boron, as in compatibility studies, there is no objection to its use for measuring boron profiles. Using 1 μ coulomb irradiations and measuring the p_q and α_o particle groups, boron concentrations have been determined in the range 0.01 - 12%. The boron profile in a nimonic alloy is shown in Figure 6. The inflexions in the profile at 8% and 3.5% boron suggest the presence of specific compounds. The boron penetration can be observed to a depth of 1000 μm. In the low concentration regions the sporadic high results suggest boron concentration at grain boundaries, which was confirmed by boron track analysis.

Boron may be measured free from nitrogen interference by the following reaction (7).

$$^{11}B(p,\alpha)^8Be \quad (E_p = 1.11 \text{ MeV})$$

While lithium and fluorine will cause interference they are unlikely to be present in steels. So far, the method has only been applied to the determination of low boron concentrations in steels, 10 - 150 ppm, and has yet to be applied to the measurement of profiles.

Table 2

Boron atom ratios in steel samples

Sample	$^{10}B/^{11}B$ atom ratio	
	Nuclear microprobe	Mass spectrometry
Enriched boron in steel	9.36 ± 1.68	9.25 ± 0.05
Natural boron in steel	0.247 ± 0.015	0.248 ± 0.007

7. CURRENT DEVELOPMENTS

The overall bulk oxygen to metal ratio of ceramic nuclear fuels is measured accurately by CO/CO_2 equilibration methods, and where less accurate oxygen measurements are needed inert or vacuum fusion techniques are employed (8). The nuclear microprobe offers the possibility of measuring the spatial variation of the oxygen to metal ratio of ceramic fuels employing nuclear reaction analysis for oxygen and simultaneous X-ray generation for the determination of the heavy metals.

Preliminary experiments show promise. The following reaction has been employed for oxygen measurement

$$^{16}O(d,p)^{17}O \quad (E_d = 1.1 \text{ MeV})$$

Emitted particles, the p_1 group, and gamma-photons at 0.87 MeV, have both been used to measure oxygen. Because of the low deuteron energy uranium has been determined employing its M X-rays. Some results are given in Figure 7. The proton and gamma-photon based results both show a reduction of O/U ratio when moving from the surface to

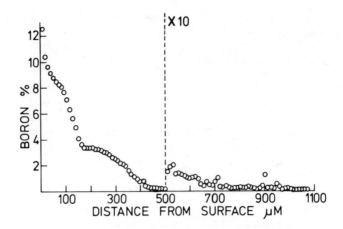

Figure 6 Boron profile in a nimonic (○, mean of 4 scans)

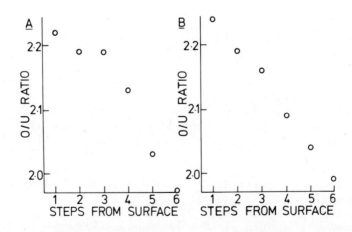

Figure 7 Profiles of O/U ratio for a ceramic fuel; A, O by proton measurement, B, O by gamma-photon measurement (○, mean of 4 scans)

the bulk, but are not strictly comparable as they were determined in separate scans.

Other profiling developments in progress are the determination of boron and carbon in boron carbide, and the measurement of silicon and beryllium in steels.

8. REFERENCES

1. Pierce, T.B., McMillan, J.W., Peck, P.F., and Jones, I.G.
   Nuclear Instruments and Methods, 118, (1974), 115

2. Olivier, C., McMillan, J.W., and Pierce, T.B.
   Nuclear Instruments and Methods, 124, (1975), 289

3. Pierce, T.B.
   Proc. Society for Analytical Chemistry, 11, (1974), 333

4. Olivier, C., McMillan, J.W., and Pierce, T.B.
   To be published.

5. Bird, J.R., Campbell, B.L., and Price, P.B.
   Atomic Energy Review, 12, (1974), 275

6. Cookson, J.A., Ferguson, A.T.G., and Pilling, F.D.
   J. Radioanalyt. Chem., 12, (1972), 39

7. Olivier, C., and Pierce, T.B.
   Radiochem. Radioanal. Letters, 17, (1974), 335

8. Taylor, B.L., Phillips, G., and Milner, G.W.C.
   'Analytical Chemistry in the Nuclear Fuel Cycle'
   Proc. Symp. Vienna, 29th Nov - 3rd Dec 1971, p.237
   Pub. IAEA Vienna, 1972

DISCUSSION

Q: (L.C. Feldman) What is the angular divergence of the micro-probe beam?

A: (J.W. McMillan) It is fairly large because the local length of the focussing system is only 25 cm.

Comment: (G. Dearnaley) In relation to the feasibility of conducting channeling experiments with the microbeam, the half-angle of convergence is approximately $3°$, but could be reduced by use of a collimator placed within the magnetic lens, at the expense of beam intensity.

Q: (M. Hufschmidt) What deuteron energies did you use in case of the (d,p) reactions on ^{12}C and ^{14}N, and what was the maximum depth (normal to the surface) you could probe by these deuterons in case of steel samples?

A: (J.W. McMillan) The deuteron energies are given in the paper and are 1.3 MeV for $^{12}C(d,p)^{13}C$ and 1.9 MeV for $^{14}N(d,p)^{15}N$ and $^{14}N(d,\alpha)^{12}C$. Because the specimens are cut to expose a cross-section for probing, the light element concentration across the whole cross-section can be measured and is only limited by the size of the sample holder and the distance the target can be moved relative to the beam; profiles of specimens as large as 30 mm across have been measured.

Q: (D. Simons) For the boron measurement have you looked into the possibility of using the $(^3He,p)$ reaction? These reactions have high Q values (for instance the $^{10}B(^3He,p)^{12}C$ reaction has Q=19.7 MeV) and should be easily separated from the Nitrogen.

A: (J.W. McMillan) No, but it could be useful reaction for exploitation in the future. Thank you for the suggestion.

ANALYSIS OF MICROGRAM QUANTITIES OF ALUMINUM IN GERMANIUM

I.V. Mitchell and W.N. Lennard

Chalk River Nuclear Laboratories

Chalk River, Ontario, Canada

ABSTRACT

We have sought a method suitable for analyzing low concentrations of Al in single crystal Ge regrowth layers. Anticipating the need for a lattice site determination for the Al, we have restricted our choice to a (non-destructive) ion-beam method.

The nuclear reaction $^{27}Al(d,p)^{28}Al$ has proved satisfactory. Yields for p_0 and p_1 groups have been used to monitor the linearity of reaction yield from prepared standards having Al concentrations from 100 $\mu gm.cm^{-2}$ to less than 1 $\mu gm.cm^{-2}$. We discuss the reasons for rejecting more familiar reactions for Al microanalysis such as $^{27}Al(p,\alpha)^{24}Mg$, $^{27}Al(p,\gamma)^{28}Si$, $^{27}Al(p,p'\gamma)^{27}Al$ and ^{27}Al (particle, X-ray).

INTRODUCTION

The demand for lattice-site determinations of impurity atoms in single crystals has been met most commonly by applying the channeling effect to an elastically scattered ion flux. In those cases where the impurity atoms are lighter than the host atoms, alternative ion-induced reactions must be sought. Numerous examples may be found in the literature. (See, for example, the contributed papers in ref. 1.)

In collaboration with G. Ottaviani and C. Canali of the University of Modena, we have undertaken a search for the lattice site symmetry of Al atoms incorporated into thin films of Ge which have been regrown as single crystals onto bulk Ge crystals by the method of solid-phase epitaxy (2). The purpose of the search is to test the conjecture that the observed p-type electrical behaviour of the film (3) is correlated with a substitutional site for the Al atoms in the Ge host.

This paper presents the first phase of this work. We describe a nuclear microanalytical method for detecting Al in Ge at concentrations below 1 at.%, typical for regrowth specimens. We also show why the better known reactions ^{27}Al(particle, X-ray), ^{27}Al(p,α)^{24}Mg and ^{27}Al(p,γ)^{28}Si are unsuccessful for this system.

METHOD

The films to be analyzed have thicknesses ranging between 2000Å and 4000Å and may contain as little as 0.1 at.% Al distributed throughout the epilayer. All were grown on thick (e.g. $\sim$ 1 mm) Ge crystals.

Anticipating the requirement for channeling effect measurements, we have chosen the ^{27}Al(d,p)^{28}Al reaction as a specific and sensitive probe for Al. (The reasons for rejecting more familiar reactions are given in the section ALTERNATIVE REACTIONS). Detailed absolute, differential and total excitation cross sections are available for the (d,p) and (d,α) reactions for deuteron energies in the range 1.5 - 3.0 MeV (4-6). The high Q-values, 5.499 and 5.467 MeV, respectively (7), lift the p_0 and p_1 group energies well away from all the common low-Z impurity lines, with the one exception of silicon (Q = 6.253 MeV). It is clear from a typical spectrum, shown in figure 1, and from the excitation functions for (d,p) shown in figure 2 - both reproduced from ref. 6 - that the ^{27}Al(d,p_{0+1})^{28}Al reaction gives an acceptable signal. The angular distributions for these two groups do not vary by more than 20% from isotropy (5).

EXPERIMENT

Targets of 0.225 µgm.cm^{-2} and 0.090 µgm.cm^{-2} Al (5 x 10^{15} atom.cm^{-2} and 2 x 10^{15} atom.cm^{-2}, respectively) were made by implantation into clean Ta and Ge samples at 40 keV energy using the CRNL 70 kV Isotope Separator. (All carbon stock contained too much trace silicon to permit their use as targets, thus precluding an independent backscattering assay of Al implant doses). Al targets of thickness 28 ± 5 µgm.cm^{-2} were made by evaporation onto Ta foil backings and the thickest Al target, 108 µgm.cm^{-2} (4000Å), was made by evaporation onto a Ge backing. The set was completed with a 2000Å thick, self-supporting Al$_2$O$_3$ film, equivalent to 33.7 µgm.cm^{-2} of Al. This was backed by clean Ta. Targets and Al-free

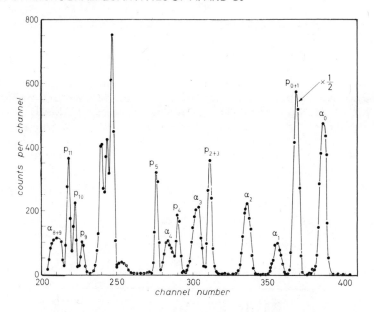

Fig. 1 α-particle and proton energy spectra at E_d = 1.8 MeV for a 220 μgm.cm^{-2} Al target. (reproduced from ref. 6).

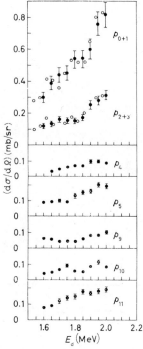

Fig. 2 Differential excitation functions at $\theta_L = 90°$ for the ^{27}Al(d,p)^{28}Al reaction. (reproduced from ref. 6).

controls were mounted on an X-Y stage with a clean Ta masking plate.

The detector was an ORTEC 300 mm^2, 300 μ depletion depth annular detector, set approximately 1 cm upstream from the targets. It was protected from elastically scattered deuterons by an Al filter 0.14 mm thick, i.e. sufficient to eliminate all protons with $E_p \lesssim 3.5$ MeV and all alpha-particles with $E_\alpha \lesssim 15$ MeV. This particle filter removed unwanted signals from $^{12}C(d,p)^{13}C$, $^{16}O(d,p)^{17}O$ and $^{14}N(d,\alpha)^{12}C$ reactions, in particular. Beam currents of 2 MeV deuterons were held to 20 nA in a 1 mm diameter spot.

RESULTS

The thickest Al target used resulted in a 38 keV loss to the 2 MeV deuteron beam, so the (d,p) cross section is constant

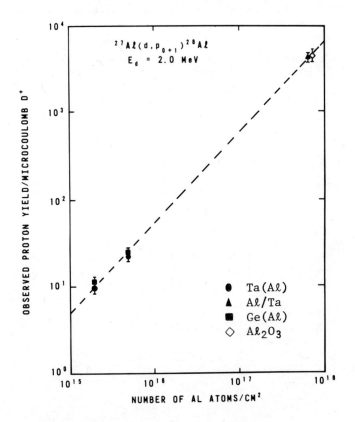

Fig. 3 Yield of $^{27}Al(d,p_{0+1})^{28}Al$ reaction per microcoulomb of 2 MeV deuterons vs. target thickness.

throughout all the Al profiles. The number of counts per microcoulomb of 2 MeV D^+ in the p_0 plus p_1 groups, plotted against Al target thickness, is shown in figure 3. The yield is linear with Al content for all targets up to the maximum thickness used (the 4000Å Al datum is not shown).

A regrowth specimen analyzed by the $^{27}Al(d,p)^{28}Al$ reaction at 2 MeV was found to contain 2.2×10^{15} Al atom.cm^{-2} in a nominal regrowth film thickness of 2500Å. This areal density corresponds to an Al atomic concentration of 0.11%.

ALTERNATIVE REACTIONS

^{27}Al(particle, X-ray)

For any light projectile e.g. protons or helium ions, the ionization cross section for Ge L (E_x = 1.2 keV) will be greater than for Al K (E_x = 1.48 keV). For either ion beam, in the range 0.5 to 2.5 MeV/amu, the BEA ratio σ_x(Ge L)/σ_x(Al K) is close to 3.2 (8). Since the Ge target is infinitely thick to the beam, the only recourse is to filter heavily on the Ge L X-rays. We make the assumption that the Al and Ge X-ray signals can be resolved, by either a proportional counter or Si(Li) spectrometer, when they have equal intensity. Adopting a figure of 0.1 at.% for the Al in Ge regrowth layer and accounting for the Ge thick target yield, we would then have to accept an attenuation of 13000 on the Al K X-ray intensity. (This is still optimistic since it takes no account of the secondary electron bremsstrahlung spectrum (9,10) in the region of 1.5 keV). This implies a calculated upper limit of 0.06 counts per microcoulomb of 0.5 MeV protons on target.

$^{27}Al(p,\alpha)^{24}Mg$

The $^{27}Al(p,\alpha)^{24}Mg$ reaction (Q = 1.595 MeV) has a strong resonance at a proton energy of 1183 keV and a resonance width of $\sim$740 eV (7). The number of Al atoms (at 0.1 at.% in Ge) that are sampled under the resonance is 6×10^{13} cm^{-2}. One microcoulomb of resonant protons in our geometry would yield 0.34 α_0 counts. There is still the problem of pulse pileup due to a massive flux of back-scattered protons (end-point energy $\sim$1.1 MeV). Filtering has the effect of decreasing the energy separation of the α_0 group and the proton continuum. Schemes to separate out the bulk of the protons by electrostatic or magnetic deflection carry a large penalty in detection solid angle. With pulse pileup rejection circuitry, we were still obliged to operate at proton currents that would limit us to a rate of 0.006 α_0 counts per second.

$^{27}Al(p,\gamma)^{28}Si$

Dunning et al. (11) have demonstrated good sensitivity for this reaction in profiling Al implants in SiC targets but they required an experimental arrangement that would not readily accommodate a goniometer for channeling work. Further, the strongest resonance in the $^{27}Al(p,\gamma)^{28}Si$ reaction for proton energies of 2 MeV or less is at 991.9 keV (7) and has a width of 80 eV. Consequently, even less of the Al in the regrowth distribution is sampled than in the case of the $^{27}Al(p,\alpha_o)^{24}Mg$ reaction, in fact as few as 6×10^{12} Al atom.cm^{-2}. We failed to detect Al at this level with beam currents of 100 nA or less. Similar remarks apply to $^{27}Al(p,p'\gamma)^{27}Al$ and $^{27}Al(p,\alpha\gamma)^{24}Mg$ reaction yields.

SUMMARY

We have found that the $^{27}Al(d,p)^{28}Al$ reaction satisfies the requirements of sensitivity and selectivity in an ion-induced reaction for detecting low concentrations ($\sim$ 0.1 at.%) of Al distributed through thin Ge regrowth layers. Detection of Al by X-ray spectrometry was frustrated by intense emission from the Ge bulk. Nuclear reaction methods were hampered by narrow resonance widths and, in the case of the $^{27}Al(p,\alpha)^{24}Mg$ reaction method, by severe pulse pileup arising from the backscattered proton flux.

Similar arguments imply that the (d,p) reaction would be the preferred choice for microanalysis of low concentrations of Si in Ge.

ACKNOWLEDGEMENTS

We acknowledge the encouragement of G. Ottaviani and C. Canali (U. of Modena) who motivated this work and supplied the trial regrowth sample. We also thank O. Westcott for making the implants, J. Gallant for making the evaporated targets and F. Brown for useful discussions.

REFERENCES

(1) Thin Solid Films 19 (1973) 1-463.
(2) V. Marrello, J.M. Caywood, J.W. Mayer and M.-A. Nicolet, Phys. Stat. Sol.(a) 13 (1972) 531.
(3) C. Canali, J.W. Mayer, G. Ottaviani, D. Sigurd and W. van der Weg, Appl. Phys. Lett. 25 (1974) 3.
(4) E. Gadioli, I. Iori, M. Mangialaio and G. Pappalardo, Nuov. Cim. 38 (1965) 1105.

(5) M. Corti, G.M. Marcazzan, L. Milazzo Colli and M. Milazzo, Nucl. Phys. 77 (1966) 625.
(6) G. Corleo and S. Sambataro, Nuov. Cim. 56B (1968) 83.
(7) P.M. Endt and C. Van der Leun, Nucl. Phys. 34 (1962) 1-324.
(8) J.D. Garcia, R.J. Fortner and T.M. Kavanagh, Rev. Mod. Phys. 45 (1973) 111.
(9) F. Folkmann, J. Phys. E 8 (1975) 429.
(10) J.F. Chemin, I.V. Mitchell and F.W. Saris, J. Appl. Phys. 45 (1974) 532.
(11) K.L. Dunning, G.K. Hubler, J. Comas, W.H. Lucke and H.L. Hughes, Thin Solid Films 19 (1973) 145.

ANALYSIS OF FLUORINE BY NUCLEAR REACTIONS AND APPLICATION TO

HUMAN DENTAL ENAMEL

> J. STROOBANTS, F. BODART, G. DECONNINCK, G. DEMORTIER
> and G. NICOLAS (x)
> L.A.R.N., Facultés Universitaire, 5000-NAMUR, Belgium
> (x) Université de Louvain, 3000-Leuven, Belgium

Abstract: Nuclear reactions induced on Fluorine by low energy protons are investigated, thick target excitation yield curves and tables for $^{19}F(p,p'\gamma)^{19}F$ and $^{19}F(p,\alpha\gamma)^{16}O$ reactions are given between 0.3 and 2.5 MeV. Interferences from other nuclear reactions, detection limits and sensitivity for Fluorine detection are investigated.

After a wide investigation of the repartition of Fluorine in tooth enamel it is concluded that there is an equilibrum of the concentrations between tooth and saliva which is rapidly restored after the perturbation introduced by the external treatments.

Prompt γ-Rays from Proton Bombardment of Fluoride Samples

The technique of Fluorine analysis by low energy nuclear reaction has often been described [1,2,3] and the application to Fluorine concentration determination is now routinely used in different laboratories. An intense γ-ray emission is observed when Fluorine is bombarded with low energy protons : a) low energy γ-rays from inelastic scattering ($E\gamma$= 110 and 197 keV), b) high energy γ-rays from (p,αγ) reactions ($E\gamma$=6.13, 6.72 and 7.12 MeV). At low energy, isolated resonances in the cross-section curves can be used for depth profile analysis, at higher energy the resonances overlap and cross sections are monotonic function of the proton energy.

The γ-ray yield from a thick sample of CaF_2 was detected at different proton energies, the data were corrected for stopping power and concentrations, in order to correspond to a sample of pure Fluorine. The number of γ-rays emitted by an homogeneous thick sample bombarded with protons of energy E_0 is then given by[4]

$$N_s = N_\gamma C_s \frac{S(\bar{E})}{S_s(\bar{E})} \quad (\mu C \times Sterad)^{-1} \tag{1}$$

where N_s is the number of γ-rays detected at energy E_0, N_γ is the number given in the table, C_s is the concentration of Fluorine in the sample, $S(\bar{E})$ and $S_s(\bar{E})$ are the stopping power parameters of Fluorine (in MeV/g cm^2) and of the sample respectively, $\bar{E}$ is the proton energy for which the yield has half the value it has at energy E_0. For bulk analysis, the sample is bombarded with 2.5 MeV protons. γ-rays are detected in a Ge-Li detector for low energy γ-rays, in a large NaI scintillator for high energy γ-rays. The concentration is obtained by comparison with a standard by application of the following formula[4]

$$C_s = C_{st} \frac{N_s}{N_{st}} \frac{S_s(\bar{E})}{S_{st}(\bar{E})} \tag{2}$$

The accuracy of the determination was tested by making calibration curves with standard samples and also by comparing the results obtained by this technique with chemical determinations on samples of unknown composition. The absolute accuracy is better than 1% for concentrations larger than 0,001, the sensitivity limit for quantitative determination is about 1 p.p.m.

1) $^{19}F(p,p'\gamma)^{19}F$ reaction. Two γ-ray lines are observed in spectra from the $^{19}F(p,p'\gamma)^{19}F$ reaction, corresponding to $E\gamma$ = 110 keV and $E\gamma$ = 197 keV. The thick sample yield is the same for both radiations for 2 MeV protons, above this energy the 197 keV γ-ray emission is more intense. A complete tabulation of the thick target excitation yield is given in Table I and the corresponding excitation curves are represented in Fig. 1-2. From these data, the number of γ-rays from bombardment of a given sample can be predicted by using formula (1). A narrow resonance is present at E_R = 938 keV (Γ_R = 4.5 keV). The corresponding plateau is clearly visible in the excitation curve. Below this energy the reaction is not very intense; above this energy the excitation yield curve reveals the presence of many resonances superposed to a continuous amplitude. A serious handicap in the utilization of these γ-rays for quantitative analysis is the presence of an important continuous spectrum from higher energy γ-rays (Compton effect). This effect can be reduced by using a high resolution Ge-Li detector (1 keV) of small thickness. The estimated sensitivity limit

ANALYSIS OF FLUORINE BY NUCLEAR REACTIONS

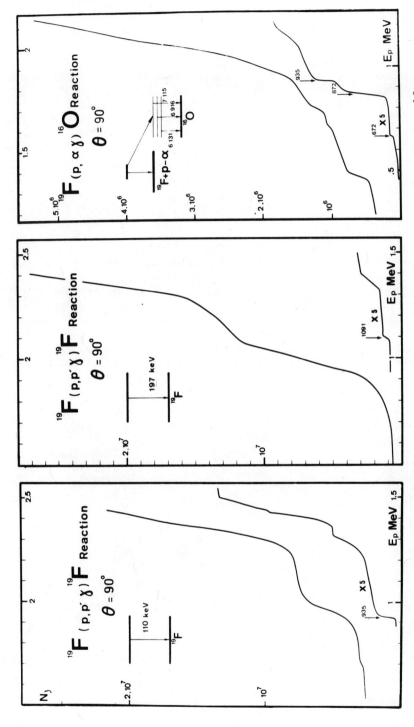

Fig. 1-2-3. Thick target excitation yield curves for (p,p'γ) and (p,αγ) reactions on ^{19}F (data are from Table I).

is then 10 p.p.m. A possible first order interference is the $^{18}O(p,\gamma)^{19}F$ reaction with the emission of 197 keV γ-rays; the intensity is very small and does not affect seriously the determinations even in presence of large concentrations of oxygen. The emission of th 110 keV γ-ray was not observed in the bombardment of oxygen with 2.5 MeV protons. This interference can be resolved in two ways :
a) by measuring the ratio Nγ(110 keV)/Nγ(197 keV) which must be identical for a Fluorine standard and for the sample; this ratio is also obtained from Tables I and II; b) by measuring the intensity of the 495 keV γ-ray yield from the $^{16}O(p,\gamma)^{17}F$ reaction for 2.2 MeV protons. The number of 197 keV γ-rays from the interfering reaction is then given by Nγ(197 keV) = 0,5 Nγ(495 keV).

2) $^{19}F(p,\alpha\gamma)^{16}O$ reaction. Several lines are seen in the spectra from the $^{19}F(p,\alpha\gamma)^{19}F$ reaction detected with a large volume (30c.c.) Ge-Li detector (fig.4), they correspond to full energy, one-escape and two-escape peaks of the 6.131, 6.916 and 7.115 MeV γ-rays. The more intense is the 6.131 MeV group γ_3, the two other groups γ_1 and γ_2 have comparable intensity and are strongly affected by Doppler effects, the life-time of the corresponding levels in ^{16}O being short enough for disintegration to take place before slowing down in the sample. A complete tabulation of the thick target yield is available while the corresponding excitation curves are represented on fig.3. Many resonances are present and can be used for depth profile analysis. The step height I_R of the corresponding plateau in the excitation curve is proportional to the product σΓ of the cross-section

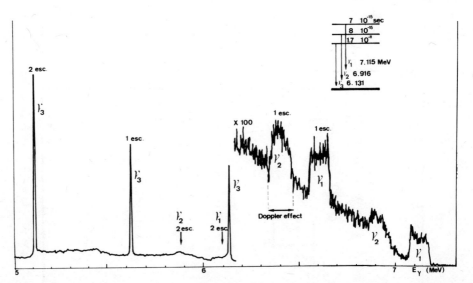

Fig.4 γ-ray spectrum from $^{19}F(p,\alpha\gamma)^{16}O$ reaction detected in a Ge-Li detector.

ANALYSIS OF FLUORINE BY NUCLEAR REACTIONS

E_p (keV)	6-7 MeV	E_0 (keV)	110 keV	197 keV	6-7 MeV	E_p (keV)	110 keV	197 keV	6-7 MeV
330	1.0	920	85	—	185	1640	2761	736	1048
340	5.5	935	317	156	234	1700	3016	944	1220
350	5.7	950	370	170	259	1730	3093	1068	1350
375	5.9	980	414	171	265	1750	3149	1152	1389
400	6.1	1030	443	171	279	1815	3253	1509	1508
450	6.5	1080	463	172	286	1880	3546	2311	1872
475	6.8	1090	488	230	289	1915	3892	3153	2153
485	8.3	1105	491	230	300	1930	4270	3559	2274
495	8.3	1130	496	259	312	1950	4508	4508	2367
520	9.0	1156	502	274	322	1980	5181	6214	2937
570	10.2	1210	580	289	376	2030	6128	8795	3651
595	13.7	1250	614	303	399	2080	6837	11782	4541
620	16.1	1263	717	304	407	2160	7377	12798	5500
645	18.2	1280	778	307	425	2210	7823	13817	—
660	20.7	1290	779	309	437	2317	7976	18088	—
670	28.5	1315	1015	319	458	2370	11420	23615	
680	28.5	1325	1015	329	460		16106		
690	30.5	1335	1015	339	463				
740	31.7	1345	1016	349	480				
780	32.0	1350	1032	368	509				
790	32.4	1365	1080	427	554				
820	35.4	1375	1111	466	661				
830	37.2	1400	1227	584	765				
835	38.5	1420	1850	592	798				
845	40.4	1430	1924	596	816				
870	120.4	1475	2384	621	868				
890	166.1	1550	2715	675	930				
900	175.7	1600	2759	705	1002				
910	179.	1610	2760	706	1047				

TABLE I. Thick target excitation yield for the (p,p'γ) and (p,αγ) reaction on ^{19}F. The yield is given in µC.ster^{-1} for a pure Fluorine sample.

at resonance by the FWHM Γ. The following narrow resonances are currently used for depth profile analysis :

E_R = 340 keV Γ_R' = 2.4 keV I_R =   6.000
E_R = 872 keV Γ_R' = 4.5 keV I_R = 126.000
E_R = 935 keV Γ_R' = 8.6 keV I_R = 70.000

None of these resonances has a pure B-W shape, a continuous background being superposed to the resonant amplitude. The γ-ray intensity from the competing $^{15}N(p,\gamma)^{16}O$ reactions is very small; for 2 MeV proton the ratio $N\gamma(^{15}N)/N\gamma(^{18}F)$ is lower than $5 \cdot 10^{-5}$. Another interference from Nitrogen is possible when low resolution detectors are used. It originates from the $^{14}N(p,\gamma)^{15}O$ reaction giving a γ-ray peak emission of 6,79 MeV without appreciable Doppler broadening. The presence of Nitrogen can be detected by the observation of 4.43 MeV γ-rays from the $^{15}N(p,\alpha\gamma)$ reaction, the intensity being 100 times higher than for the 6,79 MeV radiation above 2.5 MeV.

3) <u>Depth profile analysis</u>. The energy spread in the proton beam, the resonance width and the energy straggling in the sample determine the depth resolution in profile analysis. The depth resolution at the surface of the sample (small energy loss) must be calculated using Vavilov distributions[5]. In Fluorine analysis the resonances have large width and the distributions are strongly modified by convolution with the resonance curve ; for large depth ($x > 10^{-4}$ g cm^{-2}) the resulting distribution can be considered as <u>gaussian-like</u>, the FWHM is given for normal incidence by : $\Gamma = \sqrt{\gamma_0^2 + \Gamma_R^2 + \Gamma_S^2}$ where γ_0 is the FWHM of the beam energy distribution, Γ_R the natural width and Γ_S the straggling width given by : $\Gamma_S = 0.93 \, z \sqrt{x \, \bar{Z}/\bar{A}}$. The maximum depth and thickness that can be analysed depends on the energy of neighbouring resonances. The lower energy neighbouring resonance position determines the maximum sample thickness s_{max} which can be analysed while the high energy neighbour determines the maximum depth x_{max} which can be analysed. The following resonances can be used for depth profile analysis. The given parameters correspond to tooth enamel which is equivalent to a pure Silicon matrix.

E_R (keV)	Γ_R' (keV)	$S_{Si}(E_R)$	x_{max}	s_{max}	Resolution ($x=x_{max}$)
340	2.4	375	(3.2 10^{-4})	3.2 10^{-4}	3.4 10^{-5}
872	4.7	189	3.3 10^{-4}	3.3 10^{-4}	6.8 10^{-5}
935	8.1	183	(3.3 10^{-4})	3.3 10^{-4}	8.4 10^{-5}
1373	12.4	145	(1.8 10^{-4})	1.8 10^{-4}	1.0 10^{-4}

All these data are given in g cm^{-2}. The resonance energy and width are given in the laboratory system. From fig.3 it is concluded that the 872 keV resonance is the more intense ; the maximum depth which can be analysed without interfering with the 935 keV resonance is 6.8 10^{-5} g cm^{-2} or about 1.3 μm in tooth enamel.

Tooth Enamel Analysis

1) <u>The sample structure</u>. The important problem of dental caries following the dissolution of the enamel structure under the action of acid substances was studied for many years by means of different techniques, the Fluorine ion being supposedly playing an important action in the inhibition of enamel dental caries. The dental enamel is at 95% a layer of apatite, 500μm thick, in which the OH radical is partly substitued by Fluorine ions giving a concentration of 1 to 5°/₀₀ of Fluorine. Nuclear resonance is an unvaluable probe in dental enamel analysis since X-ray techniques (electron microprobe) are subject to errors from strong absorption of the soft X-ray radiation. Another advantage is the possibility of non destructive depth profile analysis with very small diameter beams. NMR studies indicate that the Fluorine ion would migrate along the C axis of the hydroxiapatite crystal, preventing further migration of OH⁻ ions. However the enamel structure is very different from the single crystal model, X-ray diffraction spectra show a widening of X-ray lines which indicates that the sample is a superposition of crystallites the size of which ranges between 300 and 400 Å. The intervals between these crystallites are filled by an organic network in which the -F ion can migrate. For these reasons the measurements have to be done on tooth enamel rather-than on apatite crystals and they should reproduce as close as possible the conditions of real life.

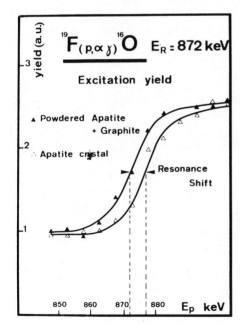

Fig.5. Energy shift observed in the bombardment of insulators.

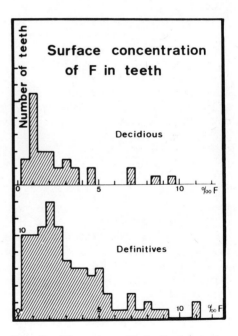

Fig.6. Fluorine distribution in tooth enamel.

2) Experimental method. Teeth are bombarded with the proton beam from the 3 MeV L.A.R.N. Van de Graaff accelerator. Very small beam currents are used in order to reduce radiation damage effects. The sample is sealed on an Aluminium frame, the tooth enamel being an excellent insulator, the distance between the beam spot and the Aluminium was kept as small as possible. Even so, important charge effects are observed. In an excitation curve taken on the 872 keV resonance, the step corresponding to the resonance is shifted toward higher proton energies; shifts as high as 15 keV are observed. This effect is due to the accumulation of electric charges in surface increasing the local potential on the target; the protons are then slowed down before reaching the surface. This effect is impossible to eliminate completely. Fortunately, the shift is constant when the samples are bombarded under identical conditions, a permanent shift of 10 keV being present in all the measurements (fig.5). The 872 keV resonance in the $^{19}F(p,\alpha\gamma)^{16}O$ reaction was chosen because of the high intensity of this reaction. It is not the best resonance for deep analysis but in the case of tooth enamel the F-concentration gradient is large at the surface and the influence of the lower energy resonance is small. On the other hand the observed diffusion region in enamel does not exceed 1μm in thickness. Excitation curves are measured on each sample in order to obtain the Fluorine concentration after various treatments. The surface of each tooth is divided into three regions: the first is left rough, the second is polished with pumice paste in order to remove a surface layer of about 100μm (P), the third region is polished and then treated with different fluoridated products (P+F). All these treatments are done in vivo. The tooth is left in mouth for a given time and then extracted before analysis. Each region is analysed for Fluorine concentration and the results are compared. The Fluorine concentration is extremely variable from tooth to tooth and only average results may be significant. For this reason, a great amount of data must be obtained before drawing conclusions.

3) Measurements. A number of teeth were first explored in order to detect systematic effects in Fluorine distributions. A beam spot of 100μm was used and a scanning was made across the different surface regions. Strong concentration variations are observed which are found to be purely random so that only average values are meaningfull. Average values are obtained by using a 500μm diameter beam spot and by moving the sample back and forth during the measurement. A systematic study of concentration in more than 150 untreated teeth of different origins has revealed that the Fluorine concentration distribution is skew with an average value of 0,25% for normal teeth and 0,15% for decidious teeth (milk-teeth).(fig.6).

For depth profile analysis the 872 keV resonance was used with detection of the high energy γ-rays in a 4"x4" NaI scintillator. In order to evaluate straggling effects, the distributions obtained

by making the convolution product of the Vavilov distributions with
the Breit Wigner distribution of 4.8 keV FWHM were calculated for different energy losses. The results of this calculation are presented
in fig.7; they are used for unfolding of the excitation yield curves.
The Fluorine distribution is obtained by using trial profile concentrations and taking the "convolution" product with the distributions
of fig.7. The results are given in fig.8-9 where the profile curves
corresponding to an average determination on 17 teeth are given. In
a first run (fig.8) teeth are treated and extracted; afterwards they
are analysed by taking excitation curves at resonance over a 50 keV
energy range on the two regions. No appreciable concentration modifications are observed when extracted teeth are kept in a dry atmosphere for a long time. The curves in fig.8 show that in the treated
region a large enrichment in Fluorine is detected to a depth of
0.1 to 0.3μm. (the first 0.1μm being uncertain, the concentration is
represented by a dashed line). In a second run of measurements,
teeth are left in mouth during 10 days before extraction, no appreciable difference is detected between polished and fluoridated regions (fig.9). A similar phenomenon was observed when teeth are kept
in distilled water.

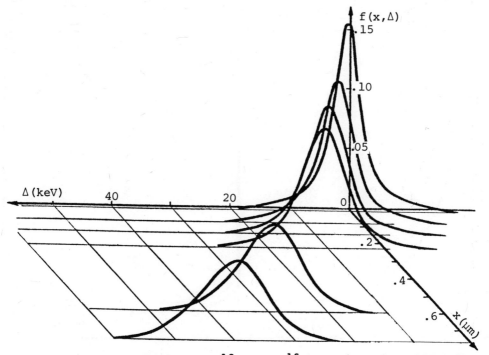

Fig.7. Resonance width of the $^{19}F(p,\alpha\gamma)^{16}O$ reaction (E_R= 872 keV
Γ_R= 4.7 keV) corresponding to different thicknesses in tooth enamel.
The distributions have been obtained by taking the convolution of
the B-W curve with Vavilov distributions.

4) <u>Conclusion</u>. Fluorine migration in teeth enamel is observed when the rough surface is eliminated by polishing and treatment with different fluoridated compounds (Fluocaril dental frost). This fluoration does not disappear when the tooth is kept in dry atmosphere. On the contrary the enrichment effect is practically absent after a few days when the tooth is kept in contact with saliva. It is then supposed that the atoms of Fluorine are eliminated by ion exchange phenomena in saliva. The hypothesis of exchange with **Chlorine ions** was made and the Cl concentration was measured by X-ray fluorescence induced by protons, no appreciable variation was observed. A similar phenomenon being observed when teeth are kept in distilled water the possibility arises of exchange with OH ions although the migration energy of a F^- ion is only 0,48 e.V. while it is 2.1 eV for an OH^- ion. Another explanation would be that the fluoride ions introduced by the treatment do not appreciably migrate into the crystallites but would simply be introduced between the crystallites through the organic matrix. A possible test of this hypothesis is to take NMR spectra from enamel powder. Unfortunately the concentration is so low that the experiment is beyond the sensitivity limits of this technique.

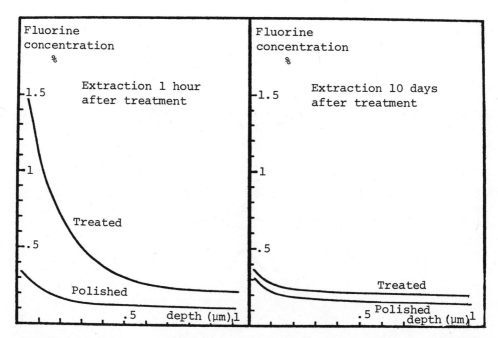

<u>Fig.8.</u> Concentration profile of Fluorine in tooth enamel. Comparison between polished and fluoridated tooth regions (average values).

<u>Fig.9.</u> Same as fig.8 but teeth are extracted 10 days after treatment (average values).

References

1. E. MOLLER and N. STARFELT. Nucl.Inst. and Methods 50 (1967) 225.
2. J.W. MANDLER, R.B.MOLER, E.RAISEN and K.S.RAJAN. Thin Solid Films 19 (1973) 165-172.
3. I.GODICHEFF, M.LOEUILLET and Ch. ENGELMAN. Journal of Radioanal. Chemistry 12 (1972) 233-250.
4. G.DECONNINCK. Journ.of Radioanal. Chem. 12 (1972) 157-169.
5. G.DECONNINCK and Y.FOUILHE. Contributed Paper n°1.10

DETERMINATION OF NITROGEN DEPTH DISTRIBUTIONS IN CEREALS USING THE $^{14}N(d,p_o)^{15}N$ REACTION

B. Sundqvist, L. Gönczi, I. Koersner,[+] R. Bergman and[+] U. Lindh

Tandem Accelerator Laboratory and [+]Gustaf Werner Institute

Box 533, S-751 21 Uppsala and [+]Box 531, S-751 21 Uppsala

ABSTRACT

Nitrogen and protein contents are correlated in agricultural products and therefore determination of nitrogen might be used for plant breeding purposes. The Kjeldahl method is time-consuming and destructive and new methods which are faster and nondestructive are searched for. It is also of great interest to extract information on the distribution of nitrogen in single seeds. In rice for example the most important part of the protein content is removed by polishing the seeds.

We have used (d,p) reactions in ^{14}N (E_d = 6 MeV) to determine the the content [2] and recently the depth distribution of nitrogen in single kernels of seed. The time needed to determine the total nitrogen content and the depth distribution with reasonable statistical accuracy is 1-5 seconds and 10-30 seconds respectively. With the deuteron energy used the analyzed depth was 250 μm and the depth resolution about 10 μm throughout the layer.

A comparison, concerning total nitrogen, of this method with the Kjeldahl method was made on individual peas and beans. The results were found to be strongly correlated. The technique to obtain depth distributions of nitrogen was used on high- and low-lysine varieties of barley for which large differences in nitrogen depth distributions were found.

1 INTRODUCTION

As nitrogen and protein are correlated in agricultural products [1], determination of nitrogen in single seeds is of interest, as well as determination of nitrogen distributions. For example, in rice, the most important part of the protein content is removed by polishing. New methods which are faster than the Kjeldahl method and, preferably nondestructive are searched for.

Recently, a fast method to measure the nitrogen content in single seeds was presented [2]. The large positive Q-values of (d,p) and (d,α) reactions in ^{14}N in comparison with competing reactions made it possible to measure the nitrogen content from a charged particle spectrum containing almost no background. High precision was achieved in a few seconds of analysis time.

The possibility to determine depth distributions of nitrogen is demonstrated in this paper. The method has been applied to two varieties of barley of low- and high-lysine content, respectively.

2 EXPERIMENTAL TECHNIQUE AND PROCEDURE

In all experiments the samples were mounted in a charged-particle scattering chamber and exposed to a beam of 6 MeV deuterons from the Uppsala Tandem van de Graaff accelerator. The charged particles emerging at a scattering angle of 165° were detected with a surface barrier detector [2].

To calibrate the depth scale for a thick target spectrum (see fig 3 ref 2) the following procedure was used.

A thin layer of melamine ($C_3H_6N_6$), about 1 μm thick, evaporated on a 0.5 μm carbon foil was exposed to the beam and gave the spectrum shown at top of fig 1 (D = 0.0 μm). With different polyethylene absorbers with thicknesses 45, 90, 180 and 225 μm, respectively, approximately equivalent to corresponding layers of tissue, put in front of the melamine foil , (see fig 1) the proton peak was shifted towards lower energies and somewhat broadened.

A peak due to the reaction $^{13}C(d,p)^{14}C$, (not seen because of a lower cut off in the logaritmic y-scale) sets a lower limit to the part of the spectrum which unambiguously can be used for extracting information on the depth distribution of ^{14}N. This limitation in depth corresponds to about 0.2 mm when 6 MeV deuterons are used.

The coat contribution should be avoided in order to derive information from the grain endosperm. This is accomplished by discriminating against the highest energies to a depth corresponding

DETERMINATION OF NITROGEN DEPTH DISTRIBUTIONS IN CEREALS

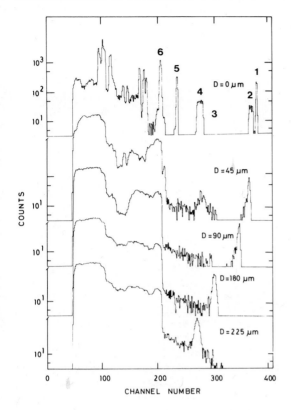

Fig 1
Charged particle yields at $\theta_{lab}=165°$ induced by 6 MeV deuterons on a thin melamine foil (D=0 μm) and added absorbers in front of the melamine foil with thickness 45, 90, 180 and 225 μm respectively. The numbers in the figure are explained below.

(1) $^{14}N(d,p)^{15}N$ (g s)
(2) $^{14}N(d,\alpha)^{12}C$ (g s)
(3) $^{13}C(d,p)^{14}C$ (g s)+
(4) $^{14}N(d,\alpha)^{12}C$ (1st e s)
(5) $^{14}N(d,p)^{15}N$ (1st and 2nd e s)
(6) $^{12}C(d,p)^{13}C$ (g s)

+ Not seen because of the lower cutoff in the y-scale.

to the layer thickness (∼ 50 μm for barley) and against low energies to avoid the interval corresponding to surface contributions from ^{13}C from the reaction $^{13}C(d,p)^{14}C$ (g s). The number obtained by integrating the signals in this selected interval is thus associated with nitrogen only. The total number of signals in the low energy part of the spectrum, i.e. from energies corresponding to deuterons elastically scattered from oxygen and carbon to a level well above the electronic noise, we have assumed to be proportional to the dry matter weight of the part of the grain analyzed.

With this (d,p) technique to measure the average nitrogen content in single seeds, eleven peas (Pisum sativum) and fourteen beans (Vicia faba) were analyzed. In addition individual Kjeldahl measurements were made afterwards.

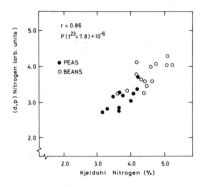

Fig 2
Individual comparison between the (d,p) and Kjeldahl methods for determination of nitrogen contents in peas and beans.

Also the nitrogen depth distribution is obtainable from the spectral information. The variation of nitrogen content with depth in a specimen relative to that in a reference material is a convenient measure we have used for the depth distribution analysis, which is illustrated by the present study of two closely related varieties of barley. Five intact grains of low-lysine barley (Boni) and five intact grains of high-lysine barley (Boni, Risø mutant 1508) were analyzed. The average nitrogen contents according to Kjeldahl analysis were 1.9 % and 2.0 % respectively. The high-lysine group was normalized to the low-lysine group of Boni seeds by the ratio of total number of counts in the summed spectra of each group. Furthermore, the spectra were divided channel by channel (fig 3). The ratios thus obtained would all be equal to one, if the two groups of seed were identical. The same analysis was made on five seeds from each group with the seed coats removed (fig 4). Using the depth scale derived from fig 1 by converting channel number to a certain depth in µm, information on the relative depth distribution of nitrogen could be obtained.

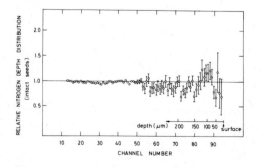

Fig 3
The relative nitrogen depth distributions obtained from a comparison between five intact grains of high- and low-lysine barley.

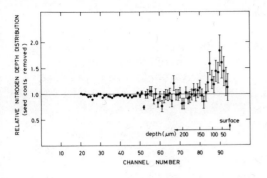

Fig 4
The relative nitrogen depth distributions obtained from a comparison between five grains of high- and low-lysine barley with coat removed.

To study the radiation damage on seeds, 227 seeds of wheat were exposed to the deuteron beam. After at least 8 days of recovery, the seeds were placed on moist sand and exposed to daylight lamps. After 7 days the germination and straw length were measured (see table I) [4].

Group	Total number of seeds	Normal seedlings No	Normal seedlings %	Abnormal seedlings No	Dead No
F1	12	0	0	6	6
F2	21	2	9±7	18	1
F3	85	59	69±11	19	7
B1	24	17	71±22	2	5
B2	26	26	100±28	0	0
B3	59	50	85±16	5	4
Control	147	138	94±11	2	7

Table I

Results from the study of radiation damage on seeds of wheat. The numbers of normal and abnormal seedlings and the number of dead seeds are shown together with the corresponding results from a (not irradiated) control group.

3 RESULTS

The correlation between individual measurements of average nitrogen content with the (d,p) and Kjeldahl methods on intact peas and beans is shown in fig 2. The results are strongly correlated as can be seen from the correlation coefficient of 0.86 derived from a statistical t-test with 23 degrees of freedom.

Nitrogen depth distributions are illustrated in figs 3 and 4. Nitrogen concentration differs by as much as 30 - 50 % in a layer 50 - 150 µm below the intact surface of the two varieties of barley. With the seed coats removed ($\sim$ 40 µm) the distribution is similar to that of intact seeds but translated a distance about equal to the coat thickness towards the surface. The large difference in nitrogen content appears at depths corresponding to the aleuron layer and possibly also the subaleuron layer observed in high-lysine barley [3].

From the study of the radiation damage on seeds of wheat, (table I) it is clear that a hit of the embryo is lethal. For seeds, hit in other parts of the endosperm, about 85 % of the seeds grew [4].

4 DISCUSSION

The nitrogen distributions presented in this paper correspond to a depth of about 0.2 mm. This depth can be extended by using a higher deuteron bombarding energy. However, the depth resolution mainly associated with the specific energy loss of the incoming deuteron will deteriorate at higher energies. The extraction of depth distributions in seeds requires somewhat better statistics in the spectra than in the determination of average nitrogen content.
To obtain the precision of the depth distributions shown in this paper it will take 10-15 seconds with proper equipment.

The techniques discussed above are useful for genetic studies. Together with prof J McKey at the Agricultural College in Uppsala, two studies are suggested. The first idea is to investigate the process of grain filling of protein in the seed up to maturity and how this process is synchronized within one inflorescence. Also a study of effects of different fertilizers is suggested. This study is already in progress in Uppsala.

The second idea is to study the localization of the genetical regulation to single chromosome pairs in wheat regarding protein content and protein distribution in different parts of the seed.

REFERENCES

1. G Ågren and R Gibson, Children's Nutrition, Unit Rept No 16 Inst Med Chem, University of Uppsala, Box 551, S-751 21 Uppsala, Sweden

2. B Sundqvist, L Gönczi, R Bergman and U Lindh, Int J Appl Radiat Isotopes 25 (1974) 277

3. L Munck, Hereditas 72:1-128 (1972)

4. B Larsson et al, Gustaf Werner Institute, Uppsala, private communication

EXPERIMENTAL MEASUREMENTS, MATHEMATICAL ANALYSIS AND PARTIAL DECONVOLUTION OF THE ASYMMETRICAL RESPONSE OF SURFACE BARRIER DETECTORS TO MeV ^{4}He, ^{12}C, ^{14}N, ^{16}O IONS [x]

G. AMSEL, C. COHEN, A. L'HOIR

Groupe de Physique des Solides de l'Ecole Normale
Supérieure, Tour 23, 2 Place Jussieu, 75221 PARIS CEDEX 05

ABSTRACT

The response of surface barrier detectors was systematically investigated for various ions (^{4}He, ^{12}C, ^{14}N, ^{16}O) at energies from 0.3 to 1.7 MeV. All the resolution curves are asymmetrical with a marked tail towards low energies, slight for ^{4}He, much stronger for the heavier ions. For ^{12}C, ^{14}N, ^{16}O ions and at all energies these curves appeared fully homothetic ; for a given incident particle the similarity factor varies with energy as $aE^{1/3} + b$. It is shown that these curves may be fitted with high accuracy by the convolution product of a gaussian with a one-sided exponential factor :

$$R(\eta) = \exp(-(\eta-\bar{\eta})^2/2\sigma^2) \ast \exp(\eta/\eta_0) Y(-\eta)$$

For C, N and O the homothetic curves yield $\sigma/\eta_0 = 0.65 \pm 0.03$. The less asymmetrical response for He yields $\sigma/\eta_0 \simeq 1.05$. A simple linear **numerical** filtering procedure is presented which leads to the deconvolution of the exponential factor in a way similar to the pole-zero cancellation techniques used in signal shaping theory. The interpretation of experimental spectra is considerably simplified by this procedure, all asymmetries of instrumental origin being removed. Applications to the measurement of the energy spread curve shape of ^{4}He, ^{12}C, ^{14}N, ^{16}O ions passing through thin self supporting Al$_2$O$_3$ films are shown as illustration.

[x] Work supported by the Centre National de la Recherche Scientifique (R.C.P. N° 157).

INTRODUCTION

The precise knowledge of the resolution curve of surface barrier detectors is of prime importance in nuclear physics for energy level determinations or in experiments such as depth profiling and as energy straggling measurements where the energy spread is of the order of instrumental resolution. Several authors [1], [2] [3] have discussed the advantage of using beams of particles of various Z_1, M_1 such as ^{4}He, ^{12}C, ^{16}O ions in backscattering experiments, an important parameter being the width of the resolution curves. The response of surface barrier detectors to energetic ions has already been studied experimentally and theoretically for light and heavy ions as a function of energy by various authors [4] [5] [6] [7] [8], the two main parameters studied being the pulse height defect (PHD) and the energy spread variance. The PHD and the energy spread depend on the energy loss in the dead layer of the detector, on the charge recombination and trapping in the depleted region, on the elastic nuclear energy losses, etc. The electronic noise may also contribute to the broadening of the resolution curves but for ions such as ^{16}O where the resolution (FWHM) is $\sim$ 50 KeV at 1 MeV the contribution of this factor may generally be neglected.

In this work a precise study of the resolution curves was undertaken both for shape and position (PHD) in the case of ^{4}He, ^{12}C, ^{14}N, ^{16}O ions between 0.3 and 1.7 MeV. The marked asymmetry of the resolution curves was measured and fitted analytically. No physical explanation of these behaviours will be presented in this paper as, in a first step, our main aim was to optimize the interpretation of experimental spectra. A numerical filtering procedure was hence developed, allowing one to compensate for these instrumental asymmetries.

EXPERIMENTAL

Monoenergetic ions were obtained by backscattering ion beams produced with the 2 MeV Van de Graaff of the Ecole Normale Supérieure. Very thin films of tantalum ($\sim$ 4 µg cm^{-2}) deposited on a vitrous carbon substrate by triode sputtering were used. The beam and pulse handling system is fully described in ref. [9]. The detector used here was a silicon surface barrier detector from ORTEC, with a resistivity of 800 Ω.cm, a barrier depth of 120µ (under 65 volt bias) and a 25 mm^2 area, located at 50 mm from the 1 mm diameter beam impact point, provided with a 3 mm diameter diaphragm and located at a laboratory angle of 150°. Amplification was achieved with ORTEC's 109 A charge sensitive preamplifier and 410 main amplifier ; the latter was used with equal differentiation and integration time constants (0.5 µS) and was followed by a 444 biased pulse stretcher. The overall electronic noise in these conditions was 5 keV (FWHM). The ADC used was a CA13 Intertechnique unit coupled to a BM96 memory. The experiments were performed in the scattering chamber described in ref. [9], in which the vacuum was kept below 10^{-6} Torr using a local liquid nitrogen trap.

The energy resolution of the incident beam was 250 eV FWHM. The thickness uniformity of the tantalum film was checked by comparing resolution curves at the same energy for various thin targets (Au, Ta) deposited on silicon or carbon. No differences in shape were observed. The calculated energy spread (FWHM) introduced by backscattering on the thin tantalum film assumed to be uniform was less than 0.2 $L_{1/2}$ in all cases. The kinematic effect due to the 3 mm diameter diaphragm was very small (less than 0.05 $L_{1/2}$ in all cases). In what follows we neglected these phenomena assuming a monoenergetic backscattered beam.

Beam currents were of the order of 50 nA for experiments with ^{4}He ions and 10 nA for experiments with heavier ions. The overall counting rate did not exceed 10^3 counts per second for all spectra, as the backscattering from the substrate was low for ^{4}He ions and absent for the others. In this way no damage was observed [2] on the detector during the experiments with heavy ions although high statistics were obtained in rather short times ($\sim 2.10^5$ counts in about 5 mn). The beam impact point on the target was systematically renewed so as to avoid any carbon build up effect. In fact no contamination influence was observed when comparing spectra corresponding to a fresh or an already irradiated impact point. The stability of the electronic set up and its linearity were checked using a mercury relay pulser and comparing spectra taken at the same energy at several hours of interval. The accelerator calibration on which we relied for PHD measurements was carried out using well known narrow resonances of nuclear reactions [9].

RESULTS

The local KeV/channel scale of the non linear amplitude response of the detector was determined in the following way. For a given incident ion, two different spectra taken at backscattered energies E_i and E_{i+1} with $\Delta E_i = E_{i+1} - E_i$ small as compared to E_i were recorded. The amplitudes $A(E_i)$ and $A(E_{i+1})$ corresponding to the peak positions expressed in channels were used to calculate the local KeV/channel figure at the mean energy $(E_i+E_{i+1})/2$, defined as : $\Delta E_i/(A(E_i)-A(E_{i+1}))$: the precise position of the maximum of each resolution curve being determined using a parabolic least square fit. For protons, this parameter is nearly a constant but for heavier ions such as ^{16}O it decreases by $\sim$ 17 % between 0.5 and 1.5 MeV. The PHD in KeV for a given ion was calculated with respect to protons of same energy. The results as a function of energy for ^{12}C, ^{14}N, ^{16}O ions are plotted in Fig. (1). For comparison the energy loss of 1.25 MeV ^{16}O ions in the rated 40 µg cm^{-2} gold electrode is $\sim$ 45 KeV according to the Northcliffe and Schilling table [10] and the PHD for the same ion at the same energy is $\sim$ 130 KeV. The PHD

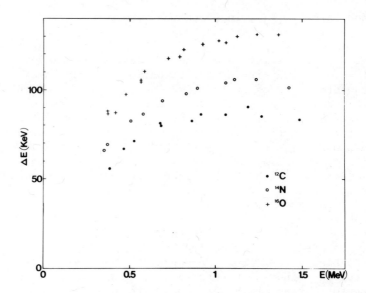

Fig.1 Pulse height defect ΔE in KeV as a function of energy for ^{12}C, ^{14}N, ^{16}O ions in an ORTEC silicon surface barrier detector.

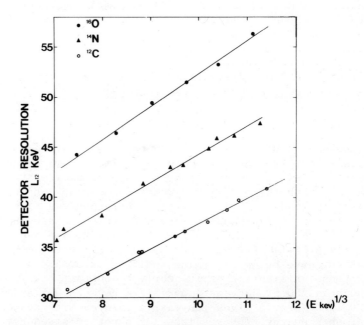

Fig. 2 Full width at half maximum $L_{1/2}$ expressed in KeV of the resolution curve of the surface barrier detector for ^{12}C, ^{14}N and ^{16}O ions as a function of the cubic root of energy.

was found to be independent of the applied bias from 40 to 65 volts.

The FWHM of all the resolution curves were always in the 10 to 20 channels range and their determination using least square fit obtained with a precision of $\sim$ 0.2 channels. The resolution $L_{1/2}$ in KeV was then obtained using the local KeV/channel curves (and not the proton energy scale). For ^{4}He ions the resolution, after correcting for a 5 KeV electronic noise using the square sum law, is found to increase slightly with energy from 10.4 KeV at 0.3 MeV to 11.2 KeV at 1.7 MeV. The resolutions $L_{1/2}$ for ^{12}C, ^{14}N and ^{16}O ions are plotted in Fig. (2) as a function of the cubic root of the detected ion energy E. In our energy range a linear variation of $L_{1/2}$ with $E^{1/3}$ is found for each ion :

$L_{1/2}$ (KeV) = 2.55 $E_{KeV}^{1/3}$ + 11.8   for ^{12}C ions

$L_{1/2}$ (KeV) = 2.87 $E_{KeV}^{1/3}$ + 15.6   for ^{14}N ions

$L_{1/2}$ (KeV) = 3.30 $E_{KeV}^{1/3}$ + 19.4   for ^{16}O ions

The extrapolation of this law for ^{16}O to the energies of 5 MeV and 10 MeV respectively leads to 76 KeV and 90.5 KeV for $L_{1/2}$, when ref [3] gives 60 KeV and 84 KeV for the same ion and energies. Thus the cubic root law seems to hold reasonably in a wide range. For a given matrix and in the energy range where the stopping power of the incident ions (^{12}C, ^{14}N, ^{16}O) increases with energy as $E^{1/2}$, i.e. for ions with energies of some MeV, the depth resolution in backscattering experiments is thus improved when increasing the bombarding energy.

As shown in Fig.(3) the resolution curves are strongly asymmetrical for ^{12}C, ^{14}N, ^{16}O ions and $L_{1/2}$ is not sufficient to specify their stage. In the next section this aspect is analysed in detail.

ASYMMETRY ANALYSIS

We found that for ^{12}C, ^{14}N, ^{16}O ions and at all energies all the resolution curves appear fully homothetic. As an illustration four resolution curves are plotted on Fig. (3) : two for ^{16}O ions for different energies, one for ^{12}C and one for ^{14}N ions, the corresponding similarity factors having been taken into account for the channel scale, the maxima being normalized to the ^{14}N curve. It appears that the long tails towards low energies do not differ significantly. The resolution curves for ^{4}He ions and at all energies are also homothetic, the asymmetry being smaller than for the heavier ions.

The resolution curves $R(\eta)$ (η being the local energy scale variable expressed in channels) appearing universal, an attempt was made to find a simple analytical representation describing correctly the asymmetry. Having in mind the possibility to apply to these

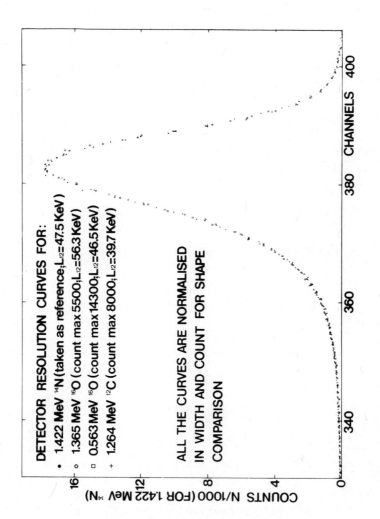

Fig. 3   Detector resolution curves for various ions and energies.

curves a deconvolution procedure, to remove the asymmetries, as will be shown below, we tried to fit the experimental curves with the convolution product of a symmetrical function with a highly asymmetrical one, the latter lending itself to easy deconvolution. It was found that a remarkable fit may be obtained using the product of a gaussian with a one-sided exponential :

$$R(\eta) = \exp(-(\eta-\bar{\eta})^2/2\sigma^2) \ast \exp(\eta/\eta_o) Y(-\eta) \qquad (1)$$

where $Y(-\eta)$ is the left-sided Heaviside unit step function. Due to the similar shape of all curves the only independent parameter in equ.(1) is the dimensionless figure σ/η_o, which is a measure of the asymmetry. The mean value corresponding to $R(\eta)$ is $\bar{\eta} - \eta_o$, the variance is $\sigma^2 + \eta_o^2$ and the third order central moment $-2\eta_o^3$ i.e. the coefficient of skewness γ_1 is :

$$\gamma_1 = -2/\left[(\sigma/\eta)^2 + 1\right]^{3/2}$$

For ^{12}C, ^{14}N, and ^{16}O ions the best fit yields $\sigma/\eta_o = 0.65 \pm 0.03$ (i.e. $\gamma_1 = -1.18$) and for ^4He $\sigma/\eta_o \simeq 1.05$ (i.E. $\gamma_1 \simeq -0.65$). Fig.(4) shows a typical fit for 1.235 MeV ^{14}N ions. Except for the small discrepancies near the high energy region the fit appears surprisingly good, down to very low values of the tail. Such a good fit might suggest that this is not share coincidence but reveals an underlying physical process in the amplitude spread phenomenon, which would be due to the sum of two independent random variables, one of them gaussian. These aspects are being investigated but their interpretation goes beyond the scope of this paper.

ASYMMETRY DECONVOLUTION

The two-sided Laplace transform of a function $s_1(t)$:

$$s_1(t) = g_1(t) \ast \exp(-at) y(t)$$

where $g_1(t)$ is an arbitrary function, is $S_1(p)$:

$$S_1(p) = G_1(p)/(p+a) \qquad (2)$$

Hence the exponential factor can easily be removed by multiplying $S_1(p)$ by $p+a$, or equivalently by applying the convolution operator $\delta' + a\delta$ to $s_1(t)$ i.e. calculating $(d/dt + a) s_1(t)$. When used in signal shaping theory this procedure is called pole-zero cancellation, and is systematically used in nuclear pulse amplifiers, the operator being approximated with well choosen passive filters. A similar principle was used by Amsel et al. [11] for shortening detector signals containing random components. The general principles of pulse tail removal by linear filtering (i.e. deconvolution) developed in Ref.[11] have been applied to the asymmetrical resolution curves considered here which bear many similarities with randomly fluctuating signals. The procedure is analogous to that of Ref. [11] with some modifications due to the particular aspects encountered here :

- filtering was carried out numerically using well choosen convolutive linear operators,
- the tail in our case is left-sided, the exponential having the transform $1/(a-p)$,
- the discrete character of the amplitude spectra was taken into account.

Due to the large number of channels in all the peaks, the spectra, although discrete, could be well fitted with continuous functions as in equ.(1). However in the numerical calculations the discrete aspect must be taken into account as follows.

The discrete equivalent of a spectrum $s(\eta)$ is characterized by the numbers s_k (content of channel k) and may be represented mathematically as a series of point masses (i.e. δ functions) located at the integer numbers of the channel scale. Hence the corresponding Laplace transform is

$$S(p) = \sum s_k \exp(-kp)$$

The discrete equivalent of the exponential of equ.(1) is a geometric series r^k, the point masses being placed at the points $k = 0, -1, -2$ etc, with $r = \exp(-1/\eta_0)$. Its Laplace transform is hence :

$$1 + r \exp(p) + r^2 \exp(2p) + \ldots = 1/(1-r \exp(p)) \qquad (3)$$

The Laplace transform of the discrete resolution spectrum being $\hat{R}(p)$, the transform of the descrete equivalent of equ.(1) is :

$$\hat{R}(p) = G(p)/(1-r \exp(p)) \qquad (4)$$

$G(p)$ being the transform of the symmetrical component of the resolution spectrum. It appears thus that multiplying $R(p)$ by $1-r \exp(p)$ or equivalently applying the convolution operator $\delta - r\delta_{-1}$ to the resolution spectrum removes the exponential type factor. Numerically this means, calling R_k the channel contents corresponding to the resolution spectrum, to replace the latter by R'_k, where

$$R'_k = R_k - r R_{k+1} \qquad (5)$$

This operation, like any deconvolution procedure, increases the count spread in each channel, the factor being here rather large. This factor was hence reduced by applying a smoothing procedure which is also carried out by numerical filtering, the whole calculation being performed in a single step. The details of the complete procedure and the corresponding count spread calculations will be presented in Ref. [12].

In Fig.(4) along with the resolution spectrum for 1.235 MeV 14N ions is plotted the deconvoluted spectrum obtained by numerical filtering. The smoothing procedure used here leads to a slight increase of the FWHM of the ideal deconvoluted spectrum i.e. the gaussian of variance σ^2 in equ.(1) ; this increase was here only about 4 %. A gaussian of same FWHM than the deconvoluted spectrum has also been plotted. The latter appears fully gaussian as expected : the tail towards low energies has been removed and replaced by channel content numbers spreading around a zero mean value. The statistical spreads of the channel counts can be easily calculated [12] and are indicated by error bars. These errors could be reduced by further smoothing, but this would lead to some broadening of the spectrum : the result of Fig.(4) is a compromise, which in general should be choosen according to the actual aims of the experiments and the available statistics. As expected, the deconvolution procedure shifts the peak of the resolution curve towards higher energies by around η_o, the mean value of the exponential factor removed. In this case this was 5.1 channels (see fig.4). As also expected, due to the reduction of variance entailed by the deconvolution, the resolution curve is also narrowed in its bulk : the FWHM is reduced by 35 % in the present case, in spite of the smoothing applied.

The deconvolution procedure developed above can readily be applied to any experimental spectrum $s(\eta)$ obtained using the surface barrier detector studied. Such a spectrum is indeed the convolution product of the instrumental response $R(\eta)$ by a function $s_o(\eta)$ which represents the ideal spectrum :

$$s(\eta) = s_o(\eta) \ast R(\eta) \qquad (6)$$

The convolution product being associative, equ.(1) leads to :

$$s(\eta) = [s_o(\eta) \ast g(\eta)] \ast \exp(\eta/\eta_o) \, Y(-\eta) \qquad (7)$$

$g(\eta)$ standing for the symmetrical factor which in equ.(1) is a gaussian. Hence $s(\eta)$ is also the convolution product of a given function by an exponential factor. So the same deconvolution procedure removes the exponential factor leading to a new spectrum $s'(\eta)$

$$s'(\eta) = s_o(\eta) \ast g(\eta) \qquad (8)$$

from which the instrumental asymmetries are removed.

As an illustration we plotted in Fig.(5) the spectrum obtained when measuring the energy straggling of ^{12}C ions in self supporting amorphous Al_2O_3 foils. This spectrum exhibits a long tail towards low energies. The corresponding deconvoluted spectrum is plotted

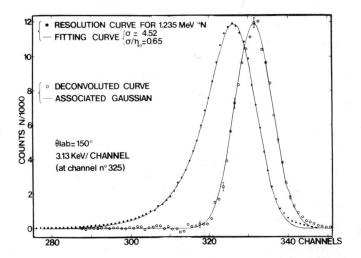

Fig. 4 Detector resolution curve for 1.235 MeV ^{14}N ions (•), along with the fitting curve $R(\eta)$ for $\sigma/\eta_0 = 0.65$ (solid line). The deconvoluted spectrum (o) is plotted along with a gaussian (solid line) of same FWHM. The shift is due to the deconvolution procedure.

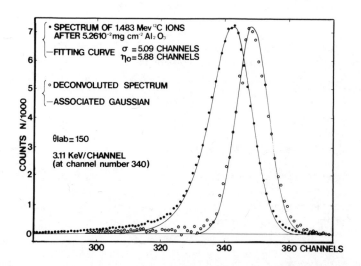

Fig. 5 Spectrum corresponding to 1.483 MeV ^{12}C ions having passed through a 52.6 µg cm^{-2} Al$_2$O$_3$ foil analyzed with a surface barrier detector (•). The deconvoluted spectrum (o) is plotted along with a gaussian of same FWHM. This spectrum exhibits a small positive tail towards low energies.

in Fig.(5) along with a gaussian of same FWHM. A tail towards low energies still remains on the deconvoluted spectrum. According to Equ.(8), this tail necessarily comes from the physical processes described by $s_o(\eta)$ and can in no way be attributed to the asymmetrical resolution curve. The corresponding results are presented in detail in the next paper at this conference [13].

REFERENCES

1. J.W. MAYER, L. ERIKSSON, J.A. DAVIES,
   Ion implantation in semiconductors, Acad.Press, London 1970, p. 140.
2. F. ABEL, G. AMSEL, M. BRUNEAUX, C. COHEN, B. MAUREL, S. RIGO, J. ROUSSEL, J.Radioanal.Chem. 16, 587 (1973).
3. S. PETERSSON, P.A. TOVE, O. MEYER, B. SUNDQVIST and A. JOHANSSON, Thin Solid Films 19, 157 (1973).
4. E.L. HAINES and A.B. WHITEHEAD,
   Rev.Sci.Instr. 37, 190 (1966).
5. B.D. WILKINS, M.J. FLUSS, S.B. KAUFMAN, C.E. GROSS, E.P. STEINBERG, Nucl.Inst.and Meth. 92, 381 (1971).
6. T. KARCHER and N. WOTHERSPOON,
   Nucl.Inst.and Meth. 93, 519 (1971).
7. J.J. GROB, A. GROB, A. PAPE, P. SIFFERT,
   Phys.Rev B, 11, n° 9, 3273 (1975).
8. J. LINDHARD, V. NIELSEN, M. SCHARFF and P.V. THOMSEN,
   Mat.Fys.Med. 33, n° 10 (1963).
9. G. AMSEL, J.P. NADAI, E. d'ARTEMARE, D. DAVID, E. GIRARD and J. MOULIN, Nucl.Inst.and Meth. 92, 481 (1971).
10. L.C. NORTHCLIFFE and R.F. SCHILLING,
    Nucl.Data Tables A7, 233 (1970).
11. G. AMSEL, R. BOSSHARD and C. ZAJDE,
    Nucl.Inst.and Meth. 71, 1 (1969).
12. G. AMSEL and A. L'HOIR, to be published.
13. A. L'HOIR, C. COHEN and G. AMSEL, this conference.

DISCUSSION

Q: (J.F. Ziegler) Did you try a lower resistivity detector to try to determine the contribution of recombination to your detector response?

A: (G. Amsel) We had no lower resistivity detectors, which would also put some problems for keeping electronic noise negligible. It is quite clear that in order to draw detailed conclusions on the phenomena responsible for the resolution asymmetry we should perform more experiments with detectors of various types. However, in practical B.S. work the type of detector investigated is the most commonly used.

Q: (R.F. Lever) How big is the <u>actual</u> energy shift in deconvolution?

A: (G. Amsel) As the mean values of convoluted curves are additive the shift is just equal to the mean value of the exponential, i.e. η_0, if you write it as $e^{-\eta/\eta_0} y(-\eta)$. This amounts to about 2/3 of the FWHM of the registered resolution curves in our experiments.

Q: (W.L. Brown) 1. What is the origin of the tail on the low energy side of the pulse height distribution?
2. Because of the exponential form of the added transform, could it be that hole-electron recombination is occurring with low probability to skew the distribution? Heavy ions should be much more sensitive to this problem than alphas.

A: (G. Amsel) 1. The results presented here are very recent and we did not yet try to interpret them in terms of physical processes. To do so we must anyhow study other types of detectors (varying gold layer thickness, silicon resistivity, applied bias voltage etc.). We may only remark that the convolutive decomposition could be interpreted stating that the collected charge spread is due to the sum of two independent random variables, one with gaussian distribution, the other with an exponential probability density. The $E^{1/3}$ behaviour is yet not understood neither.
2. Such types of phenomena might be responsible for the observed results.

EXPERIMENTAL STUDY OF THE STOPPING POWER AND ENERGY STRAGGLING OF
MeV ^{4}He, ^{12}C, ^{14}N and ^{16}O IONS IN AMORPHOUS ALUMINIUM OXIDE[x]

A. L'HOIR, C. COHEN, G. AMSEL

Groupe de Physique des Solides de l'Ecole Normale

Supérieure, Tour 23, 2 Place Jussieu, 75221 PARIS CEDEX 05

ABSTRACT

Stopping power and energy straggling have been measured in transmission type experiments with self-supporting amorphous Al_2O_3 foils for ^{4}He, ^{12}C, ^{14}N, ^{16}O ions in the 0.3 - 1.7 MeV range. Results on the stopping power of ^{4}He ions are compared to various experimental measurements on ^{27}Al and ^{16}O, the Bragg rule being assumed : our results are systematically $\sim$ 12% lower. For ^{12}C, ^{14}N, ^{16}O ions the stopping power varies linearly with the ion velocity in the 2 to 4.5 10^8 cm s^{-1} range ; our experimental slopes are systematically lower (8 to 35 %) than the various predicted slopes (Firsov, Lindhard and Scharff, Northcliffe and Schilling), the Bragg rule being assumed for this comparison. The energy straggling standard deviation Ω for ^{4}He, ^{12}C, ^{14}N, ^{16}O ions exhibits a nearly linear variation with the ion velocity in the 2.5 to 6.10^8 cm s^{-1} range. For higher velocities, Ω which is $\sim$ 20 % higher than the Chu and Mayer and Lindhard and Scharff predictions approaches asymptotically a constant value $\sim$ 10 % higher than Bohr's theoretical prediction.

INTRODUCTION

Transmission type experiments for energy loss and straggling measurements are by principle the simplest to interpret and more precise, especially for heavy ions, for which the kinematic factor is markedly smaller than one, even for backscattering on high Z materials. However such experiments put severe technical problems as for the preparation of microscopically uniform, thin self-suppor-

[x] Work supported by the Centre National de la Recherche Scientifique (R.C.P. N° 157).

ting films. In the present work we took advantage for carrying out such experiments of a technique developed in our laboratory [1] for preparing self-supporting Al_2O_3 anodic films in the 200 to 5000 Å thickness range ; these films are known to be of well defined stoechiometry, amorphous and of very well controlled uniform thickness in the anodic oxidation conditions. Our results on stopping power for ^4He ions are compared with the Ziegler and Chu tables [2] and experimental measurements of Feng [3] in aluminium, the Bragg rule [4] being assumed. The validity of empirical corrections for the energy loss of ^4He in oxides recently proposed by Ziegler et al. [5] is also checked. For ^{12}C, ^{14}N and ^{16}O ions we measured the stopping power in the 2 to 4.5 10^8 cm s^{-1} velocity range i.e. in a domain where the electronic stopping cross section is velocity proportional. Our results are compared with Firsov [6], Lindhard and Scharff [7] theoretical predictions and with the Northcliffe and Schilling tables [8] assuming Bragg's rule. In straggling measurements comparisons were made with Bohr's theory [9] and with more recent calculations by Hvelplund using Firsov theory [10], Lindhard and Scharff [11] and Chu and Mayer [12] using the formalism given by Bonderup and Hvelplund [13]. A numerical filtering procedure described in an other paper [14] presented by the authors at this conference and referred to hereafter as A was used to calculate the standard deviation Ω of the energy loss processes under study.

EXPERIMENTAL

Principle

Monoenergetic ions were produced by backscattering beams from the 2 MeV Van de Graaff accelerator of the Ecole Normale Supérieure on a very thin film of tantalum (4µg cm^{-2}) deposited on a vitrous carbon substrate by triode sputtering. Backscattering spectra were registered, the experimental conditions being the same as in A. The self-supporting Al_2O_3 foils could be interposed on the path of the backscattered particles, parallel to the detector surface, at 1 cm of the latter. Spectra were recorded at various energies of the incident beam with and without absorber. The energy steps were choosen in order to obtain spectra with and without absorber having very close mean detected energies. In this way a direct measurement of the energy loss in the foils relying almost only on the accelerator calibration was obtained. Moreover as seen in A, the resolution curves of the surface barrier detector depend on energy ; hence the energy straggling in the Al_2O_3 foils can only be calculated if the resolution function at the mean detected energy corresponding to the spectrum with absorber is known.

Foils preparation

Self-supporting Al_2O_3 foils were prepared as described in detail in Ref.(1). They were obtained by anodic oxidation of industrial laminated annealed aluminium sheets 20 μm thick of purity 99.8 %, that not too high purity being necessary to obtain a quick selective dissolution of the metal by H Cl [15]. Oxidations were carried out in aqueous buffer solutions of 0.5 % by weight ammonium citrate (pH6), the current density being $\sim$ 5 mA cm^{-2}. The thickness of the anodic oxide films is then about 12 Å per volt [16] assuming a 3.17 g cm^{-2} density. The Al_2O_3 self-supporting targets are stretched on a metallic frame, the free area of these targets being $\sim$ 2 cm^2.

Determination of the foils thickness and stoechiometry

Although known from the litterature [16] the oxygen content of the foils was directly determined using the $^{16}O(d,p)^{17}O^x$ nuclear reaction at 900 KeV [17]. The standard references used were anodic Ta_2O_5 for which the oxygen content is known by coulometry with a 3 % precision [18], [17]. ^4He backscattering experiments at $\theta_{lab} = 150°$ were also performed for stoechiometry measurements on the Al_2O_3 foils at various beam energies E_i between 0.5 and 1.9 MeV. Fig. 1 represents a typical spectrum taken at Ei = 1.9 MeV. These backscattering spectra, assuming that the Rutherford law is valid for ^4He ions on oxygen and aluminium in our energy domain, allow us to determine the stoechiometry by direct comparison of the counts corresponding to the particles backscattered on oxygen and aluminium (the center of mass correction being carried out). Moreover the oxygen content of the foils was also measured in these backscattering experiments assuming Rutherford cross sections and using as a reference sputtered films of tantalum deposited on a silicon substrate, the amount of deposited tantalum being determined by weight with a 2 % precision [19]. The oxygen contents measured in this way and using the $^{16}O(d,p)^{17}O^x$ nuclear reaction were found to be equal within 3 %. Finally, we found for all the aluminium oxide foils examined a stoechiometry $Al_2O(3.05 \pm 0.05)$; the slight oxygen excess with respect to Al_2O_3 might be attributed to water adsorption at the surface of the foils. The thickness of the foils used were in the 50 to 85 μg cm^{-2} range i.e. in the 1600 to 2800 Å range assuming a density of 3.17 g cm^{-3}. It can be seen in Fig.(1) that carbon build up was negligible. The macroscopic lateral uniformity of the foils was also checked comparing spectra obtained for different beam impact points on the foils. The samples were found to be uniform within the experimental precision ($\sim$1%).

Measurement conditions

All measurements were performed in the scattering chamber described in Ref.[17]. Experimental conditions were exactly the same as in A, the only difference being the use of Al_2O_3 foils.

For ^{4}He ions, the resolution function FWHM which increases slightly with energy was about 12 KeV. For heavier ions, the dependence with energy of this function was systematically studied (see A) ; in the present experiments the FWHM was in the 30 KeV to 55 KeV range.

DATA REDUCTION

Targets of different thicknesses were used in our experiments. The thicknesses were choosen so as to have the best compromise between several conditions. i) the energy loss through the absorber must be great as compared to the FWHM of the detector resolution in order to obtain a precise measurement. ii) the energy loss must be small as compared to the energy of the slowed down particles. If these conditions are respected the fact that the stopping power and energy straggling vary with energy can easily be taken into account in the data reduction. iii) the spread of the energy loss through the absorber should be large enough as compared to the FWHM of the detector resolution so as to get measurements of the energy straggling with reasonable precision. These conditions could be easily fulfilled for ^{4}He ions : typically the energy losses through the absorbers were five to ten times the FWHM of the energy resolution and never exceeded 20 % of the energy of the slowed down particles. The FWHM of the energy straggling through the absorbers was practically equal to the FWHM of the energy resolution. For heavier ions the above conditions were not so easily fulfilled. The energy losses through the absorber were about five times the FWHM of the energy resolution, this representing already 20 to 30 % of the energy of the slowed down particles. However a simple calculation assuming a velocity proportional law for the stopping power shows that even for such losses the error introduced by considering that one measures the stopping power at the mean energy $\overline{E} = E_b - \Delta E/2$ where E_b is the energy of the backscattered beam and ΔE the energy loss, is small as compared to the overall measurement uncertainties. The FWHM of the energy straggling was ~ 35 % lower than the FWHM of the corresponding energy resolutions leading to poorer precision than for ^{4}He ions. However as the deconvolution technique described in A and used here leads to a reduction of about 35 % of the FWHM of the resolution curves the situation with the numerically filtered curve was finally similar to that of the ^{4}He case.

It has been shown in A that the resolution curve at a mean detected energy E_1 can be fitted with high accuracy by

$$R_1(\eta) = \exp(-(\eta-\overline{\eta}_1)^2 / 2\sigma_1^2) \ast \exp(\eta/\eta_o^1) \, Y(-\eta) \qquad (1)$$

with $\sigma_1/\eta_o^1 = 0.65$ for ^{12}C, ^{14}N and ^{16}O ions and $\sigma_1/\eta_o^1 = 1.05$ for ^{4}He ions, the value of σ_1 depending on the particle and the mean detected energy. A spectrum obtained when an absorber is used, with the same mean detected energy E_1, the energy before the absorber

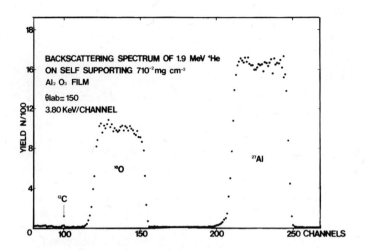

Fig. 1 Typical ^{4}He backscattering spectrum obtained from one of the self supporting Al_2O_3 foils used.

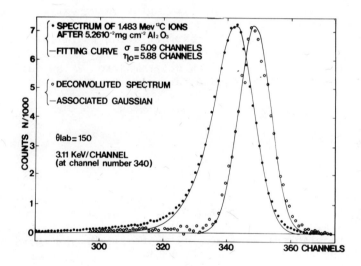

Fig. 2 Typical spectrum of ^{12}C ions having passed through an Al_2O_3 foil (●) fitted with a function given by equ.(2). The deconvoluted spectrum (O), plotted along with a gaussian of same FWHM exhibits a small positive tail towards low energies.

being E_2, and assuming a gaussian energy straggling of variance Ω^2 must hence necessarily follow the law

$$S(\eta) = \exp(-(\eta-\bar{\eta}_1)^2 / 2\sigma_2^2) \ast \exp(\eta/n_o^1) \, Y(-\eta) \qquad (2)$$

with $\sigma_2^2 = \sigma_1^2 + \Omega^2$.

Fig. 2 shows an energy spectrum of ^{12}C ions after passing through a 52.6 g cm^{-2} Al_2O_3 absorber and its fitting curve using equation (2). It can be seen that the fit is satisfactory but for a long tail of the spectrum towards low energies, the integral of this tail representing about 3 % of the overall integrated spectrum. Fig. (2) also compares the deconvoluted spectrum (see A) with a gaussian function (solid line) of variance σ_2^2. Here again the fit is satisfactory but for the tail towards low energies of the deconvoluted spectrum. The origin of this tail is under study. There might be some contribution of high losses by nuclear collisions but the main contributions are likely to be local, microscopic target over-thicknesses. The fact that on the high energy part of the spectrum the fit is as satisfactory as for the resolution curve (see A) shows that there are no noticeable porosities or holes in the Al_2O_3 films. In the present interpretation we have neglected the low energy tail and assumed in a first approximation a gaussian energy straggling of variance Ω^2 . Ω is given in KeV using the local KeV/channel scale (see A), its value being extracted from both fitting procedures reported in Fig.(2). The energy loss distribution function through the absorber being assimilated to a gaussian, the measure of the mean energy loss E_2-E_1 was easily extracted from the peak positions determination . The latter was carried out using a least square parabolic fit.

RESULTS AND DISCUSSION

A) RESULTS ON STOPPING POWER

^{4}He ions

The results are plotted in Fig.(3). The stopping power $S_{exp}(E)$ was measured between 0.3 and 1.7 MeV ; the agreement between the results obtained from four different Al_2O_3 foils is very good. Plotted along with the data are i) a stopping power curve $S_1(E)$ obtained using the Ziegler and Chu table [2] for gazeous oxygen and the recent measurement by Feng [3] for aluminium, assuming the Bragg rule (solid line) ii) a stopping power curve $S_2(E)$ obtained from $S_1(E)$ by applying the empirical corrective factor on the oxygen stopping power proposed by Ziegler et al. [5] for "solid state effect" (dotted line). Our results are systematically lower than $S_1(E)$, the descrepancies varying from 15 to 9 % between 0.4 and 1.7 MeV ; the maximum value of $S_{exp}(E)$ is between 650 and 700 KeV in good agreement with $S_1(E)$, the two curves having hence a rather

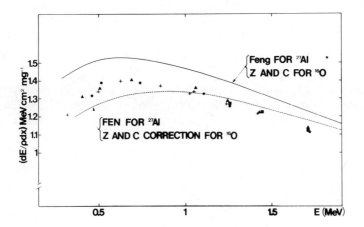

Fig. 3 Stopping power for ^{4}He in Al_2O_3 versus energy measured using Al_2O_3 self supporting foils of various thicknesses (■ 52.6μg cm^{-2}, ● and ▲ 70μg cm^{-2}, + 82.6μg cm^{-2}). The solid ($S_1(E)$) and dotted ($S_2(E)$) lines are obtained from data on stopping power in aluminium and oxygen (see text).

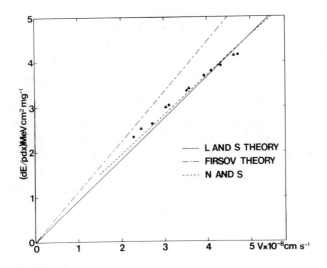

Fig. 4 Stopping power for ^{12}C in Al_2O_3 versus ion velocity, obtained with a 52.6 μg cm^{-2} (●) and a 70μg cm^{-2} (▲) Al_2O_3 foil. Results from Firsov (—·—), Lindhard and Scharff (———) and Northcliffe and Schilling (-----) are also plotted.

similar shape. The dotted curve $S_2(E)$ gives a better absolute agreement with $S_{exp}(E)$ than $S_1(E)$ while its general shape is markedly different. The proposed corrections on oxygen seem too high below 1 MeV and too low at higher energies. Typically we found that, for respecting the Bragg rule, the gazeous oxygen stopping power should be divided by 1.30 at 0.4 MeV and 1.17 at 1.6 MeV while for the same energies Ref. (5) proposed respectively 1.5 and 1.08. Our corrective factor on $S_1(E)$ relies obviously on the values of the stopping power for aluminium and oxygen taken respectively in Ref.(3) and (2).

^{12}C, ^{14}N, ^{16}O ions

For each type of particles, measurements were performed with two different Al_2O_3 foils (52.6 and 70 μg cm^{-2}), leading to results in good agreement. The stopping power for ^{12}C, ^{14}N, ^{16}O ions are respectively presented in Fig.(4,5,6) as a function of the ion velocity. On these figures our results are compared to various theories and tables, the Bragg rule being assumed. The Lindhard and Scharff theory [7] which gives for a monoatomic target an electronic stopping power $Se_{LS}(v)$:

$$Se_{LS}(v) \simeq \xi_e \; 8\pi e^2 \; a_o \; \frac{Z_1 Z_2}{Z} \; v/v_o \qquad (3)$$

where $Z^{2/3} = Z_1^{2/3} + Z_2^{2/3}$ and $\xi_e \simeq Z_1^{1/6}$, leads to the solid line. The Firsov theory [6] which gives for a monoatomic target an electronic stopping power $Se_F(v)$:

$$Se_F(v) = 5.15 \; (Z_1 + Z_2) \; 10^{-15} \; v/v_o \quad eV \; cm^2/Atom \qquad (4)$$

leads to the dot-and-dash line. The dotted line $S_{NS}(v)$ is obtained from the Northcliffe and Schilling table [8]. For the three ions the experimental stopping power $S_{exp}(v)$ appears to vary linearly with the ion velocity v in the 2 to 4 10^8 cm s^{-1} range. It is remarkable to notice that the difference between $S_{exp}(v)$ and $S_{NS}(v)$ is almost always smaller than 3 %, the greatest difference measured being 5 %. However a difference in slopes of $S_{exp}(v)$ and $S_{NS}(v)$ is clearly noticeable, the slope of $S_{exp}(v)$ being smaller in all cases by 10 to 15 %. The agreement between $S_{exp}(v)$ and the two theoretical electronic stopping powers $Se_{LS}(v)$ and $Se_F(v)$ is much poorer.

B) RESULTS ON STRAGGLING

The R.M.S. deviations Ω extracted from the experimental results were systematically normalized with respect to Bohr's theoretical prediction [9], which is :

$$\Omega_B = (4\pi Z_1^2 \; e^4 \; n\Delta R)^{1/2} \qquad (5)$$

where $n\Delta R$ is the number of electrons of the target per cm^2, e

STOPPING POWER AND ENERGY STRAGGLING IN AMORPHOUS AlO 973

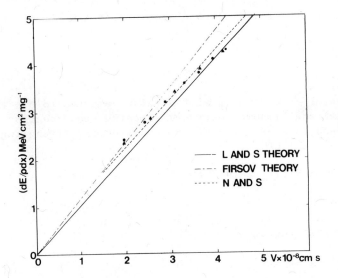

Fig. 5 Stopping power for ^{14}N in Al_2O_3 versus ion velocity, obtained with a 52.6μg cm^{-2} (●) and a 70μg cm^{-2} (▲) Al_2O_3 foil. Results from Firsov (—·—), Lindhard and Scharff (———) and Northcliffe and Schilling (-- -- --) are also plotted.

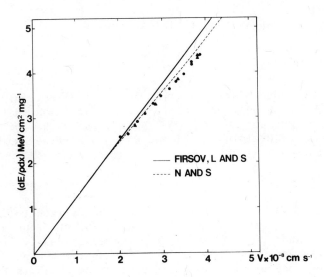

Fig. 6 Stopping power for ^{16}O in Al_2O_3 versus ion velocity, obtained with a 52.6μg cm^{-2} (●) and a 70μg cm^{-2} (▲) Al_2O_3 foil. Results from Firsov (—·—) , Lindhard and Scharff (———·) and Northcliffe and Schilling (------) are also plotted.

the electron charge. For our diatomic targets $n \Delta R$ is given by :

$$n \Delta R = N_{Al} Z_{Al} + N_{ox} Z_{ox} \qquad (6)$$

where N_{Al} and N_{ox} are respectively the number of aluminium and oxygen atoms per cm^3 and Z_{Al} and Z_{ox} respectively 13 and 8. The experimental results Ω/Ω_B will be also compared to two theoretical models : i) the Firsov electronic stopping cross section formalism, which leads to a velocity proportional energy straggling R.M.S. deviation $\Omega_F(v)$ [10] :

$$\Omega_F^2 = 8 \ N\Delta R(Z_1+Z_2)8/3 \ 10^{-15} \ (\frac{v}{v_o})^2 \ eV^2 \ cm^2/At \qquad (7)$$

This model holds for velocities markedly smaller than $v_o Z_1^{2/3}$ ii) The Lindhard and Scharff theory, [11], which gives an energy straggling characterized by Ω_{LS} with

$$\Omega^2_{LS}/\Omega_B^2 = 1 \qquad \text{for } x \gtrsim 3$$
$$\Omega^2_{LS}/\Omega_B^2 = \frac{L(x)}{2} \qquad \text{for } x \lesssim 3 \qquad (8)$$

where $x = v^2/(v_o^2 Z_2)$ and $L(x)$ is the stopping number given in terms of the stopping power dE/dR by :

$$\frac{dE}{dR} = \frac{4\pi Z_1^2 e^4}{m \ v^2} \ N \ Z_2 \ L(x) \qquad (9)$$

An appropriate expression of $L(x)$ in equ.(8) is according to ref. [11] :

$$L(x) = 1.36 \ x^{1/2} - 0.016 \ x^{3/2} \qquad 10)$$

dE/dR varies hence nearly as $1/\sqrt{E}$. Formulas (8) and (9) hold in our case for MeV ^{4}He ions. Our results were also compared to the calculations of Chu and Mayer [12] $\Omega_{CM}(E)$, using the formalism given by Bonderup and Hvelplund [13] and using a Hartree-Fock-Slater atomic electronic density.

In Fig. 7 are plotted the results for energy straggling measurements of ^{4}He, ^{12}C, ^{14}N and ^{16}O ions in Al$_2$O$_3$. All the energy straggling R.M.S. deviations Ω are normalized to the corresponding value Ω_B given by Bohr's theory. For ^{4}He ions a clear velocity dependence is observed : Ω increases nearly linearly with the velocity in the 4 to 6 10^8 cm s^{-1} range. For higher velocities Ω approaches asymptotically a constant value which is found to be slightly higher (less than 10 %) than Ω_B. The shape of the variation of $\Omega(v)$ is in good agreement with Chu and Mayer or Lindhard and Scharff, but our experimental results are systematically by $\sim$ 20 % higher than the two corresponding curves. For ^{12}C, ^{14}N and ^{16}O ions in the 0.3 to 1.5 MeV range i.e. in the 2.5 to 4.5 10^8 cm s^{-1} velo-

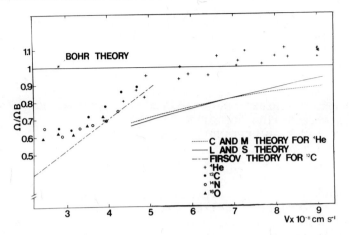

Fig. 7 Energy straggling R.M.S. deviation normalized to Bohr's prediction versus velocity for ^{4}He, ^{12}C, ^{14}N and ^{16}O in Al$_2$O$_3$. Chu and Mayer calculations for ^{4}He (----), Firsov theoretical prediction for ^{12}C (—·—) and Lindhard and Scharff theoretical prediction (———) are also plotted.

city range all our results Ω/Ω_B are lower than 1 ; the variation of Ω/Ω_B is, within experimental error, linear, in agreement with the Firsov theory. The slope for ^{12}C ions is found higher than for the others, also in agreement with the Firsov theory. However all our results are systematically higher than the Firsov predictions, the discrepancies never exceeding 20 %.

REFERENCES

1. G. AMSEL, Annales de Physique 9, 297 (1964).
2. J.F. ZIEGLER, W.K. CHU, Atomic data and nuclear data tables 13, 463 (1974).
3. J.S.-Y. FENG, J.Appl.Phys. 46, 444 (1975).
4. W.H. BRAGG and R. KLEEMAN, Phil.Mag. 10, 5318 (1905).
5. J.F. ZIEGLER, W.K. CHU and J.S.-Y. FENG, this conference.
6. O.B. FIRSOV, Soviet Phys. JETP, 9, 1076 (1959).
7. J. LINDHARD and M. SCHARFF, Phys.Rev. 124, 128 (1961).
8. L.C. NORTHCLIFFE and R.F. SCHILLING, Nuclear data tables A7, 233 (1970).
9. N. BOHR, Mat.Fys.Medd. 18, n° 8 (1948).
10. P. HVELPLUND, Mat.Fys.Medd. 38, n° 4 (1971).
11. J. LINDHARD and M. SCHARFF, Mat.Fys.Medd., 27, n° 15 (1953).
12. W.K. CHU and J. MAYER, Catania working data (1974).
13. E. BONDERUP and P. HVELPLUND, Phys.Rev. A4, 562 (1971).

14. G. AMSEL, C. COHEN and A. L'HOIR, this conference.
15. A.M. LANE, Rev.Mod.Phys. $\underline{32}$, 519 (1960).
16. J. SIEJKA, J.P. NADAI and G. AMSEL,
    J.Electrochem.Soc. $\underline{118}$, 727 (1971).
17. G. AMSEL, J.P. NADAI, E. D'ARTEMARE, D. DAVID, E. GIRARD and
    J. MOULIN, Nucl.Inst. and Meth. $\underline{92}$, 481 (1971).
18. G. AMSEL, C. CHERKI, G. FEUILLAGE and J.P. NADAI,
    J.Phys.Chem.Solids 30, 2117 (1969).
19. S. RIGO and E. BACKELANDT, to be published.

DISCUSSION

Q: (B.R. Appleton) Were nuclear stopping corrections made to the data? Also, a comment: Charlie Moak at ORNL has made an extensive set of measurements for heavy ions which show that Se is $\underline{not}$ velocity proportional but his results all have a different $\underline{intercept}$ than yours (i.e. the extrapolated line falls below the origin).

A: (A. L'Hoir) No nuclear stopping corrections were made because $\Omega_n^2 \ll \Omega_e^2$ in our energy range and because the experimental errors are much higher than these corrections.

Q: (J.F. Ziegler) Anodic Al_2O_3 is sometimes considered porous. Can you comment on your target material?

A: (A. L'Hoir) No attempt was made to investigate the quality of the targets but we plan to do it.

AUTHOR INDEX

Aldape, F	65	Cachard, A.	425
Allen, C	901	Cairns, J.A.	773
Amsel, G.	953, 965	Calvillo, J.	65
Amtén, L.	795	Campisano, S.U.	397, 585, 597
Appleton, B.R.	607	Carstanjen, H.D.	497
Armitage, B.H.	55, 281	Chadderton, L.T.	255
Arnold, G.W.	415	Chaudhri, M.A.	873
		Chi, A.	65
Baeri, P.	597	Chinese Delegation	635
Baglin, J.E.E.	313, 385, 447	Li Te-kuang	
Baragiola, R.A.	29	Teng Hsien-tsan(f.)	
Barcz, A.	407	Tsou Shih-chang	
Barrett, J.H.	607	Cheng Hsiao-wu	
Bauer, W.	575	Liu Ching-tsien	
Bayerl, P.	363	Chang Tung-ho	
Bayly, A.R.	483	Hsieh Yu-fa	
Behrisch, R.	47, 539, 821	Wu Cheng-ming	
Betuel, H.	785	Yuan Shu-hsun	
Bergman, R.	945	Chu, W.K.	15, 125, 149
Bister, M.	75	Cohen, C.	953, 965
Blewer, R.S.	185	Cooke, B.E.	727
Bodart, F.	933	Cookson, J.	773
Bonnet, M.C.	785	Corsi, C.	585
Borders, J.A.	415	Costanzo, E.	397
Borgesen, P.	33		
Bøttiger, J.	811	Das, S.K.	567
Brennan, J.G.	851, 863	Deconninck, G.	87, 933
Brown, M.D.	851, 863	Demortier, G.	933
Burns, G.	873	Dearnaley, G.	885, 901
Butler, J.W.	3	Dodds, D.	235

Eckardt, J.C.	29	Inada, T.	375
Eckstein, W.	47,821	Ischenko, G.	265
Eichinger, P.	353,363	Iwami, M.	665
Ewan, G.T.	727		
		Jaccard, C.	627
Fallavier, M.	293,425,785	Jenkins, L.H.	607
Feldman, L.C.	735	Johansen, A.	255
Feng, J.S.-Y.	15		
Feuerstein, A.	245,471	Kalbitzer, S.	245,471
Finstad, T.G.	437	Kaminski, M.	567
Folkmann, F.	695,747	Kamke, D.	831,841
Fontell, A.	75	Käppeler, F.	803
Foti, G.	397,585,597	Kawatsura, K.	719
Fouilhe, Y.	87	Koersner, I.	945
Fujimoto, F.	719	Komaki, K.-I.	517
		Komiya, S.	665
		Kotai, G.	303
Gabler, F.	265		
Gamo, K.	375	Land, D.J.	851,863
Geerk, J.	273	Langguth, K.G.	273
Glantz, L.	795	Langley, R.A.	201,337
Gönczi, L.	945	Lee, C.P.	375
Grahmann, H.	245,471	Lennard, W.N.	925
Grant, W.A.	235	Lever, R.E.	111,125,149,163
Gyulai, J.	303	L'Hoir, A.	953,965
		Liebl, H.	659
Hammer, W.N.	447	Lindh, U.	945
Hartley, N.E.W.	901	Linker, G.	273
Heck, D.	803	Luomajärvi, M.	75
Heintze, V.	831	Lurio, A.	773
d'Heurle, F.M.	385		
Henrich, E.	675	Mac Donald, R.J.	483
Hiraki, A.	665	Mackenzie, C.D.	281
Hirvonen, J.K.	163,457	Manuaba, A.	303
Hofer, W.O.	659	Marsaud, S.	293,425,785
Holloway, D.F.	773	Martin, P.J.	483
Howard, J.K.	125	Mayer, J.W.	33,375
Huber, H.	627	Mc Millan, J.W.	913
Hübler, G.K.	457	Mezey, G.	303
Hufschmidt, M.	831	Miller, J.W.	607

AUTHOR INDEX

Mitchell, I.V.	925	Schou, J.	255
Möller, W.	831,841	Schow, O.E.	607
Morenius, B.	795	Shimizu, A.	665
Müller, P.	265	Simons, D.G.	851,863
		Silverman, P.J.	735
Nagy, T.	303	Sobiesiak, H.	803
Narusawa, T.	665	Spicer, B.M.	873
Nicolas, G.	933	Steenstrup, S.	255
Nicolet, M.-A.	33	Stroobants, J.	933
Noggle, T.S.	607	Sundqvist, B.	795,945
Oetzmann, H.	245,471	Tanoue, H.	685
Olmos, D.	65	Tardy, J.	425
Olsen, T.	437	Terasawa, M.	719
Ozawa, K.	719	Thieme, G.	99
		Thomas, G.J.	575
Pabst, W.	211,363	Thomas, J.P.	293,425,785
Pfeiffer, Th.	841	Trehan, P.N.	55
Pierce, T.B.	913	Tsurushima, T.	685
Picraux, S.T.	527,597,811	Turos, A.	407
Pihl, J.	795		
Poate, J.M.	317	Vitali, G.	585
Price, P.B.	727		
		Wahl, M.	353
Reistad, R.	437	Whitton, J.L.	727
Rickards, J.	65	Wieluński, L.	407
Rimini, E.	397,585,597	Williams, J.S.	223,235
Romero, S.	65	Wittmaack, K.	649
Rossing, T.	567		
Roth, J.	47,539,821	Zehner, D.M.	607
Roulet, M.	627	Ziegler, J.F.	15,163,759,773
Rouse, J.	873		
Rud, N.	811		
Samid, I.	375		
Satake, T.	665		
Sauermann, H.	353		
Scherzer, B.M.U.	33,47		
Schmidt, H.J.	675		

SUBJECT INDEX

Aluminium, 75, 245, 363, 487, 506, 567, 597, 777
 - in GaAs, 737
 - in Germanium, 925
 - amorphous Oxide, 965
Aluminium-Chromium, 117, 141
Aluminium-Copper, 128
Aluminium-Gold, 327, 480
Aluminium-Nickel, 329, 391
Aluminium-Oxide, 287, 425
Angular Distribution - see channeling
Arsenic-Tellurium, 295
Asymmetrical Response of Surface Barrier Detectors, 953
Atom Location - see Lattice Location
Auger Electron Spectroscopy, 125, 326, 609, 667, 780

Background, 196
Backscattering, Analysis of
 - Alloys, 128, 323
 - Compound Formation, 117, 141, 152, 319, 327
 - Damage Distribution, 539, 597
 - Elements, 169, 170
 - Foils, 185, 266
 - Interdiffusion, 143, 171, 293, 322, 397, 407, 539, 597
 - Insulators, 415
 - Ion Implanted Layers, 229, 505
 - Pore Size, 281
 - Random Spectra, 273
 - Layered Structures, 166, 215, 269
 - Organic Materials, 437
 - Thin Films, 149, 166, 204, 215, 227, 268, 317, 337, 357, 365, 375, 397, 480
Backscattering with Protons,
 - 185, 265, 281, 313, 375
 - ^{4}He ions, 21, 39, 50, 125, 149, 163, 211, 255, 293, 313, 340, 365, 386, 397, 407, 416, 553
 - Deuterons, 47, 247
 - Lithium, 425
 - Carbon, 265
 - Nitrogen, 211, 553
 - Oxygen, 265
Backscattering, 37, 103
 - Computer Analysis,
 - High Resolution, 168, 211, 223, 245, 480

- Resonant, 93, 284, 303
- Standards Project, 313
- Transmission technique, 186, 282

Binary Encounter Approximation, 697
Biological Samples, 437, 785, 795, 933, 945
Bismuth-Copper, 390
Blistering, 567, 841, 846
Blocking, 511
Boron, 920
- K-Shell Ionization, 719
Bragg's Rule, 49, 129
Bubbles, 575

Carbon, 65, 901, 913
Carbon-Niobium, 273
Catalysts, 733
Channeling, 531
- angular distribution, 501, 510, 520, 531,
- damage analysis, 539, 597
- impurity analysis, 505, 727
- lattice location, 505, 527
- low angle, 229
- principles of, 489, 517
Charge Build-up, 422
Chromium, 532
Chromium-Copper, 329, 386
Chromium-Gold, 407
Chromium-Nickel-Gold, 407
Cobalt, 778
Cobalt- Gadolinium, 306
Combined techniques, 125, 337, 363, 397, 407, 425, 735, 779, 795, 889
Compound Formation, 117, 141, 152, 319, 327
Continuum Potential, 529, 542
Copper, 75, 498, 501, 540
Copper-Lead, 397

Copper-Silicon, 668
Corrosion, 567, 575, 885, 901

Damage
- Depth Distribution, 539, 551, 597
- Gettering, 635
Dechanneling, 547, 597, 627
Deconvolution, 174, 293, 953
Depth Profiling, 83, 111, 185, 201, 273, 656
- of Deuterons, 207, 831
- of Helium, 47, 207
- of Nitrogen, 864
- by Backscattering, 419
- by Nuclear Reactions, 466, 811
- by X-Ray Analysis, 735
Depth Resolution, 212, 223, 255, 266, 428, 821
- Backscattering, 463
- Detector Resolution, 213, 257
- Electrostatic Analyzer, 471
- Magnetic Spectrometer, 457
Detection Limits, 464, 710
Deuterium, 527
Deuterium-Erbium, 203
Diffusion, 143, 171, 298, 322, 385 397, 407, 419, 539, 597
Disorder
- Thermal Disorder, 627
Dissolution Enhancement, 685
Distribution Profiles - see Depth Profiling
Double Alignment, 512

Energy Analysis Method, 814
Energy Loss Distribution, 89, 476, 965
- modification of theory, 3
- energy dependence of, 65, 81
- transmission technique, 56

SUBJECT INDEX

    – experimental values for
      protons in Al, 60
      protons in Ag, 60
      protons in Au, 60
      protons in C, 65
      protons in N, 12
      protons in Ni, 60
      protons in Mo, 60
      protons in Ta, 60
      protons in V, 60
      protons in Xe, 13
      ^{4}He in Al, 80
      ^{4}He in Cu, 80
      heavy ions in Si, 242, 250
      heavy ions in Ge, 250
    – Calculation, 87
Energy Loss – see Stopping Cross Sections
Energy deposition, 684
Exfoliation, 575

Fluorine – Detection in Matrices, 873, 933
Flux Density, 500, 528
Flux Distribution, 502, 504
Flux Peaking, 498, 502, 517, 528

Gallium-Aluminium-Arsenide, 363, 375
Gallium Arsenide, 460, 728, 735
Gallium-Gadolinium Oxide, 685
Gallium Phosphide, 519
Gas Reemission, 575, 841
Germanium, 169, 245, 275, 731, 925
Germanium-Niobium, 318
Gold, 295, 326, 462, 471, 559, 607, 675
Gold-Nickel, 143, 293
Gold-Silicon, 667
Gold-Silver, 171
Gold-Tellurium, 300
Gold-Titanum-Palladium, 323
Grazing Angle Incidence, 211, 223, 237

Helium, 47, 207, 811
High Resolution Backscattering, 168, 211, 223, 245
Human Lymphocytes, 785
    –Dental Enamel, 933
Hydrogen, 527, 811, 901

Implantation Profiles, 831
Insulators, 415
Interdiffusion – see Diffusion
Ion Implantation, 229, 236, 419
Ion Induced X-Rays, 364, 510, 695, 699, 795
    – Analysis, 762
    – Absorption, 751
    – Angular Distribution, 751
    – Calibration, 759
    – Experimental Equipment, 763
Ionization Cross Sections, 724
Iron, 519

Kinematic Recoil Factor, 188

Lateral Range – see Range
Lattice Location, 497
    – Backscattering, 505
    – Blocking, 511
    – Ion Induced X-Rays, 510
    – Nuclear Reactions, 507, 527, 534
Lattice Vibration, 500
LEED, 609
Lead-Tin-Telluride, 585
Light Atom Profiles, 47, 185, 201, 811, 821, 831, 901, 915, 938, 945
Lithia-Aluminia-Silicate-Glass, 417
Lithium, 887, 901

Magnesium, 170
Magnesium-Alloy, 891
Magnetic Spectrometer, 457
Manganese, 506

Marker, 332
Mass Resolution, 224, 428, 460 487
Matrix X-Ray Yield, 755
Metallization, 535
Microhardness, 908
Microprobe, see Nuclear Microprobe
Molecular Ions, 99
Molecular Orbits, 703
Molybdenum, 534
Moments Analysis, 149, 856
Multielemental Analysis, 773, 785, 797
Multiple Scattering, 214, 473

Nickel, 232, 675, 777, 841
Nickel-Silicon, 330
Niobium, 50, 499, 507, 575,
Niobium-Tin, 268
Niobium-Titanium, 268
Nitrogen, 10, 864, 901, 916, 945
Nuclear Microprobe, 896, 901, 913
Nuclear Reactions
- $^1H(^{11}B,\alpha)$, 901
- $^1H(^{19}F,\alpha\gamma)^{16}O$, 811
- $^2H(d,p)T$, 831
- $^3He(d,\alpha)H$, 47, 821
- $^3He(d,p)^4He$, 811
- $^4He(^{10}B,n)^{13}N$, 811
- $^7Li(p,\alpha)$ ^4He, 901
- $^9Be(p,\alpha)^6Li$, 466
- $^{10}B(\alpha,p)^{13}C$, 920
- $^{10}B(d,p)^{11}B$, 920
- $^{11}B(\alpha,p)^{14}C$, 920
- $^{11}B(p,\alpha)^8Be$, 921
- $^{12}C(d,p)^{13}C$, 798, 891, 901, 915
- $^{12}C(p,p)^{12}C$, 67
- $^{13}C(d,p)^{14}C$, 798
- $^{14}N(d,\alpha)^{12}C$, 798, 901, 918
- $^{14}N(d,p)^{15}N$, 798, 918, 945
- $^{14}N(p,\gamma)^{15}O$, 851, 863
- $^{16}O(d,p)^{17}O$, 798, 891, 901, 921
- $^{16}O(d,\alpha)^{14}N$, 407, 798
- $^{16}O(\alpha,\alpha)^{16}O$, 303
- $^{18}O(p,\alpha)^{15}N$, 891, 901
- $^{19}F(p,\alpha\gamma)^{16}O$, 202, 873, 936
- $^{19}F(p,\alpha)^{16}O$, 891
- $^{19}F(p,p'\gamma)^{19}F$, 934
- $^{20}Ne(\alpha,\gamma)^{24}Mg$, 76
- $^{27}Al(d,p)^{28}Al$, 925
- $^{27}Al(p,\gamma)^{28}Si$, 10, 368, 425
- $^{32}S(d,p)^{33}S$, 798, 984
- $^{32}S(d,\alpha)^{30}P$, 798

Oligo-Elements, 785
Oxidation, 425
Oxidation Enhancement, 308
Oxygen, 306, 508, 901
Oxygen-Silicon, 227, 304, 471

Palladium-Silicon, 152
Photon Emission, 483
Platinum, 313, 471, 740
Plane Wave Born Approximation, 698
Pore Size, 281
Porous Samples, 281, 775
Potential
- Continuum, 529, 542
- String, 518
Pulsed Proton Beam, 803

Quadrupole Mass Filter, 650

Radioactive Background, 803
Range
- Projected, 235
- Lateral, 235
for
- Heavy Ions in Si, 238, 245
- Heavy Ions in Ge, 245

Reactor Materials, 913

SUBJECT INDEX

Reactions — see Nuclear Reactions
Resolving Power — see Depth Resolution
Resolution — see Depth Resolution
Resonance Method, 814
Resonance Reactions, 67, 93, 303, 368, 425, 851
Rutherford Scattering — see Backscattering

Scanning Electron Microscope, 393, 559
Scattering Cross Sections, Enhancement, 190, 201
 for Molecular Ions, 99
 for H^+ on D, 203
 for H^+ on ^4He, 203
Secondary Electron Bremsstrahlung, 704, 749
Sensitivity, 194, 226, 303, 464
 in Ion Induced X-Ray Analysis, 708, 747
 in Nuclear Reaction Analysis, 812
Silica, 288
Silicides, 152, 331, 357
Silicon, 245, 275, 353, 463, 511, 522, 635, 665
Si(Li)-Time Resolution, 805
Silver, 506
Silver-Bromide 627
Silver-Phosphate-Glass, 416
Silver-Silicon, 668
SIMS, 659, 665, 895
Sputtering, 134, 659, 670, 676
Standards Project, 313
Standard Samples, 785,
Stopping Cross Sections, 471
 Solid/Gas differences, 15
 exper./th. value ratios, 15
 empirical formula, 29
 determination by RBS, 33, 82, 213
 experimental values for protons in C, 69

^4He in Al, 82, 383
 Al_2O_3, 965
 Au, 41
 Cu, 82
 Ge, 252, 383
 Pt, 41
 Si, 252
 SiO_2, 42
 Ta_2O_5, 42
^{12}C in Al_2O_3, 965
^{14}N in Various Materials, 863, 965
^{16}O in Al_2O_3, 965
Stopping Powers — see Stopping Cross Sections
Straggling — see Energy Loss Distribution
Sulphur, 728
Superconducting Layers, 268, 318

Tantalum-Nitride, 337
Thin Film Analysis — see Backscattering
Titanium, 663
Trace Analysis, 759
Transmission Electron Microscopy, 397
Transverse Energy, 545
Tungsten, 215, 353, 519, 533, 540
Tunneling Elements, 269

Unfolding Techniques, 851

Vavilov Distributions, 89
Vanadium, 266

Xenon, 13
X-Ray see Ion Induced X-Rays
X-Ray Photo-Electron-Spectroscopy, 675

Yield
 aligned, 544, 559
 minimum, 501
 Reaction-, 500, 852
Yttrium, 787